A MOUSETRAP FOR DARWIN

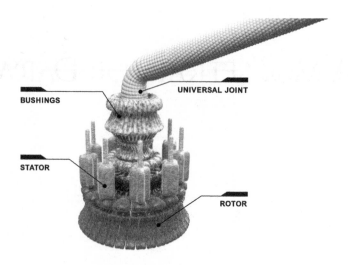

BUSHINGS

UNIVERSAL JOINT

STATOR

ROTOR

The Bacterial Flagellum Motor

A MOUSETRAP
FOR DARWIN

MICHAEL J. BEHE

SEATTLE   DISCOVERY INSTITUTE PRESS   2020

Description

In 1996 *Darwin's Black Box* thrust Lehigh University biochemist Michael Behe into the national spotlight. The book, and his subsequent two, sparked a firestorm of criticism, and his responses appeared in everything from the *New York Times* to science blogs and the journal *Science*. His replies, along with a handful of brand-new essays, are now collected in *A Mousetrap for Darwin*. In engaging his critics, Behe extends his argument that much recent evidence, from the study of evolving microbes to mutations in dogs and polar bears, shows that blind evolution cannot build the complex machinery essential to life. Rather, evolution works principally by breaking things for short-term benefit. It can't construct anything fundamentally new. What can? Behe's money is on intelligent design.

Library Cataloging Data

A Mousetrap for Darwin by Michael J. Behe

556 pages, 6 x 9 x 1.1 inches & 1.6 lb, 229 x 152 x 29 mm. & 0.74 kg

Library of Congress Control Number: 2020948398

ISBN-13: 978-1-936599-90-5 (hardback); 978-1-936599-91-2 (paperback); 978-1-936599-92-9 (EPUB); 978-1-936599-93-6 (Kindle)

BISAC: SCI027000 SCIENCE / Life Sciences / Evolution

BISAC: SCI049000 SCIENCE / Life Sciences / Molecular Biology

BISAC: SCI007000   SCIENCE / Life Sciences / Biochemistry

Publisher Information

Discovery Institute Press, 208 Columbia Street, Seattle, WA 98104

Internet: http://www.discoveryinstitutepress.com/

Published in the United States of America on acid-free paper.

First Edition, November 2020.

Praise for A Mousetrap for Darwin

"Wow, what a book! Michael Behe's *A Mousetrap for Darwin* is a compelling read for anyone following the Darwin vs. ID debate. It not only is a magnificent testimony to his own massive contributions in so many areas of ID over more than two decades since he first published *Darwin's Black Box*, but also provides perhaps the most comprehensive and incisive critique of Neo-Darwinism currently in print. In the more than one hundred articles and posts in the book, Behe revisits key arguments for ID which he initially developed and that have since become foundational to the defense of ID, such as the argument from irreducible complexity and the argument from waiting times. The book represents a devastatingly brilliant unanswerable response to his Darwinian critics and to the whole Darwinian worldview. Behe brings out more forcibly than any other author I have recently read just how vacuous and biased are the criticisms of his work and of the ID position in general by so many mainstream academic defenders of Darwinism. And what is so telling about his many wonderfully crafted responses to his Darwinian critics is that it is Behe who is putting the facts before theory while his many detractors—Kenneth Miller, Jerry Coyne, Larry Moran, Richard Lenski, and others—are putting theory before the facts. In short, this volume shows that it is Behe rather than his detractors who is carefully following the evidence. Of all the fine essays in this volume, I think his responses to Lenski in Parts 4 and 7 are particularly outstanding and unanswerable. Lenski's inability to undermine Behe's critique gives the lie to the notion that Darwinism provides anything resembling a convincing account of the biological world."
—Michael Denton, PhD, MD, former Senior Research Fellow in the Biochemistry Department at the University of Otago in Dunedin, New Zealand, author of *Evolution: A Theory in Crisis* and *Nature's Destiny*

"Ever since the release of *Darwin's Black Box* in 1996, I have been impressed with Mike Behe's ability to respond to his critics quickly, respectfully, and with clarity and patience. This collection of many of those responses over a twenty-four-year period is a gift to all who promote intelligent design, and indeed to anyone passionately interested in the science of biological origins and the contemporary debate over Darwinism and design. Both critics and champions of intelligent design will find much to ponder here. Behe repeatedly meets his critics head on and with no apologies. This collection is a treasure."
—Raymond G. Bohlin, PhD, co-author of *The Natural Limits to Biological Change*

"Michael Behe's *Darwin's Black Box*, *The Edge of Evolution*, and *Darwin Devolves* clearly describe the problems and limits of Darwinism as well as what the mechanism of random mutation and natural selection actually does. Over the years Behe has received a mountain of criticism, all of which has been answered in detail by him in letters to the editors of various journals, newspapers, and blogs. Now, in *A Mousetrap for Darwin*, Behe treats his readers to his compelling and thorough responses to his critics. Anyone reading this book will become better informed of the powerful arguments for design in biology and better educated regarding the Design vs. Darwin debate. I greatly enjoyed *Mousetrap* and highly recommend it."
—Russell W. Carlson, PhD, Emeritus Professor of Biochemistry & Molecular Biology and the Complex Carbohydrate Research Center University of Georgia

"The humorous Mosquito Bite Scratcher illustration in Michael Behe's 1996 book *Darwin's Black Box* opened my eyes to the irreducibility of biological systems. Ever since, I have followed Behe's tireless defense of his position in numerous well-argued articles and responses. His style is respectful; he carefully studies the arguments of his critics and explains why they err. His mousetrap example, the bacterial flagellum motor and other irreducibly complex biological systems, the mounting laboratory evidence of the strict limits of evolution, and his devolution argument should drive the message home: the idea that life's diversity emerged through the Darwinian evolutionary mechanism is a dead idea. This book is a welcome and valuable collection of these brilliantly argued articles."
—Matti Leisola, DSc, Professor Emeritus of Bioprocess Engineering, Aalto University, Finland

"Over the years I have followed Michael Behe's work in building an arsenal of arguments for intelligent design. And I have followed the desperate attempts of mainstream evolutionists to discredit that work. I've found that in their attacks, they have used fallacious logic and zombie science at every turn. A few of the critiques are superficially persuasive, but they hold up best if you don't think too hard about the biochemical details of their evolutionary scenarios. If you fear to doubt Darwinism, read further at your own peril. Behe's devastating rebuttals are here in spades. If, however, you are ready and willing to follow the evidence, take heart: Behe guides us into state-of-the-art biochemistry—and into the case for intelligent design—with elegance, clarity, and good grace. This collection is a delight."
—Marcos Eberlin, PhD, member of the Brazilian Academy of Sciences and winner of the prestigious Thomas Medal (2016)

CONTENTS

INTRODUCTION

SINCE THE TURN OF THE MILLENNIUM A RAFT OF DISTINGUISHED biologists have written books critically evaluating evolutionary theory.[1] *None* of them think that Darwin's mechanism is the main driver of life. It may surprise people who get their information about the state of science from gee-whiz puff pieces in the mainstream media, but, although strong partisans still hold out, the eclipse of Darwinism in the scientific community is well-advanced. A few years ago the journal *Nature*[2] published an exchange between two groups of scientists, one defending Darwin and the other saying it's time to move on. It's nice to have defenders, but when an idea has been around for 150 years—wished well by all right-thinking people, investigated to death by the scientific community—and a piece appears in the world's leading science journal saying it's time to move on, then it's time to move on.

The question of course is, move on to what? Those books by scientists dissing Darwin offer their own clever ideas, but so far the scientific community isn't buying any of them. All the new ideas—self-organization, facilitated variation, symbiosis, complexity theory, and more—are quickly concluded to be nonstarters, to have the same problems as Darwin's theory, or both. In the absence of an acceptable replacement—and because of its usefulness as a defensive talking point in fending off skepticism from the public—intellectual inertia maintains Darwinism as textbook orthodoxy.

How can Darwin's theory be on the ropes and yet presidential candidates still be roasted for mealy-mouthed answers about evolution? How can new fossils be paraded monthly across the front pages of major newspapers as fresh and powerful evidence of evolution, but distinguished scientists still argue against the sufficiency of simple random mutation and natural selection? To understand the seeming conundrum one has to make a critical distinction between evolution itself—call it *mere* evolution—and the *mechanism* or *cause* of evolution. Mere evolution is the bare proposition that organisms living today are descended with modification from organisms that lived in the distant past. There is terrific evidence consistent with that. On the other hand we can ask, well, what *caused* such astounding changes to take place? What is the *reason* or *mechanism* for evolution? That's the sticking point.

Darwin's claim to fame is not that he proposed mere evolution. The idea that modern organisms descended from ancient ones had been discussed by other scientists well before Charles Darwin came on the scene. Those pre-Darwinian scientists, however, all proposed teleological reasons for evolution—that is, reasons above and beyond nature itself, such as a guiding force or some inborn drive to improve. Darwin's claim to fame, rather, was to propose a cause for evolution that involved only mechanical considerations that science could readily study, such as natural selection acting on random mutation. (Darwin actually wrote of unspecified random "variation" because he and all the other scientists of his day were in the dark about the mechanism of heritable changes to life—mutations in DNA).

Although he wasn't the first to propose it, Darwin popularized the idea of evolution by envisioning it as a wholly mechanized phenomenon that fell fully within the purview of science. Darwin's proffered mechanism of random variation plus selection actually fell out of favor quickly in the nineteenth century, but the residual notion of mere evolution continued to provide a framework for scientists to classify life—comparing similarities between different organisms, noting differences between similar ones, and building a pattern of relationships. It wasn't until the 1940s, long after the publication of *The Origin of Species*, when increasing understanding of the genetics of inheritance was seen to be at least compatible with Darwin's mechanism, that scientific leaders met and proclaimed what was dubbed the "neo-Darwinian synthesis"—a fusion of Darwin's theory with modern genetics. At that point Darwin's mechanism became the presumptive explanation in the scientific community and popular press for the cause of evolution.

At least for a few decades.

1. The Problem for Darwin

After World War II new equipment and new techniques allowed biologists to probe ever more deeply into life, until at last they hit its foundation—molecules. From today's perspective it seems odd but, until remarkably recently in history, scientists had little idea of exactly how life worked or what its underlying basis looked like. The big breakthroughs arguably came in the 1950s with the discovery of the shapes of DNA and myoglobin (a protein that stores oxygen in muscles), as well as the cracking of the genetic code. The structure of myoglobin showed that life's molecules were not built to look pretty, like some New Age crystals. Rather, they were *machines*, and they were built to do specific, concrete jobs. The genetic code demonstrated that DNA carried *information*—an unheard-of property for a natural chemical.

Since that auspicious beginning biological knowledge has increased by leaps and bounds. The most profound insight of modern science is that life is based literally on *machines made of molecules*. Much like for the alien, half-humanoid, half-machine Borg in Star Trek, tiny nanomachines perform the necessary tasks of cells. Among their many roles, machines in cells act as taxis and trucks, shuttling passengers and supplies across vast distances (relative to the size of the molecules), along cellular highways marked by traffic signs, both also made of molecules. Cellular computer programs of bewildering sophistication control the assembly of the machinery. Elegant genetic regulatory networks express the information in DNA to produce the right molecules at the right times in the right places, building the intricate bodies of animals. What scientists of earlier times took to be a primitive abacus has turned out to be a futuristic supercomputer. What biologists of Darwin's day thought was a "simple little lump of albuminous combination of carbon" (the cell)[3]—pretty much just a microscopic piece of Jell-O—has turned out to be a fully automated, nanoscale factory, sophisticated beyond human imagining.

And that's the problem for Darwin: the molecular foundation of life has turned out to be astoundingly, gobsmackingly sophisticated, elegant beyond words. It's not only that the information in life is stored in a genetic code—a feature that's strongly evocative of intelligence and the likes of which had never been anticipated by chemical principles. Rather, it's that the more researchers look, the more and more levels of codes, programs, and controls are found. But don't take my word for it. Listen to the prominent embryologists Michael Levine and Eric Davidson describe control systems—at the time newly discovered—that help coordinate animal development:

> Gene regulatory networks (GRNs) are logic maps that state in detail the inputs into each *cis*-regulatory module, so that one can see how a given gene is fired off at a given time and place.... The architecture reveals features that can never be appreciated at any other level of analysis but that turn out to embody distinguishing and deeply significant properties of each control system. These properties are composed of linkages of multiple genes that together perform specific operations, such as positive feedback loops, which drive stable circuits of cell differentiation.[4]

And read then-president of the National Academy of Science Bruce Alberts marveling about the completely unexpected molecular machinery of the cell:

> We have always underestimated cells. Undoubtedly we still do today. But at least we are no longer as naive as we were when I was a graduate student in the 1960s....
>
> ... As it turns out, we can walk and we can talk because the chemistry that makes life possible is much more elaborate and sophisticated than anything we students had ever considered.... Indeed, the entire cell can be viewed as a factory that contains an elaborate network of interlocking assembly lines, each of which is composed of a set of large protein machines....
>
> Why do we call the large protein assemblies that underlie cell function protein *machines*? Precisely because, like the machines invented by humans to deal efficiently with the macroscopic world, these protein assemblies contain highly coordinated moving parts.[5]

Sure, the details of life also show dents and scratches, just like a fine automobile will show dents and scratches after long service on the road. But, as with a car, it's not the dents that require much explanation; it's the highly functional arrangement of parts that does.

How could Darwin's clunky mechanism—one tiny, random change at a time, each followed by a long, fitful, and uncertain period of natural selection, with no ability to anticipate future needs—account for the molecular marvels that modern biology had uncovered? Increasingly the answer became, it couldn't. The more science advanced and the more elegance and complexity was uncovered, the more biologists drifted away from Darwinism. New proposals have offered ideas about how the cell might engineer itself, or arise mysteriously from raw complexity, or how different kinds of cells might join forces. Yet, while each new idea usually can explain at least one area of biology reasonably well, none seem any better than Darwinism itself in accounting for the whole. So even as biologists continue at breakneck speed to discover how life actually works in the present, and despite cheerleaders in the popular press and aggressive proselytizers, origins science is in a funk. The question that Darwin was supposed to have answered—what is the *reason* for the unfolding of life?—stands unanswered.

Yet, for those willing to see, the solution is blazingly obvious. Whenever we come across "logic maps" in our everyday world, or "a factory that contains an elaborate network of interlocking assembly lines," we *immediately* conclude that they were purposely put together. Such arrangements simply *reek* of design. And why in such cases do we so confidently conclude design? Why—whenever we observe a purposeful arrangement of parts in our everyday world, such as, say, in an ordinary mechanical mousetrap—do we routinely conclude that intentional intelligent design is the explanation? The reason is that only *minds* can have purposes, so the discovery of parts arranged for a purpose allows us to infer that other minds besides our own exist. Now that we have unexpectedly stumbled over such very purposeful arrangements at the very foundation of life (the cell), there is no reason to withhold an identical judgment.

2. INTO THE RING

I FIRST entered the fray with the 1996 publication of *Darwin's Black Box: The Biochemical Challenge to Evolution.* As a biochemist I had become quite disenchanted with the Darwinian theory I had been taught throughout my education, and had concluded that tweaks to the theory were woefully inadequate. Enough was enough. It was time to take the bull by the horns, to stop pussyfooting around, to wake up and smell the coffee. It was time, I decided, for science to acknowledge that pre-Darwinian biologists were right—life had indeed been purposely, intelligently designed.

Initial reactions in the scientific community to my first book were muted (in private, some scientists were intrigued). At the beginning, even a couple of semi-positive book reviews were published, and the idea was quite well received among the general public. But soon criticisms came in force. Apparently, the popularity of the idea with the public increased to the point where the scientific community actually felt threatened, and initiated political steps to fight it. Science organizations published official pamphlets denigrating the concept of design. Popular science magazines warned of the impending end of western civilization. Governments—influenced more by science organizations than by parents—issued edicts against discussing design in schools. Even a committee of the Council of Europe condemned it.

All that seemed a rather excessive reaction to me, especially since the idea—agree with it or not—is spectacularly obvious. Design was quite easy for biologists to spot before Darwin based on plant and animal anatomy. And now that his once-promising idea hasn't panned out, it should be exceedingly easy for biologists after Darwin to see design based on the elaborate molecular machinery of the cell. After all, accounting for the strong appearance of design in life was the triumph of Darwin's theory, we were frequently told. Random mutation and natural selection mimicked design, we were oft assured. But now that Darwin is in decline we are left with the overwhelming appearance of design in life and no realistically plausible theory to explain it.

I followed *Darwin's Black Box* with *The Edge of Evolution: The Search for the Limits of Darwinism* in 2007, and extended my case in 2019's *Darwin Devolves: The New Science about DNA That Challenges Evolution*. Each book drew heavily upon recent advances in molecular biology, and confirmed and deepened the conclusions I had come to in my first book. No matter whether new results and a new mechanical theory may come along in the future to explain what Darwin's theory couldn't (I very much doubt it), the current state of our knowledge intellectually justifies a firm conclusion of the intelligent design of life.

Each of the three books drew responses from places high and low—some civil, some not. These responses gave me the opportunity to further extend my analysis, clarify points, and rebut fresh objections—or, as was often case, find fresh illustrations to rebut old objections I had already addressed previously.

The present book is a selection of my writings and talks over several action-packed, fun-filled decades in many battles over Darwinism and intelligent design. Although I hope that the reader will be persuaded about the merit of design both by the force of the arguments in its favor and the weakness (in my view) of those against, the book offers more than that. In the back-and-forth, point-counterpoint of the arguments it provides an insight into how people who think about these things reason. It vividly shows that even the smartest of scientists and intellectuals are human and fallible. We all are influenced by our own interests and priorities, come from our own peculiar backgrounds, hope for our own visions of the future. All of those factors enter into our reasoning. Nobody on earth is the reincarnation of Mr. Spock, either singly or collectively.

In an ideal world this book would include all of my critics' articles as well as my responses, but that would have made for an unwieldy and expensive book, even if I could have gotten permission to reprint them all. However, many of their essays are freely available online. Wherever possible I provide readers with the information needed to track down the written criticisms I am responding to.

Since I initially wrote the essays in this book for audiences that hadn't read much on the subject, most of them can stand alone. For ones that could use a bit of background knowledge, I add explanatory comments as needed. In a few, I tweak the prose a bit to make it read more easily for this collection, or to make a point clearer. A few of the pieces were written for publications that edited them rather severely. Here I include the original, unedited essays. A few were intended as letters to the editors of various science journals or newspapers but were not published, either because they were refused or because they missed a deadline.

The original publications ranged widely, from formal science journals to newspaper op-eds to popular magazines to blog posts. So, since one has to write for the intended audience, the style and level of technical detail ranges widely, too. Some are stiffly formal, others breezy; some give an overview, others have lots of specifics; some are longer, some shorter. Nonetheless, even in the technical articles the gist of the argument is easy to follow.

PART ONE: A BLACK BOX UNDER THE MICROSCOPE

DARWIN'S ELEGANT IDEA OF EVOLUTION BY RANDOM CHANGES AND natural selection provided a good jumping-off point for research in the nineteenth century. But it was severely hobbled by an utter lack of knowledge of the underlying mechanisms of biology. Yes, organisms come in astonishing varieties, and, yes, they have amazing abilities. But how exactly do they do what they do? How does even a single cell live, metabolize, reproduce, and transmit information to its offspring? In 1859, when Darwin's *The Origin of Species* first appeared, even rudimentary answers to such questions were decades down the road. In the meantime many scientists grew far too comfortable with Darwin's simplistic explanation and were oblivious to the growing disconnect between theory and facts. Darwinian orthodoxy chugged along pretty much by dint of intellectual inertia, and still does. The startling challenge that modern working science threw down for Darwin's hypothesis was the discovery of *machines* and *information* at the foundation of life. Quite literally, life—like the Borg in *Star Trek*—depends on tiny machines and computers made of molecules.

Not long after *Darwin's Black Box* was published in 1996, criticisms came flying thick and fast, and grew ever more intense over the years. Fair enough—if you write about a controversial topic, you have to expect controversy. Yet, as you'll see in my responses, surprisingly little of the initial fire was directed at what I was actually arguing. Rather, much of it seemed aimed at tangential associations that popped into critics' minds when they heard the phrase "intelligent design." Put another way, they often were battling straw men, not the actual case I had set before them. Now, one might imagine that such behavior from your intellectual opponents would prove frustrating, but it was actually a lot of fun for me to discover what side issues made critics stumble and to show why the objections were irrelevant.

The first two pieces in this section distill key arguments I made in *Darwin's Black Box*. If you have read that book and those arguments remain fresh in your mind, you may wish to skip the two pieces and move right into my *New York Times* op-ed and, from there, on to my earliest responses to critics of *Darwin's Black Box*.

1. EVIDENCE FOR INTELLIGENT DESIGN FROM BIOCHEMISTRY

This talk was given shortly after *Darwin's Black Box* was published and summarizes its argument.

"Evidence for Intelligent Design from Biochemistry," lecture, Discovery Institute's God & Culture Conference, Seattle, August 10, 1996.

HOW DO WE SEE? IN THE NINETEENTH CENTURY THE ANATOMY OF the eye was known in great detail, and its sophisticated features astounded everyone who was familiar with them. Scientists of the time correctly observed that if a person were so unfortunate as to be missing one of the eye's many integrated features, such as the lens, or iris, or ocular muscles, the inevitable result would be a severe loss of vision or outright blindness. So it was concluded that the eye could only function if it were nearly intact.

A SERIES OF EYES

CHARLES DARWIN knew about the eye too. In *The Origin of Species*, Darwin dealt with many objections to his theory of evolution by natural selection. He discussed the problem of the eye in a section of the book appropriately entitled "Organs of Extreme Perfection and Complication." For evolution to be believable, Darwin had to somehow convince the public that complex organs could be formed gradually, in a step-by-step process.

He succeeded brilliantly. Cleverly, Darwin didn't try to discover a real pathway that evolution might have used to make the eye. Instead, he pointed to modern animals with different kinds of eyes, ranging from the simple to the complex, and suggested that the evolution of the human eye might have involved similar organs as intermediates.

Here is a paraphrase of Darwin's argument: Although humans have complex camera-type eyes, many animals get by with less. Some tiny creatures have just a simple group of pigmented cells, or not much more than a light-sensitive spot. That simple arrangement can hardly be said to confer vision, but it can sense light and dark, and so it meets the creature's needs. The light-sensing organ of some starfishes is somewhat more sophisticated. Their eye is located in a depressed region. This allows the animal to sense which direction the light is coming from, since the curvature of the depression blocks off light from some directions. If the curvature becomes more pronounced, the directional sense of the eye improves. But more curvature lessens the amount of light that enters the eye, decreasing its sensitivity. The sensitivity can be increased by placement of gelatinous material in the cavity to act as a lens. Some modern animals have eyes with such crude lenses. Gradual improvements in the lens could then provide an image of increasing sharpness, as the requirements of the animal's environment dictated.

Using reasoning like this, Darwin convinced many of his readers that an evolutionary pathway leads from the simplest light-sensitive spot to the sophisticated camera-eye of man. But the question remains, how did vision begin? Darwin persuaded much of the world that a modern eye evolved gradually from a simpler structure, but he did not even try to explain where his starting point for the simple light-sensitive spot came from. On the contrary, Darwin dismissed the question of the eye's ultimate origin: "How a nerve comes to be sensitive to light hardly concerns us more than how life itself originated."

He had an excellent reason for declining the question: it was completely beyond nineteenth century science. How the eye works; that is, what happens when a photon of light first hits the retina, simply could not be answered at that time. As a matter of fact, no question about the underlying mechanisms of life could be answered. How did animal muscles cause movement? How did photosynthesis work? How was energy extracted from food? How did the body fight infection? No one knew.

To Darwin vision was a black box, but today, after the hard, cumulative work of many biochemists, we are approaching answers to the question of sight. Here is a brief overview of the biochemistry of vision. When light first strikes the retina, a photon interacts with a molecule called 11-*cis*-retinal, which rearranges within picoseconds to *trans*-retinal. The change in the shape of retinal forces a change in the shape of the protein, rhodopsin, to which the retinal is tightly bound. The protein's metamorphosis alters its behavior, making it stick to another protein called transducin. Before bumping into activated rhodopsin, transducin has tightly bound a small molecule called GDP. But when transducin interacts with activated rhodopsin, the GDP falls off and a molecule called GTP binds to transducin. (GTP is closely related to, but critically different from, GDP.)

GTP-transducin-activated rhodopsin now binds to a protein called phosphodiesterase, located in the inner membrane of the cell. When attached to activated rhodopsin and its entourage, the phosphodiesterase acquires the ability to chemically cut a molecule called cGMP (a chemical relative of both GDP and GTP). Initially there are a lot of cGMP molecules in the cell, but the phosphodiesterase lowers its concentration, like a pulled plug lowers the water level in a bathtub.

Another membrane protein that binds cGMP is called an ion channel. It acts as a gateway that regulates the number of sodium ions in the cell. Normally the ion channel allows sodium ions to flow into the cell, while a separate protein actively pumps them out again. The dual action of the ion channel and pump keeps the level of sodium ions in the cell within a narrow range. When the amount of cGMP is reduced because of cleavage by the phosphodiesterase, the ion channel closes, causing the cellular concentration of positively charged sodium ions to be reduced. This causes an imbalance of charge across the cell membrane which, finally, causes a current to be transmitted down the optic nerve to the brain. The result, when interpreted by the brain, is vision.

My explanation is just a sketchy overview of the biochemistry of vision. Ultimately, though, this is what it means to "explain" vision. This is the level of explanation for which biological science must aim. In order to truly understand a function, one must understand in detail every relevant step in the process. The relevant steps in biological processes occur ultimately at the molecular level, so a satisfactory explanation of a biological phenomenon such as vision, or digestion, or immunity must include its molecular explanation.

Now that the black box of vision has been opened, it is no longer enough for an "evolutionary explanation" of that power to consider only the anatomical structures of whole eyes, as Darwin did in the nineteenth century, and as popularizers of evolution continue to do today. Each of the anatomical steps and structures that Darwin thought were so simple actually involves staggeringly complicated biochemical processes that cannot be papered over with rhetoric. Darwin's simple steps are now revealed to be huge leaps between carefully tailored machines. Thus biochemistry offers a Lilliputian challenge to Darwin. Now the black box of the cell has been opened and a Lilliputian world of staggering complexity stands revealed. It must be explained.

IRREDUCIBLE COMPLEXITY

How CAN we decide if Darwin's theory can account for the complexity of molecular life? It turns out that Darwin himself set the standard. He acknowledged that "if it could be demonstrated that any complex organ existed which could not possibly have been formed by numerous, successive, slight modifications, my theory would absolutely break down."[1]

But what type of biological system could not be formed by "numerous, successive, slight modifications"? Well, for starters, a system that is irreducibly complex. Irreducible complexity is just a fancy phrase I use to mean a single system which is composed of several interacting parts, and where the removal of any one of the parts causes the system to cease functioning.

Let's consider an everyday example of irreducible complexity: the humble mousetrap. The mousetraps that my family uses consist of a number of parts. There are: 1) a flat wooden platform to act as a base; 2) a metal hammer, which does the actual job of crushing the mouse; 3) a spring with extended ends to press against the platform and the hammer when the trap is charged; 4) a sensitive catch which releases when slight pressure is applied, and 5) a metal bar which connects to the catch and holds the hammer back when the trap is charged. Now you can't catch a few mice with just a platform, add a spring and catch a few more mice, add a holding bar and catch a few more. All the pieces of the mousetrap have to be in place before you catch any mice. Therefore, the mousetrap is irreducibly complex.

An irreducibly complex system cannot be produced directly by numerous, successive, slight modifications of a precursor system, with each stage a functioning system, because any precursor to an irreducibly com-

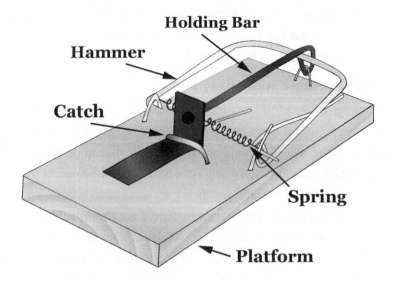

Figure 1.1. A mousetrap—irreducibly complex.

plex system that is missing a part is by definition nonfunctional. An irreducibly complex biological system, if there is such a thing, would be a powerful challenge to Darwinian evolution. Since natural selection can only choose systems that are already working, then a biological system would have to arise not gradually but as an integrated unit, in one fell swoop, for natural selection to have anything to act on.

Demonstration that a system is irreducibly complex is not a proof that there is absolutely no gradual route to its production. Although an irreducibly complex system can't be produced directly, one can't definitively rule out the possibility of an indirect, circuitous route. However, as the complexity of an interacting system increases, the likelihood of such an indirect route drops precipitously. And as the number of unexplained, irreducibly complex biological systems increases, our confidence that Darwin's criterion of failure has been met skyrockets toward the maximum that science allows.

THE CILIUM

Now, ARE any biochemical systems irreducibly complex? Yes, it turns out that many are. A good example is the cilium. Cilia are hairlike structures on the surfaces of many animal and lower plant cells that can move fluid over the cell's surface or "row" single cells through a fluid. In humans, for example, cells lining the respiratory tract each have about 200 cilia that beat in synchrony to sweep mucus towards the throat for elimination. What is the structure of a cilium? A cilium consists of a bundle of fibers called an axoneme. An axoneme contains a ring of nine double "microtubules" surrounding two central single microtubules. Each outer doublet consists of a ring of thirteen filaments (subfiber A) fused to an assembly of ten filaments (subfiber B). The filaments of the microtubules are composed of two proteins called alpha and beta tubulin. The eleven microtubules forming an axoneme are held together by three types of connectors: subfibers A are joined to the central microtubules by radial spokes; adjacent outer doublets are joined by linkers of a highly elastic protein called nexin; and the central microtubules are joined by a con-

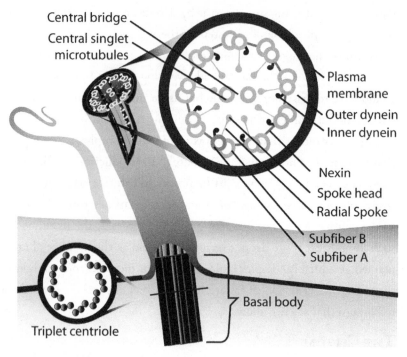

Figure 1.2. The cilium.

necting bridge. Finally, every subfiber A bears two arms, an inner arm and an outer arm, both containing a protein called dynein.

But how does a cilium work? Experiments have shown that ciliary motion results from the chemically powered "walking" of the dynein arms on one microtubule up a second microtubule so that the two microtubules slide past each other. The protein cross-links between microtubules in a cilium prevent neighboring microtubules from sliding past each other by more than a short distance. These cross-links, therefore, convert the dynein-induced sliding motion to a bending motion of the entire axoneme.

Now, let us consider what this implies. What components are needed for a cilium to work? Ciliary motion certainly requires microtubules; otherwise, there would be no strands to slide. Additionally, we require a motor, or else the microtubules of the cilium would lie stiff and motionless. Furthermore, we require linkers to tug on neighboring strands, converting the sliding motion into a bending motion, and preventing the structure from falling apart. All of these parts are required to perform one function: ciliary motion. Just as a mousetrap does not work unless all of its constituent parts are present, ciliary motion simply does not exist in the absence of microtubules, connectors, and motors. Therefore, we can conclude that the cilium is irreducibly complex, an enormous monkey wrench thrown into its presumed gradual, Darwinian evolution.

BLOOD CLOTTING

Now LET's talk about a different biochemical system, that of blood clotting. Amusingly, the way the blood clotting system works is reminiscent of a Rube Goldberg machine.

The name of Rube Goldberg, the great cartoonist who entertained America with his silly machines, lives on in our culture, but the man himself has pretty much faded from view. Here's a typical example of his humor. In one cartoon (see *Darwin's Black Box*, page 75) Goldberg depicts an elaborate "Mosquito Bite Scratcher" where water from a drainpipe fills a flask, causing a cork with attached needle to rise and puncture a paper cup containing beer, which sprinkles on a bird. The intoxicated bird falls onto a spring, bounces up to a platform, and pulls a string thinking it's a worm. The string triggers a cannon which frightens a dog. The dog flips over and lands in just the right spot such that his rapid breathing raises and lowers a scratcher over a mosquito bite on the back of a gentleman's neck, relief the man happily accepts without embarrassment as he chats with an elegantly dressed lady.

When you think about it for a moment you realize that the Rube Goldberg machine is irreducibly complex. It is a single system which is composed of several interacting parts, and where the removal of any one of the parts causes the system to break down. If the dog is missing, the machine doesn't work; if the needle hasn't been put on the cork, the whole system is useless.

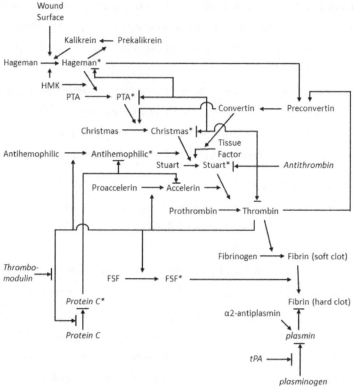

Figure 1.3. The blood clotting cascade.

It turns out that we all have Rube Goldberg in our blood. When blood clots, the cells get trapped in a meshwork structure that resembles a fisherman's net. The meshwork is formed from a protein called fibrin. But what controls blood clotting? Why does blood clot when you cut yourself, but not at other times when a clot would cause a stroke or heart attack? Take a look at the complexity of Figure 1.3, at what's called the blood clotting cascade.

Let's go through just some of the reactions of clotting. When an animal is cut, a protein called Hageman factor sticks to the surface of cells near the wound. Bound Hageman factor is then cleaved by a protein called HMK to yield activated Hageman factor. Immediately the activated Hageman factor converts another protein, called prekallikrein, to its active form, kallikrein. Kallikrein helps HMK speed up the conversion of more Hageman factor to its active form. Activated Hageman factor and HMK then together transform another protein, called PTA, to its active form. Activated PTA in turn, together with the activated form of another protein (discussed below) called convertin, switch a protein called Christmas factor to its active form. Activated Christmas factor, together with antihemophilic factor (which is itself activated by thrombin in a manner similar to that of proaccelerin) changes Stuart factor to its active form. Stuart factor, working with accelerin, converts prothrombin to thrombin. Finally thrombin cuts fibrinogen to give fibrin, which aggregates with other fibrin molecules to form the meshwork clot you saw in the last picture.

Blood clotting requires extreme precision. When a pressurized blood circulation system is punctured, a clot must form quickly or the animal will bleed to death. On the other hand, if blood congeals at the wrong time or place, then the clot may block circulation as it does in heart attacks and strokes. Furthermore, a clot has to stop bleeding all along the length of the cut, sealing it completely. Yet blood clotting must be confined to the cut or the entire blood system of the animal might solidify, killing it. Consequently, clotting requires this enormously complex system so that the clot forms only when and only where it is required. Blood clotting is the ultimate Rube Goldberg machine.[2]

THE PROFESSIONAL LITERATURE

OTHER EXAMPLES of irreducible complexity abound in the cell, including aspects of protein transport, the bacterial flagellum, electron transport, telomeres, photosynthesis, transcription regulation, and much more. Examples of irreducible complexity can be found on virtually every page of a biochemistry textbook. But if these things cannot be explained by Darwinian evolution, how has the scientific community regarded these phenomena of the past forty years? A good place to look for an answer to that question is in the *Journal of Molecular Evolution*. *JME* is a journal that was begun specifically to deal with the topic of how evolution occurs on the molecular level. It has high scientific standards, and is edited by prominent figures in the field. In the October 1994 issue of *JME*, there were published eleven articles; of these, all eleven were concerned with the comparison of protein or DNA sequences. A sequence comparison is an amino acid-by-amino acid comparison of two different proteins, or a nucleotide-by-nucleotide comparison of two different pieces of DNA, noting the positions at which they are identical or similar, and the places where they are not. Although useful for determining possible lines of descent, which is an interesting question in its own right, comparing sequences cannot show how a complex biochemical system achieved its function—the question that most concerns us here.

By way of analogy, the instruction manuals for two different models of computer put out by the same company might have many identical words, sentences, and even paragraphs, suggesting a common ancestry (perhaps the same author wrote both manuals), but comparing the sequences of letters in the instruction manuals will never tell us if a computer can be produced step by step starting from a typewriter.

None of the papers discussed detailed models for intermediates in the development of complex biomolecular structures. In the past ten years *JME* has published over a thousand papers. Of these, about one hundred discussed the chemical synthesis of molecules thought to be necessary for the origin of life, about fifty proposed mathematical models to improve sequence analysis, and about 800 were analyses of sequences. There were ZERO papers discussing detailed models for intermediates in the development of complex biomolecular structures. This is not a peculiarity of *JME*. No papers are to be found that discuss detailed models for intermediates in the development of complex biomolecular structures in the *Proceedings of the National Academy of Science, Nature, Science,* the *Journal of Molecular Biology* or, to my knowledge, any science journal whatsoever.

"Publish or perish" is a proverb that academicians take seriously. If you do not publish your work for the rest of the community to evaluate, then you have no business in academia and, if you don't already have tenure, you will be banished. But the saying can be applied to theories as well. If a theory claims to be able to explain some phenomenon but does not generate even an attempt at an explanation, then it should be banished. Despite comparing sequences, molecular evolution has never addressed the question of how complex structures came to be. In effect, the theory of Darwinian molecular evolution has not published, and so it should perish.

DETECTION OF DESIGN

WHAT'S GOING on? Imagine a room in which a body lies crushed, flat as a pancake. A dozen detectives crawl around, examining the floor with magnifying glasses for any clue to the identity of the perpetrator. In the middle of the room, next to the body, stands a large, gray elephant. The detectives carefully avoid bumping into the pachyderm's legs as they crawl, and never even glance at it. Over time the detectives get frustrated with their lack of progress but resolutely press on, looking even more closely at the floor. You see, textbooks say detectives must "get their man," so they never consider elephants.

There is an elephant in the roomful of scientists who are trying to explain the development of life. The elephant is intelligent design. To a person who does not feel obliged to restrict his search to unintelligent causes, the straightforward conclusion is that many biochemical systems were designed. They were designed not by the laws of nature, not by chance and necessity. Rather, they were planned. The designer knew what the systems would look like when they were completed; the designer took steps to bring the systems about. Life on earth at its most fundamental level, in its most critical components, is the product of intelligent activity.

The conclusion of intelligent design flows naturally from the data itself, not from sacred books or sectarian beliefs. Inferring that biochemical systems were designed by an intelligent agent is a humdrum process that requires no new principles of logic or science. It comes simply from the hard work that biochemistry has done over the past forty years, combined with consideration of the way we reach conclusions of design every day.

What is "design"? Design is simply the purposeful arrangement of parts. The scientific question is how we detect design. This can be done in various ways, but design can most easily be inferred for mechanical objects. While walking through a junkyard you might observe separated bolts and screws and bits of plastic and glass, most scattered, some piled on top of each other, some wedged together. Suppose you saw a pile that seemed particularly compact, and when you picked up a bar sticking out of the pile, the whole pile came along with it. When you pushed on the bar it slid smoothly to one side of the pile and pulled an attached chain along with it. The chain in turn yanked a gear which turned three other gears which turned a red-and-white striped rod, spinning it like a barber pole. You quickly conclude that the pile was not a chance accumulation of junk, but was designed, was put together in that order by an intelligent agent, because you see that the components of the system interact with great specificity to do something.

It is not only artificial mechanical systems for which design can easily be concluded. Systems made entirely from natural components can also evince design. For example, suppose you are walking with a friend in the woods. All of a sudden, your friend is pulled high in the air and left dangling by his foot from a vine attached to a tree branch. After cutting him down you reconstruct the trap. You see that the vine was wrapped around the tree branch, and the end pulled tightly down to the ground. It was securely anchored to the ground by a forked branch. The branch was attached to another vine, hidden by leaves so that, when the trigger-vine was disturbed, it would pull down the forked stick, releasing the spring-vine. The end of the vine formed a loop with a slipknot to grab an appendage and snap it up into the air. Even though the trap was made completely of natural materials you would quickly conclude that it was the product of intelligent design.

A Complicated World

A word of caution: intelligent design theory has to be seen in context; it does not try to explain everything. We live in a complex world where lots of different things can happen. When deciding how various rocks came to be shaped the way they are, a geologist might consider a whole range of factors: rain, wind, the movement of glaciers, the activity of moss and lichens, volcanic action, nuclear explosions, asteroid impact, or the hand of a sculptor. The shape of one rock might have been determined primarily by one mechanism, the shape of another rock by another mechanism. The possibility of a meteor's impact does not mean that volcanos can be ignored; the existence of sculptors does not mean that many rocks are not shaped by weather.

Similarly, evolutionary biologists have recognized that a number of factors might have affected the development of life: common descent, natural selection, migration, population size, founder effects (effects that may be due to the limited number of organisms that begin a new species), genetic drift (spread of neutral, nonselective mutations), gene flow (the incorporation of genes into a population from a separate population), linkage (occurrence of two genes on the same chromosome), meiotic drive (the preferential selection during sex cell production of one of the two copies of a gene inherited from an organism's parents), transposition (the transfer of a gene between widely separated species by nonsexual means), and much more. The fact that some biochemical systems were designed by an intelligent agent does not mean that any of the other factors are not operative, common, or important.

Curiouser and Curiouser

So as this talk concludes we are left with what many people feel to be a strange conclusion: life was designed by an intelligent agent. In a way, though, all of the progress of science over the last several hundred years has been a steady march toward the strange. People up until the middle ages lived in a natural world. The stable earth was at the center of things; the sun, moon, and stars circled endlessly to give light by day and night;

the same plants and animals had been known since antiquity. Surprises were few.

Then it was proposed, absurdly, that the earth itself moved, spinning while it circled the sun. No one could feel the earth spinning; no one could see it. But spin it did. From our modern vantage it's hard to realize what an assault on the senses was perpetrated by Copernicus and Galileo; they said in effect that people could no longer rely on even the evidence of their eyes.

Things got steadily worse over the years. With the discovery of fossils it became apparent that the familiar animals of field and forest had not always been on earth; the world had once been inhabited by huge, alien creatures who were now gone. Sometime later Darwin shook the world by arguing that the familiar biota was derived from the bizarre, vanished life over lengths of time incomprehensible to human minds. Einstein told us that space is curved and time is relative. Modern physics says that solid objects are mostly space, that subatomic particles have no definite position, that the universe had a beginning.

Now it's the turn of the fundamental science of life, modern biochemistry, to disturb. The simplicity that was once expected to be at the foundation of life has proven to be a phantom. Instead, systems of astounding, irreducible complexity inhabit the cell. The resulting realization that life was designed by an intelligence is a shock to us in the twentieth century who have gotten used to thinking of life as the result of simple natural laws. But other centuries have had their shocks and there is no reason to suppose that we should escape them. Humanity has endured as the center of the heavens moved from the earth to beyond the sun, as the history of life expanded to encompass long-dead reptiles, as the eternal universe proved mortal. We will endure the opening of Darwin's black box.

2. BLIND EVOLUTION OR INTELLIGENT DESIGN?

The following is a talk I presented at a debate sponsored by the American Museum of Natural History, and moderated by the then-director of the National Center for Science Education, Eugenie Scott. Bill Dembski and I represented the intelligent design side, and Kenneth Miller and Robert Pennock represented the Darwinian evolution side. The structure of the debate was peculiar—Dembski and I gave presentations, and then were questioned by Pennock and Miller. They didn't get to give talks, and we didn't get to ask questions. The rationale was that intelligent design was the topic under scrutiny, not Darwinism. Despite the odd format I thought the evening went very well indeed.[3]

"Blind Evolution or Intelligent Design?," debate, American
Museum of Natural History, New York, April 2002.

It's great to be back in New York City. I taught at Queens College and City University for three years in the early 1980s; my wife grew up on Cambreleng Avenue near 187th Street in the Bronx; and our first child was born here, so New York holds many happy memories for our family.

My talk will be divided into four parts: first, a sketch of the argument for design; second, common misconceptions about the mode of design; third, misconceptions about biochemical design; and finally, discussion of the future prospects of design. Before I begin, however, I'd like to emphasize that the focus of my argument will not be descent with modification, with which I agree. Rather, the focus will be the mechanism of evolution—how did all this happen, by natural selection or intelligent design? My conclusion will not be that natural selection doesn't explain anything; rather, the conclusion will be that natural selection doesn't explain everything.

So, let's begin with a sketch of the design argument. In *The Origin of Species*, Darwin emphasized that his was a very gradual theory; natural selection had to work by "numerous, successive, slight modifications" to pre-existing structures. However, "irreducibly complex" systems seem quite difficult to explain in gradual terms. What is irreducible complexity? I've defined the term in various places, but it's easier to illustrate what I mean with the following example: the common mousetrap. A common mechanical mousetrap has a number of interacting parts that all contribute to its function, and if any parts are taken away, the mousetrap doesn't work half as well as it used to, or a quarter as well—the mousetrap is broken. Thus it is irreducibly complex.

Suppose we wanted to evolve a mousetrap by something like a Darwinian process. What would we start with? Would we start with a wooden platform and hope to catch mice inefficiently? Perhaps tripping them? And then add, say, the holding bar, hoping to improve efficiency? No, of course not, because irreducibly complex systems only acquire their function when the system is essentially completed. Thus irreducibly complex systems are real headaches for natural selection because it is very difficult to envision how they could be put together—that is, without the help of a directing intelligence—by the "numerous, successive,

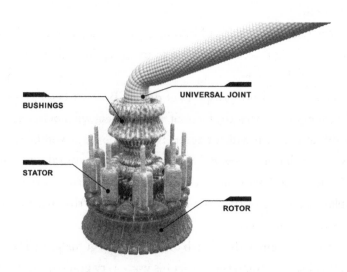

Figure 1.4. The bacterial flagellum.

slight modifications" that Darwin insisted upon. Irreducibly complex biological systems would thus be real challenges to Darwinian evolution.

Yet modern science has discovered irreducibly complex systems in the cell. An excellent example is the bacterial flagellum (Figure 1.4), which is literally an outboard motor that bacteria use to swim. The flagellum has a large number of parts that are necessary for its function—a propeller, hook, drive shaft, and more. Thorough studies show it requires 30 to 40 protein parts. And in the absence of virtually any of those parts, the flagellum doesn't work, or doesn't even get built in the cell. Its gradual evolution by unguided natural selection therefore is a real headache for Darwinian theory. I like to show audiences a picture of the flagellum from a biochemistry textbook because, when they see it, they quickly grasp that *this is a machine*. It is not *like* a machine, it is a *real* molecular machine. Perhaps that will help us think about its origin.

I have written that the flagellum is not only a problem for Darwinism, but is better explained as the result of design—deliberate design by an intelligent agent. Some of my critics have said that design is a religious conclusion, but I disagree. I think it is wholly empirical, that is, the conclusion of design is based on the physical evidence along with an appreciation for how we come to a conclusion of design. To illustrate how we come to a conclusion of design, consider a *Far Side* cartoon by Gary Larson showing a troop of jungle explorers walking in single file, and the lead explorer has just a moment before been strung up and skewered. The third fellow in line comments to the fourth, "That's why I never walk in front." Now, everyone who looks at this cartoon immediately realizes that the trap that skewered the lead explorer was designed. But how do you know that? That is, if you encountered this situation in the real world, how is it that you really would know the trap was designed? Is it a religious conclusion? Probably not. You know it's designed because you see a number of very specific parts acting together to perform a function; you see something like irreducible complexity or specified complexity.

Now I will address common misconceptions about the mode of design, that is, how design may have happened.

My book, *Darwin's Black Box*, in which I flesh out the design argument, has been widely discussed in many publications. What have other scientists said about it? Well, they've said many things—not all flattering—but the general reaction is well summarized in a recent book, *The Way of the Cell*, published last year by Oxford University Press, and authored by Colorado State University biochemist Franklin Harold, who writes, "We should reject, as a matter of principle, the substitution of intelligent design for the dialogue of chance and necessity (Behe 1996); but we must concede that there are presently no detailed Darwinian accounts of the evolution of any biochemical system, only a variety of wishful speculations."[4]

Let me take a moment to emphasize Harold's two points. First, he acknowledges that Darwinists have no real explanations for the enormous complexity of the cell, only hand-waving speculations, more colloquially known as "just-so stories"—how the rhinoceros got its horn; how the bacterium got its flagellum. I find this an astonishing admission for a theory that has dominated biology for so long. Second, apparently he thinks that there is some principle that forbids us from investigating the idea of intelligent design, even though design is an obvious idea that quickly pops into your mind when you see a drawing of the flagellum or other complex biochemical systems. But what principle is that?

I think the principle boils down to this: Design appears to point strongly beyond nature. It has philosophical and theological implications, and that makes many people uncomfortable.

But any theory that purports to explain how life occurred will have philosophical and theological implications. For example, the Oxford biologist Richard Dawkins has famously said that "Darwin made it possible to be an intellectually fulfilled atheist." Ken Miller has written that God "used evolution as the tool to set us free." Stuart Kauffman, a leading complexity theorist, thinks Darwinism cannot explain all of biology, and thinks that Kauffman's own theory will somehow show that we are "at home in the universe." So all theories of origins carry philosophical and theological implications.

But how could biochemical systems have been designed? Did they have to be created from scratch in a puff of smoke? No. The design process may have been much more subtle. It may have involved no contravening of natural laws. Let's consider just one possibility. Suppose the designer is God, as most people would suspect. Well, then, as Ken Miller points out in his book, *Finding Darwin's God*, a subtle God could cause mutations by influencing quantum events such as radioactive decay, something that I would call guided evolution. That seems perfectly possible to me. I would only add, however, that that process would amount to intelligent design, not Darwinian evolution.

Now let's look at common misconceptions about biochemical design.

Some Darwinists have proposed that a way around the problem of irreducible complexity could be found if the individual components of a system first had other functions in the cell. For example, consider a hypothetical example such as pictured here, where all of the parts are supposed to be necessary for the function of the system. Might the system have been put together from individual components that originally worked on their own? Unfortunately this picture greatly oversimplifies the difficulty, as I discussed in *Darwin's Black Box*.

Here analogies to mousetraps break down somewhat, because the parts of the system have to automatically find each other in the cell. They can't be arranged by an intelligent agent, as a mousetrap is. To find each other in the cell, interacting parts have to have their surfaces shaped so that they are very closely matched to each other. Originally, however, the individually acting components would not have had complementary surfaces. So all of the interacting surfaces of all of the components would first have to be adjusted before they could function together, and only then would the new function of the composite system appear. Thus the problem of irreducibility remains, even if individual components separately have their own functions.

Another area where one has to be careful is in noticing that some systems with extra or redundant components may have an irreducibly complex core. For example, a car with four spark plugs might get by with three or two, but it certainly can't get by with none. Rat traps often have two springs, to give them extra strength. They can still work if one spring is removed, but they can't work if both springs are removed. Thus in trying to imagine the origin of a rat trap by Darwinian means, we still have all the problems we had with a mousetrap.

A cellular example of redundancy is the hugely complex eukaryotic cilium, which has multiple copies of a number of components, yet needs at least one copy of each to work, as I pictured in *Darwin's Black Box*.

Many other criticisms have been made against intelligent design. I have responded to a number of them in book chapters and journal articles.

I will now discuss how I view the future prospects of a theory of intelligent design. I see them as very bright indeed. Why? Because the idea of intelligent design has advanced, not primarily because of anything I or any individual has done. Rather, it's been the very progress of science itself that has made intelligent design plausible. Fifty years ago much less was known about the cell, and it was much easier then to think that Darwinian evolution was true. But with the discovery of more and more complexity at the foundation of life, the idea of intelligent design has gained strength.

That trend is continuing. As science pushes on, our sense of the complexity of the cell is not getting any less; on the contrary, it is getting much greater. For example, a recent issue of the journal *Nature* carried the most detailed analysis yet of the total protein complement of yeast—the so-called yeast "proteome." The authors point out that most proteins they investigated in the cell function as multiprotein complexes—not as solitary proteins as scientists had long thought. In fact they showed that almost fifty percent of the proteins in the cell function as complexes of a half dozen or more.[5] To me, this implies that irreducible molecular machinery is very likely going to be the rule in the cell, not the exception. We will probably not have to wait too long to see.

Another example comes from a paper published in the *Journal of Molecular Biology* two years ago, which showed that some enzymes have only a limited ability to undergo multiple changes in their amino acid sequence, even when the enzymes function alone, as single proteins, and even when the changes are very conservative ones. This led the author to caution that "homologues sharing less than about two-thirds sequence identity should probably be viewed as distinctive designs with their own optimizing features."[6] The author pictured such proteins as near-islands of function, virtually isolated from neighboring protein sequences. This may mean that even individual proteins from separate species that are similar but not identical in their amino acid sequence might not have been produced by Darwinian processes, as most scientists thought and as even I was willing to concede. Perhaps even I give too much unearned credit to Darwinian theory.

Finally, to show what research questions might be asked by a theory of intelligent design, I'd like to briefly describe some of my own recent work. The title slide of a seminar I gave six weeks ago to the biotechnology group at Sandia National Laboratory, "Modeling the Evolution of Protein Binding Sites: Probing the Dividing Line between Natural Selection and Intelligent Design," points to a question I'm very interested in exploring. If you are someone like myself who thinks that some things in biology are indeed purposely designed, but that not all things are designed, then a question which quickly arises is, where is the broad dividing line between design and unintelligent processes? I think that question has to be answered at the molecular level, particularly in terms of protein structure.

Drawings of the bacterial flagellum picture proteins as bland spheres or ovals, but each protein in the cell is actually itself very complex. Proteins are polymers of amino acid residues, and some structural features of proteins require the participation of multiple residues. For example, a disulfide bond requires two cysteine residues—just one cysteine residue can't form such a bond. Thus, in order for a protein that did not have a disulfide bond to evolve one, several changes in the same gene first have to occur. Thus in a real sense the disulfide bond is irreducibly complex, although not nearly to the same degree of complexity as systems made of multiple proteins.[7]

The problem of irreducibility in protein features is a general one. Whenever a protein interacts with another molecule, as all proteins do, it does so through a binding site, whose shape and chemical properties closely match the other molecule. Binding sites, however, are composed of perhaps a dozen amino acid residues, and binding is generally lost if any of the positions are changed. One can then ask the question, how long would it take for two proteins that originally did not interact to evolve the ability to bind each other by random mutation and natural selection, if binding only occurs when all positions have the correct residue in place?

Although it would be difficult to experimentally investigate this question, the process can be simulated on a computer, which I did and described at a 2001 meeting of the Protein Society in Philadelphia. The log of the expected time to generate what I call "irreducibly complex" protein features was shown as a function of the log of the population size and the log of the probability of the feature. In the presentation a yellow dot represented the time expected to generate a new disulfide bond in a protein that did not have one if the population size is a hundred million organisms. The expected time, I showed, is roughly a million generations. A red dot showed that the expected time needed to generate a new protein binding site would be a hundred million generations.

Using data from these simulations as well as Bill Dembski's concept of probabilistic resources, we can come to several broad, tentative conclusions: 1) that undirected irreducibly complex mutations cannot have been regularly involved in the evolution of large animals—the time frame required for this is too long; and 2) that undirected IC systems of the complexity of two or more protein binding sites cannot have been regularly involved in the evolution of vertebrates. This work assumed that all mutations were neutral. Future work could investigate such questions as, what if intermediate mutations are selected against? And what happens if there is competition between IC mutations and single-site mutations?

The broad motivation behind this work is to start getting some good numbers to plug into Bill Dembski's explanatory filter, to try to come to a reasoned conclusion about where in nature design leaves off.

In summary, I want to leave you with four take-home points: 1) the question is open: no other scientific theory has yet explained the data; 2) intelligent design is an empirical hypothesis that flows easily from the data, as you can tell by looking at a drawing of the flagellum; 3) there is no non-question-begging principle that forbids our considering design; and best of all, 4) there are exciting research questions that can be asked within a design framework.

3. DARWIN UNDER THE MICROSCOPE

After a meeting in 1996 of the Pontifical Academy of Sciences on evolution, Pope St. John Paul II wrote the usual letter of thanks. Although very nuanced (as papal statements usually are), the letter's statement that evolution is "more than a theory," grabbed newspaper headlines. This was just months after *Darwin's Black Box* had been published. Gail Collins, later the editorial page editor of the *New York Times*, had seen my book and thought an op-ed by me would be timely. I can still remember the reaction when I was late to a Lehigh faculty meeting and apologized, explaining I had been on the phone with the *New York Times*.

"Darwin Under the Microscope," *New York Times*, October 29, 1996.

POPE JOHN PAUL II'S STATEMENT LAST WEEK THAT EVOLUTION IS "more than just a theory" is old news to a Roman Catholic scientist like myself.

I grew up in a Catholic family and have always believed in God. But beginning in parochial school I was taught that He could use natural processes to produce life. Contrary to conventional wisdom, religion has made room for science for a long time. But as biology uncovers startling complexity in life, the question becomes, can science make room for religion?

In his statement, the Pope was careful to point out that it is better to talk about "theories of evolution" rather than a single theory. The distinction is crucial. Indeed, until I completed my doctoral studies in biochemistry, I believed that Darwin's mechanism—random mutation paired with natural selection—was the correct explanation for the diversity of life. Yet I now find that theory incomplete.

In fact, the complex design of the cell has provoked me to stake out a distinctly minority view among scientists on the question of what caused evolution. I believe that Darwin's mechanism for evolution doesn't explain much of what is seen under a microscope. Cells are simply too complex to have evolved randomly; intelligence was required to produce them.

I want to be explicit about what I am, and am not, questioning. The word "evolution" carries many associations. Usually it means common descent—the idea that all organisms living and dead are related by common ancestry. I have no quarrel with the idea of common descent, and continue to think it explains similarities among species. By itself, however, common descent doesn't explain the vast differences among species.

That's where Darwin's mechanism comes in. "Evolution" also sometimes implies that random mutation and natural selection powered the changes in life. The idea is that just by chance an animal was born that was slightly faster or stronger than its siblings. Its descendants inherited the change and eventually won the contest of survival over the descendants of other members of the species. Over time, repetition of the process resulted in great changes—and, indeed, wholly different animals. That's the theory. A practical difficulty, however, is that one can't test the theory from fossils. To really test the theory, one has to observe contemporary change in the wild or the laboratory, or at least reconstruct a detailed pathway that might have led to a certain adaptation.

Darwinian theory successfully accounts for a variety of modern changes. Scientists have shown that the average beak size of Galapagos finches changed in response to altered weather patterns. Likewise, the ratio of dark- to light-colored moths in England shifted when pollution made light-colored moths more visible to predators. Mutant bacteria survive when they become resistant to antibiotics. These are all clear examples of natural selection in action.

But these examples involve only one or a few mutations, and the mutant organism is not much different from its ancestor. Yet to account for all of life, a series of mutations would have to produce very different types of creatures. That has not yet been demonstrated.

Darwin's theory encounters its greatest difficulties when it comes to explaining the development of the cell. Many cellular systems are what I term "irreducibly complex." That means the system needs several components before it can work properly. An everyday example of irreducible complexity is a mousetrap, built of several pieces (platform, hammer, spring, and so on). Such a system probably cannot be put together in a Darwinian manner, gradually improving its function. You can't catch a mouse with just the platform and then catch a few more by adding the spring. All the pieces have to be in place before you catch any mice.

An example of an irreducibly complex cellular system is the bacterial flagellum: a rotary propeller, powered by a flow of acid, that bacteria use to swim. The flagellum requires a number of parts before it works—a rotor, stator, and motor.

Furthermore, genetic studies have shown that about forty different kinds of proteins are needed to produce a working flagellum.

The intracellular transport system is also quite complex. Plant and animal cells are divided into many discrete compartments; supplies, including enzymes and proteins, have to be shipped between these compartments. Some supplies are packaged into molecular trucks, and each truck has a key that will fit only the lock of its particular cellular destination. Other proteins act as loading docks, opening the truck and letting the contents into the destination compartment.

Many other examples could be cited. The bottom line is that the cell—the very basis of life—is staggeringly complex. But doesn't science

already have answers, or partial answers, for how these systems originated? No. As James Shapiro, a biochemist at the University of Chicago,
wrote, "There are no detailed Darwinian accounts for the evolution of
any fundamental biochemical or cellular system, only a variety of wishful speculations."[8] A few scientists have suggested non-Darwinian theories to account for the cell, but I don't find them persuasive. Instead, I
think that the complex systems were designed—purposely arranged by
an intelligent agent.

Whenever we see interactive systems (such as a mousetrap) in the
everyday world, we assume that they are the products of intelligent activity. We should extend the reasoning to cellular systems. We know of no
other mechanism, including Darwin's, which produces such complexity.
Only intelligence does.

Of course, I could be proved wrong. If someone demonstrated that,
say, a type of bacterium without a flagellum could gradually produce
such a system, or produce any new, comparably complex structure, my
idea would be neatly disproved. But I don't expect that to happen.

Intelligent design may mean that the ultimate explanation for life is
beyond scientific explanation. That assessment is premature. But even if
it is true, I would not be troubled. I don't want the best scientific explanation for the origins of life; I want the correct explanation.

Pope John Paul II spoke of "theories of evolution." Right now it
looks as if one of those theories involves intelligent design.

4. THE STERILITY OF DARWINISM

Darwin's Black Box provoked a lot more reaction than I had
anticipated. (Before it came out, I was just hoping that at least a
few people in the field would read it.) But it eventually gathered
over a hundred reviews in various publications. One of the most
interesting was in a magazine called *Boston Review*, published by
MIT. The initial review was written by a prominent evolutionary
biologist named H. Allen Orr, who didn't care much for it.[9] The
magazine then invited a number of academics to comment on

the review, and was overwhelmed with responses. Participants included not only myself, but also University of Chicago evolutionary biologist Jerry Coyne, UC San Diego biochemist Russell Doolittle, SUNY Stonybrook evolutionary biologist Douglas Futuyma, University of Guelph philosopher Michael Ruse, and University of Chicago microbiologist James Shapiro, as well as one friendly voice, UC Berkeley law professor Phillip Johnson. Richard Dawkins responded in the same section to a review of his own recently published book, *Climbing Mount Improbable*, but would not deign to comment on mine.

"The Sterility of Darwinism: A Reply to H. Allen Orr's Review of *Darwin's Black Box,*" *Boston Review*, February/March 1997.

A S IT STRUGGLES TO COMPREHEND NATURE, SCIENCE SOMETIMES has to completely rethink how the world works. For example, Newton's laws apply to everyday objects, but they can't handle nature's tiny building blocks. Propelled by this discovery, quantum mechanics overthrew Newton's theory. Revolutions in biology have included the cell theory of life in the nineteenth century, as well as the slow realization in this century that cells are composites of enormously complex molecular systems.

Newton's theory remains very useful, and we can still learn many things by studying whole animals or cells. When explaining the nuts and bolts of the world, however, those views must yield to more basic descriptions. A mechanical engineer can't contradict a physicist on fundamental principles of matter. And evolutionary biology can't overrule biochemistry on fundamental principles of life.[10] It's not a question of pride—that's just the way the world works.

Curiously, some people seem offended by the way the world works. In his review of my book, *Darwin's Black Box: The Biochemical Challenge to Evolution*, evolutionary biologist H. Allen Orr unexpectedly attempts to claim priority for his field. He grouses that pre-med students are required to take biochemistry but not evolutionary biology. He plaintively asks, "Why is everyone an expert witness when the topic is Darwinism but not when it's biochemistry?" The obvious reply is that the evolution of biochemical systems *is itself biochemistry.* When a protein sequence changes, when DNA mutates, those are *biochemical* changes. Since inherited changes are caused by molecular changes, it is biochemists—not evolutionary biologists—who will ultimately decide whether Darwin's mechanism of natural selection can explain life. No offense—that's just the way the world works.

Orr hankers for the respect accorded physicists, and thinks evolutionary biologists can finally lay aside their "physics envy" because "we biologists have discovered the structure of DNA, broken the genetic code, sequenced the entire genome of some species." Orr is like a podiatrist claiming credit for progress in brain surgery. *Biochemistry* made those dramatic advances; evolutionary biology played no part. I mean no disrespect, but this is not a minor academic turf war—the point is crucial. Anyone who wants to address questions about life's basic mechanisms has to do so from a molecular perspective. Orr does not.

Declining the opportunity to address my biochemical arguments, Orr questions the concept of irreducible complexity on *logical* grounds. He agrees with me that "You cannot... gradually improve a mousetrap by adding one part and then the next. A trap having half its parts doesn't function half as well as a real trap; it doesn't function at all." So Orr understands the point of my mousetrap analogy—but then mysteriously forgets it. He later writes, "Some part (A) initially does some job (and not very well, perhaps). Another part (B) later gets added, because it helps A." *Some* part initially does *some* job? Which part of the mousetrap is he talking about? A mouse has nothing to fear from a "trap" that consists of just an unattached holding bar, or spring, or platform, with no other parts.

I do sympathize with Professor Orr's muddling of the analysis. The concept of irreducible complexity is new, and can be difficult to grasp for people who have always assumed without demonstration that small, continuous changes could produce virtually any biological structure. Perhaps in the future that assumption will not have such a strong hold on the minds of evolutionary biologists.

Having completed his logical analysis, Orr turns to the topic of gene duplication: "So how does Behe explain duplicate genes? He doesn't." But I do. I discuss them on pages 89–90 of my book, concluding, "The sequence similarities are there for all to see.... By itself, however, the hypothesis of gene duplication... says nothing about how any particular protein or protein system was first produced." For example, the DNA in each of the antibody-producing cells of your body is very similar to that of the others, but not identical. The similarities are due to common descent; that is, all the cells in your body descended from one fertilized egg cell. The differences, however, are not due to Darwinian natural selection. Rather, there is a very clever, built-in program to rearrange antibody genes. Billions of different kinds of antibody genes are "intentionally" produced by your body from a pre-existing stock of just a few hundred gene pieces. Perhaps because of his unfamiliarity with molecular systems, Orr has trouble seeing that similarity in gene sequences may indicate common ancestry, but is not itself evidence that a system was constructed by natural selection.

To test natural selection requires much more evidence than mere sequence similarity: it requires *experimentation*. In all of the scientific literature, however, no experimental evidence can be found that natural selection can produce irreducibly complex biochemical systems. To rebut my arguments Orr could simply have cited papers in the science literature where the systems I discuss have been explained. He didn't do that because explanations are nowhere to be found.

What has biochemistry found that must be explained? Machines— literally, machines made of molecules. Let's look at just one example. The flagellum is an outboard motor that many bacteria use to swim. It consists of a rotary propeller, motor, and stationary framework. Yet this short description can't do justice to the machine's full complexity. Writing of the flagellum in *Cell*, Lucy Shapiro of Stanford University marvels, "To carry out the feat of coordinating the ordered expression of about 50 genes, delivering the protein products of these genes to the construction site, and moving the correct parts to the upper floors while adhering to the design specification with a high degree of accuracy, the cell requires impressive organizational skills."[11] Without any one of a number of parts, the flagellum does not merely work less efficiently; it does not work at all. Like a mousetrap, it is irreducibly complex and therefore cannot have arisen gradually.

The rotary nature of the flagellum has been recognized for about twenty-five years. During that time not a single paper has been published in the biochemical literature even attempting to show how such a machine might have developed by natural selection. Darwin's theory is completely barren when it comes to explaining the origin of the flagellum or any other complex biochemical system.

The sterility of Darwinism indicates that it is the wrong framework for understanding the basis of life. As I argue in my book, an alternative hypothesis is both natural and obvious: systems such as the flagellum were intentionally designed by an intelligent agent. Just as in the everyday world we immediately conclude design when we see a complex, interactive system such as a mousetrap, there is no reason to withhold the same conclusion from interactive molecular systems. This conclusion may have theological implications that make some people uncomfortable; nonetheless it is the job of science to follow the data wherever it leads, no matter how disturbing.

One last charge must be met: Professor Orr maintains that the theory of intelligent design is not falsifiable. He's wrong. To falsify design theory a scientist need only experimentally demonstrate that a bacterial flagellum, or any other comparably complex system, could arise by natural selection. If that happened I would conclude that neither flagella nor any system of similar or lesser complexity had to have been designed. In short, biochemical design would be neatly disproved.

Let's turn the tables on Orr. Is natural selection falsifiable? He writes, "We have no guarantee that we can reconstruct the history of a biochemical pathway. But even if we can't, its irreducible complexity cannot count against its gradual evolution." This is a dangerously anti-scientific attitude. In effect he is saying, "I just know that phenomenally complex biochemical systems arose gradually by natural selection, but don't ask me how." With such an outlook, Orr runs the risk of clinging to ideas that are forever insulated from contact with the outside world.

BERWICK ON DAWKINS

AFTER READING Robert C. Berwick's criticism of *Climbing Mount Improbable*, I find myself in the odd position of sympathizing with Richard Dawkins. Although his book is a juicy target for debunking, Berwick chides Dawkins for all the wrong reasons. Berwick points out that natural selection is sometimes not a complete explanation for some biological feature. For example, he writes that polioviruses have shapes like geodesic domes not because selection made them that way, but because the symmetrical shape is required by physical law. Well, fine. But there are many tasks a virus faces that are not explained at all by simple physical laws: the virus has to attach to a cell surface, inject its genetic material into the cell, hijack the cell's machinery, make copies of the poliovirus DNA, and re-package the genetic material. In response the body's immune system launches a counterattack to ferret out and destroy the virus. *None* of these processes is explained by simple physical constraints. Berwick seems mesmerized by the simple crystal that covers nature's watch, and ignores the complex ticking gears of the mechanism within. Dawkins's writing should be roundly criticized for failing to answer the question he has set himself: what is the origin of biological complexity? But, to his credit, Dawkins at least knows the important question. Berwick doesn't.

5. BOSTON REVIEW ON THE WEB

Discussion from the many academic commenters about Orr's review of *Darwin's Black Box* threatened to overwhelm the physical pages of *Boston Review*. So the magazine moved discussion to a then-new venue—its website. My comments below were posted there.[12] None of the other print participants bothered.

Continued web discussion, *Boston Review*, 1997.

M Y THANKS TO THE *BOSTON REVIEW* FOR PUBLISHING MY REPLY
to Allen Orr's review of my book. I would like to address the main
points of several critics in the symposium who, I believe, have mistaken
notions of what I am arguing.

ALLEN ORR

PROFESSOR ORR has a mistaken notion of irreducible complexity. I
thought I made that clear in my reply, but from his response I suppose
I did not, so let me try again. I define irreducible complexity in *Darwin's
Black Box* as "a single system composed of several well-matched, interacting parts that contribute to the basic function, wherein the removal of
any one of the parts causes the system to effectively cease functioning."[13]
Orr, however, uses the term loosely to mean something like "if you remove a part, the organism will die." In his review he talks about lungs,
saying "we grew thoroughly terrestrial and lungs, consequently, are no
longer luxuries—they are essential." The problem is, if you quickly dis-
sect lungs from an animal, many parts of it *will continue to work*. The liver
will work for a while, muscles will twitch, cells will metabolize until they
run out of oxygen. Thus lungs are not absolutely required for the function of those other parts, not in the way that a spring is absolutely required for a standard mousetrap or nexin linkers are required for ciliary
function. That's the problem with using poorly chosen examples, especially at the whole-organ level. I am careful in my book[14] to say that you
must look at molecular systems to see if Darwinism can explain their development. When you look at irreducibly complex molecular examples,
it is clear that Darwinism has not and, I believe, cannot explain them.
Orr's main line of argument, therefore, simply misses the point.

I should also point out that, contrary to Professor Orr's assertion,
we do not know that swim bladders evolved into lungs *by natural selection*. There is absolutely no evidence for it. It may be likely that lungs are
descended from swim bladders, but no experiment has indicated that
natural selection can do the trick. In fact, no one even knows at the nuts-
and-bolts molecular level what it would take. Orr simply assumes it is

possible because he is not bothering with the myriad molecular difficulties that would face such a transformation.

Orr says that the parts of a mousetrap might have started out as something else, and then were changed into their current parts. I address this type of argument in *Darwin's Black Box*.[15] Essentially this approach doesn't help. The parts still have to be adjusted to each other at some stage, and they still don't work until all the parts have been so adjusted. That requires intelligent activity.

Orr says we know mousetraps are designed because we have seen them being designed by humans, but we have not seen irreducibly complex biochemical systems being designed, so we can't conclude they were. I discuss this in the book.[16] We apprehend design from the system itself, even if we don't know who the designer is. For example, the SETI project (Search for Extraterrestrial Intelligence) scans space for radio waves that might have been sent by aliens. However, we have never observed aliens sending radio messages; we have never observed aliens at all. Nonetheless, SETI workers are confident, and I agree, that they can detect intelligently produced phenomena, even if they don't know who produced them. Orr's criterion is also subject to a *reductio ad absurdum*. Suppose we flew to an alien planet and observed a deserted city. Orr's position would hold that we can't conclude the city was designed, because we have never seen aliens producing cities, and he would oblige us to search for an unintelligent cause for the manifestly designed city.

Orr finds it "extremely curious" that I think some systems could evolve by natural selection, but that others couldn't. I discuss this in the book.[17] Simply put, some systems are more complex than others, irreducibly so. If one biochemical system looks pretty much like the other to Orr, then he isn't going to see any problems. However, if you attend to the details of each system, as I tried to do, difficulties for Darwinism loom at many places.

Phillip Johnson can fight his own battles, but I'd just like to say I think it odd that Orr jumps on Johnson for an understandable confusion of terms. Orr seems to think that the essence of explanation is in knowing the labels that evolutionists have put on concepts, rather than on whether the concepts actually explain how life got here. It is especially odd when Orr gives no indication of understanding much about the molecular basis of life, where all the inheritable action necessarily takes place.

JERRY A. COYNE

PROFESSOR COYNE seems really to have been traumatized by being quoted in my book. He should relax. My purpose in quoting him and others was to show that many thoughtful biologists found Darwinism to be an incomplete theory of life. I did not say that Coyne or the others agreed with intelligent design. Indeed, for several of the people I quoted (e.g., Stuart Kauffman and Lynn Margulis) I specifically discuss their alternative theories to Darwinism. I start off the section by saying, "A raft of evolutionary biologists examining whole organisms wonder just how Darwinism can account for their observations." After a few other people, I quote Coyne as saying, "We conclude—unexpectedly—that there is little evidence for the neo-Darwinian view: its theoretical foundations and the experimental evidence supporting it are weak."[18] In Coyne's paper, the sentence did not stop there; it continued with "and there is no doubt that mutations of large effect are sometimes important in adaptation." I do not see, however, where that changes the sense of the sentence at all. In my manuscript I had his quote ending with an ellipsis, but the copy editor took out all ellipses in this section and put in periods, so I assume that it is in keeping with standard editorial practices. It is extremely difficult for me to understand why Coyne thinks his idea is anything other than a doubt about the efficacy of Darwinism, or what context could possibly change its plain meaning. Coyne goes on to quote the entire paragraph in which the sentence appeared, but that changes nothing of the basic thrust as far as I can see.

Coyne says my book bears the four marks of "crank science," which I will address in turn:

1) Coyne says I did not present my views "directly to the scientific community." Free Press sent my book out to a number of scientists for their review. One angrily told Free Press not to publish because he viewed intelligent design theory as "giving up." Three said they thought the book meritorious and worthy of publication, although they did not agree with my conclusion of intelligent design. One scientist thought the book worthy of publication, and thought that intelligent design was possibly true. Additionally, my book was put up for competitive bidding, and several university presses were interested. They were outbid by Free Press.

2) Coyne complains that if one biochemical pathway is explained by natural selection, intelligent design advocates can just move on to another, and so ID is not falsifiable. This complaint would have some merit if Darwinists had explained *any* complex biochemical system. I can't speak for others, but for myself if I were convinced that natural selection could explain a system of a certain degree of complexity, then I would assume it could explain other systems of a similar or lesser degree of complexity. However, to date it has not been able to explain the origins of functional systems of much complexity at all.

3) Coyne says Behe "likens himself to Newton, Einstein, and Pasteur." I do not. I clearly acknowledge that the credit belongs to the scientific community as a whole, whose cumulative work makes design apparent. Here are some relevant sentences:

> The result of these cumulative efforts to investigate the cell—to investigate life at the molecular level—is a loud, clear, piercing cry of "*design!*" The result is so unambiguous and so significant that it must be ranked as one of the greatest achievements in the history of science. The discovery rivals those of Newton and Einstein, Lavoisier and Schrödinger, Pasteur and Darwin.... The magnitude of the victory, gained at such great cost through sustained effort over the course of decades, would be expected to send champagne corks flying in labs around the world. This triumph of science should evoke cries of "Eureka" from ten thousand throats, should occasion much hand-slapping and high-fiving, and

perhaps even be an excuse to take a day off.... Why does the scientific community not greedily embrace its startling discovery?[19]

It does not take a rocket scientist to see design; the hard work was in the day-in, day-out elucidation of the molecular workings of the cell. For someone who is as touchy as he is to possible misinterpretation, Coyne seems not to mind putting a strained interpretation on other people's writing.

4) Coyne complains the book is "heavily larded" with quotations from evolutionists. This leads into his being upset with being quoted himself, as discussed above. That aside, however, I don't know what to make of this statement. What is a book concerning evolution supposed to contain if not quotes from evolutionists? Quotes from accountants?

Russell F. Doolittle

Professor Doolittle is a prominent scientist, a member of the National Academy of Sciences who has worked hard on many aspects of protein structure over the course of a distinguished career. He knows more about the process of blood clotting, and more about the relationships among the protein members of the clotting cascade, than perhaps anyone else on earth. He does not, however, know how natural selection could have produced the clotting cascade. In fact, he has never tried to explain how it could have. Nonetheless, as reflected in his comments in *Boston Review*, he clearly thinks he has addressed the question. This results from a basic confusion, which I will try to clarify.

As Professor Doolittle points out, the sequence of amino acids in one protein might be strikingly like that in a second protein. A good example is the one he gives us—the different subunits of hemoglobin. This gave rise to the idea that the similar proteins might have descended from a common gene, when in the past the gene was duplicated. Virtually all biochemists accept this, and so do I. Many proteins of the clotting cascade are similar to each other, and similar to other non-cascade proteins, so they also appear to have arisen by some process of gene duplication. I think this is a very good hypothesis too. The critical point, however, is that the duplicated gene is simply a copy of the old one, with the same properties as the old one—it does not acquire sophisticated new properties simply by being duplicated. In order to understand how the present-day system got here, a scientist has to explain how the duplicated genes acquired their new, sophisticated properties.

With hemoglobin the task of getting from a simple protein with one chain to a complex of four chains does not appear to present problems, as I discussed in *Darwin's Black Box*.[20] In both cases the proteins simply bind oxygen, with more or less affinity, and they don't have to interact critically with other proteins in a complex protein system. There is a fairly obvious pathway leading from a simple hemoglobin to a more complex one.

With the proteins of blood clotting, however, the task of adding proteins to the cascade appears to be horrendously problematic. With one protein acting on the next, which acts on the next, and so forth, duplicating a given protein doesn't give you a new step in the cascade. Both copies of the duplicated protein will have the same target protein which they activate, and will themselves be activated by the same protein as before. In order to explain how the cascade arose, therefore, a scientist has to propose a detailed route whereby a duplicated protein turns into a new step in the cascade, with a new target, and a new activator. Furthermore, because clotting can easily go awry and cause severe problems when it is uncontrolled, a serious model for the evolution of blood clotting has to include quantitative factors, such as how much of a clot forms, what pressure it can resist, how frequent inappropriate clots would be, and many, many more such questions.

Professor Doolittle has addressed none of these questions. He has confined his work to the question of what proteins appear to be descended from what other proteins, and is content to wave his hands and assert that, well, those systems must have been put together by natural selection somehow. The title of the reference to his work that he cites here says it all: "Reconstructing the History of Vertebrate Blood Coagulation from a Consideration of the Amino Acid Sequences of Clotting

Proteins." His work concerns sequence comparisons. Doolittle has no idea of whether the clotting cascade could have been built up by natural selection.

An illustration of this fact is shown in his citing Bugge et al.[21] Professor Doolittle writes:

> Recently the gene for plaminogen [*sic*] was knocked out of mice, and, predictably, those mice had thrombotic complications because fibrin clots could not be cleared away. Not long after that, the same workers knocked out the gene for fibrinogen in another line of mice. Again, predictably, these mice were ailing, although in this case hemorrhage was the problem. And what do you think happened when these two lines of mice were crossed? For all practical purposes, the mice lacking both genes were normal! Contrary to claims about irreducible complexity, the entire ensemble of proteins is not needed. Music and harmony can arise from a smaller orchestra.

However, if one goes back and looks at Bugge et al., one sees that Professor Doolittle misread the paper. The mice that have had both genes knocked out do not have a functioning clotting system: they can't form clots; they hemorrhage; females die during pregnancy. They are certainly not candidates for evolutionary intermediates.

The lesson here is not that Doolittle misread a paper, which can happen to anyone. Rather, there are two points. First, a Darwinian mindset can tend to make one glide over problems that would occur in the real world. And second, sequence information is not sufficient to conclude that a system evolved by natural selection. The sequence information that Professor Doolittle had did not stop him from mistakenly pointing to a nonviable situation as a potential evolutionary intermediate. One can go further to say that, if a scientist as prominent as Russell Doolittle does not know of a detailed route by which natural selection could produce the clotting cascade, nobody knows.

I argue that each of the steps of the clotting cascade is irreducibly complex (see Chapter 4 of *Darwin's Black Box*)—requiring the rearrangement of several components simultaneously before a viable, controlled clotting system could be in place, and that is why I conclude that the cascade is a product of design. Clotting factors may be related by common descent, but the clotting cascade was not produced by natural selection.

On a different note, I'm glad Professor Doolittle likes Rube Goldberg too, but unfortunately it supplies what I think is his rock-bottom reason for deciding that natural selection produced the clotting cascade: "no Creator would have designed such a circuitous and contrived system." Well, Doolittle is a good scientist, but he's no theologian, and he doesn't serve science well when he lets his theological presuppositions influence his scientific judgment.

DOUGLAS J. FUTUYMA

PROFESSOR FUTUYMA advances arguments that I have dealt with in my replies to Orr and Doolittle: a mistaken notion of what constitutes irreducible complexity, and a confusion of what gene duplication is able to explain. He also offers a polemic, saying that I "claim a miracle in every molecule" and "Behe invites us to give up." I do not, however, claim to see miracles—I see design. Design and intelligence are two phenomena of which we have direct experience; they are parts of the world we see every day. On the other hand, the ability of natural selection to produce complex molecular systems resides only in the mind of Futuyma and others. Far from giving up, intelligent design theory takes the world as we see it, without philosophical preconceptions. Futuyma concludes, however, that the world must have behaved in a way we have never seen it behave, all to fit his extrascientific views.

MICHAEL RUSE

PROFESSOR RUSE asks if I have the right to appeal to design as a scientist. Well, many scientists already appeal to design. I mentioned the SETI program earlier; clearly those scientists think they can detect design (and nonhuman design at that). Forensic scientists routinely make decisions of whether a death was designed (murder) or an unfortunate accident. Archaeologists decide whether a stone is a designed artifact or just a chance shape. And in Ruse's own example of the downed airliner, investigators spent large amounts of effort to determine if the crash was designed. As I explain in my book,[22] it can be easy to determine that a system was designed, but extremely difficult to determine who the designer was. I do not argue that the designer was God, although it could have been and certainly many people will think so. I argue simply for the conclusion of design itself.

JAMES SHAPIRO

I APPRECIATE Professor Shapiro giving everyone a dose of reality in his descriptions of the enormous and interactive complexities of the cell, and certainly agree with him that they are beyond Darwinian explanation. I can't quite understand, however, what he envisions as an explanation for the *origin* of those systems. He seems to imply that they somehow assemble themselves, which strikes me as having a big chicken-and-egg problem. His analogies to computers and information theory are quite congenial to intelligent design ideas. However, he draws back from that conclusion for reasons I fail to grasp. Still, I look forward to reading more of his ideas when he fleshes them out.

6. DARWINISM AND DESIGN

One of the earliest reviews of *Darwin's Black Box* was by Oxford evolutionary biologist Tom Cavalier-Smith, writing in *Trends in Ecology and Evolution*.[23] I had discussed some of his musings rather unfavorably in the book, and he returned the favor in the review. What follows is my response to Cavalier-Smith's review.

"Darwinism and Design," letter to the editor, *Trends in Ecology and Evolution* 12, no. 6 (June 1997): 229.

IN HIS REVIEW OF MY BOOK *DARWIN'S BLACK BOX*, WHICH IS CRITI-cal of Darwinian theory, Tom Cavalier-Smith alternates between calling me ignorant and calling me deceitful, but finally seems to conclude that I am intentionally dishonest because of my religious views. I do not wish to descend into acrimony, so let me state plainly that my religious convictions can easily accommodate a Darwinian explanation for life, and I said so in my book.[24] I have no motive, religious or otherwise, to be dishonest. I wrote the book (which I knew would be controversial) out of a straightforward, professional conviction that many complex biochemical systems are beyond Darwinian explanation.

Well, I am not dishonest, but am I ignorant? Perhaps so. No one can be completely in command of the literature, and I would be very happy to be shown citations to published work explaining in detail how complex biochemical systems evolved by natural selection. However, Professor Cavalier-Smith says there is no such work: "For none of the cases mentioned by Behe is there yet a comprehensive and detailed explanation of the probable steps in the evolution of the observed complexity. The problems have indeed been sorely neglected." Yet he thinks that, even if detailed explanations are unavailable, general explanations are in hand. He cites ten references in his review. Of the ten, half refer to his own work, only four are published in this decade, and only two are reports of original research. From my point of view, in all of the cited papers the evolutionary explanation takes the form "system X developed because it would help the cell to do Y," without noticing the difficulties of making X by a blind process. It's like saying, "Air conditioners developed to enable more people to work indoors in the summertime."

Much of the difficulty arises in the differing standards that different disciplines have for what constitutes an "explanation." Biochemists require molecular detail. Cavalier-Smith, however, does not. (Indeed, he even castigates *Trends in Biochemical Sciences* for noticing engineering design in biochemical systems.) Darwinian evolution, though, would necessarily have to take place at the nut-and-bolt molecular level, the domain of biochemistry. A Darwinian evolutionary explanation, therefore, has to be a detailed biochemical explanation. None currently exist. By itself this fact doesn't justify the conclusion of intelligent design that I reach. (I also advance other arguments for design in my book.) But by itself the absence of detailed Darwinian explanations should provoke more thoughtfulness than was shown in Cavalier-Smith's review.

7. Intelligent Design is Not Creationism

For over twenty years Eugenie ("Genie") Scott was head of something called the National Center for Science Education, which sounds like a generalist, do-good organization but whose mission is sharply focused on beating back challenges to evolution in general and Darwinism in particular, mostly in schools but also in broader public life. Whenever and wherever the topic of evolution is broached with less than full assent, the NCSE swoops in to supply reporters with handouts, talking points, and interviewees from a list of scientists they have on speed dial.

"Intelligent Design Is Not Creationism," letter to
the editor, eLetters, *Science*, July 7, 2000.

SCOTT REFERS TO ME AS AN INTELLIGENT DESIGN "CREATIONIST,"[25] even though I clearly write in my book *Darwin's Black Box* (which Scott cites) that I am not a creationist and have no reason to doubt common descent. In fact, my own views fit quite comfortably with the forty percent of scientists Scott acknowledges think "evolution occurred, but was guided by God." Where I and others run afoul of Scott and the National Center for Science Education (NCSE) is simply in arguing that intelligent design in biology is not invisible, it is empirically detectable.

The biological literature is replete with statements like David DeRosier's in the journal *Cell*: "More so than other motors, the flagellum resembles a machine designed by a human."[26] Exactly why is it a thoughtcrime to make the case that such observations may be on to something objectively correct?

Scott blames "frontier," "nonhierarchical" religions for the controversy in biology education in the United States. As a member of the decidedly hierarchical, mainstream Roman Catholic Church, I think a better candidate for blame is the policing of orthodoxy by the NCSE and others—abetting lawsuits to suppress discussion of truly open questions and decrying academic advocates of intelligent design for "organiz[ing] conferences" and "writ[ing] op-ed pieces and books." Among a lot of religious citizens, who aren't quite the yahoos evolutionists often seem to think they are, such activities raise doubts that the issues are being fairly presented, which might then cause some people to doubt the veracity of scientists in other areas too. Ironically, the activity of Scott and the NCSE might itself be promoting the mistrust of science they deplore.

8. REPLY TO "OF MOUSETRAPS AND MEN"

"Behe Responds to Shanks and Joplin," letter to the editor, *Reports of the National Center for Science Education* 21, no. 3-4 (November 14, 2008).

IN THEIR ARTICLE "OF MOUSETRAPS AND MEN: BEHE ON Biochemistry,"[27] which has just come to my attention, Shanks and Joplin appear to mistakenly attribute to me the contention that irreducibly complex biochemical systems must have been created *ex nihilo*. I have never claimed that. I have no reason to think that a designer could not have used suitably modified pre-existent material. My argument in *Darwin's Black Box* is directed merely toward the conclusion of design. How the design was effected is a separate and much more difficult question to address. Although creation *ex nihilo* is a formal possibility, design might have been produced by some other means which involved no discontinuities in natural law, even if the designer is a supernatural being. One possibility is directed mutations. As noted by Brown University biologist Kenneth Miller in *Finding Darwin's God*, "The indeterminate nature of quantum events would allow a clever and subtle God to influence events in ways that are profound, but scientifically undetectable to us. Those events could include the appearance of mutations." I have no reason to object to that as a route to irreducibly complex systems. I would just note further that such a process amounts to intelligent design, and that while we may be unable to discern the means by which the design is effected, the resultant design itself may be detected in the structure of the irreducibly complex system.

The core claim of intelligent design theory is quite limited. It says nothing directly about how biological design was produced, who the designer was, common descent, or other such questions. Those can be argued separately. It says only that design can be empirically detected in observable features of physical systems. As an important corollary, it also predicts that mindless processes—such as natural selection or the self-organization scenarios favored by Shanks and Joplin—will not be demonstrated to be able to produce irreducible systems of the complexity found in cells.

9. A Review of Kenneth Miller's Finding Darwin's God

"Book Review: *Finding Darwin's God: A Scientist's Search for Common Ground between God and Evolution* by Kenneth R. Miller," *National Catholic Bioethics Quarterly* (Summer 2001): 277–278.

Finding Darwin's God, by Brown University professor of biology Kenneth Miller, is a full-throated defense of Darwinian evolution against skeptics. Several things make the book more interesting than others in the genre (*Science on Trial* by Douglas Futuyma, *Scientists Confront Creationism* edited by Laurie Godfrey, *Darwinism Defended* by Michael Ruse, *Evolution and the Myth of Creationism* by Tim Berra). First, Miller knows how to write in a lively style (he is also a successful textbook author). Second, he takes on advocates of "intelligent design theory," including University of California-Berkeley law professor Phillip Johnson and—reader be warned, I'm an interested party—myself, who are relatively recent apostates from Darwinism and have gotten hearings for their ideas in a number of academic settings. (Both Johnson and I have responded to Miller's criticisms of us, Johnson in his latest book *The Wedge of Truth* and I by a series of essays on the internet.) But, third, and the most interesting facet of the book: Miller is a practicing Roman Catholic and aims to show that Darwinian evolution is at least compatible with faith in God, and perhaps it is even the best way for God to have made material life.

Miller's theological view of evolution is that God "made the world today contingent upon the events of the past. He made our choices matter, our actions genuine, our lives important. In the final analysis, He used evolution as the tool to set us free." I will focus on Miller's attempt to baptize Darwinism. Briefly, although well-intentioned, it founders on his idiosyncratic use of the concept of "chance."

Keenly aware that prominent voices in biology equate Darwinism with atheism, Miller contends it is reasonable to think that God purposefully created the universe. He points to the so-called anthropic co-incidences—physical features of the universe (such as the strength of the gravitational constant, the charge of the electron, and so on) that seem suspiciously fine-tuned for our existence. He explicitly stands against those, such as the atheistic philosopher Daniel Dennett, who try to explain them away as evidence for an infinite number of universes. Fine and good, and I agree with Miller on this point. But he seems oblivious to the fact that he is making a classic argument for intelligent design—he has concluded, based on scientific data, that a viable explanation for the laws and fundamental constants of the universe is that they were the choice of an intelligent agent.

Other reviewers have also drawn attention to Miller's design reasoning. Barry Palevitz, a determinedly atheistic biology professor, writes: "Wait! Is Ken Miller, irreducible complexity's worst nightmare, using the exact same arguments as Behe, except that instead of designing biochemical pathways, Miller's deity plays dice with quarks?"

Miller tries to justify treating physics differently from biology. "Let's keep in mind that evolution is a biological theory, not a cosmic one," he writes. "The success or failure of the anthropic principle may be relevant to whether or not we can find God in the stars, but it does not tell us whether or not we can find him in the evolution of species." But one suspects that if we do find God in the stars we are likely to find Him in a lot of other places in nature too, such as in life.

In the end, Kenneth Miller, trying to avoid deism, does allow God a place in biology, but at the cost of hopelessly confusing his argument. In a remarkable passage he writes:

> The indeterminate nature of quantum events would allow a clever and subtle God to influence events in ways that are profound, but scientifically undetectable to us. Those events could include the appearance of mutations, the activation of individual neurons in the brain, and even the survival of individual cells and organisms affected by the chance processes of radioactive decay. Chaos theory emphasizes the fact that enormous changes in physical systems can be brought about by unimaginably small changes in initial conditions; and this, too, could serve as an undetectable amplifier of divine action.

So, Miller concludes, God could act in nature without being obvious. He could cause mutations that would change species, kill selected organisms, and so forth, all without tripping scientific detectors. But the glaring problem for a self-professed "orthodox Darwinist" such as Miller is that those mutations would have been caused by an intelligent agent, and so they would be the result of intelligent design. Let me strongly emphasize that any involvement by an intelligence in the production of life is inimical to Darwinism, as Darwin himself made clear. If God is "influencing" events, then to that extent they are not "chance" events. And if they are not "chance" events, then we have left the realms of Darwin's theory and entered into intelligent design.

It seems, then, that Miller's entire theological argument rests on an equivocation. Rather than using the terms "random" and "chance" to mean in part "not intended or directed by anyone," Miller uses them to mean something like "has an undetectable cause." But that meaning is untenable. For example, suppose by influencing undetectable quantum events God caused a meteor shower to strike the moon, and the resulting craters spelled out "John 3:16." Certainly we could rationally conclude that event was intended, even if we couldn't pinpoint its cause. And if we can do it for astronomical events, why not for biological ones as well? A conclusion of intelligent design is not based on knowledge of the means by which the intelligent agent achieved the result; rather, it is based on the specified complexity of the result (see William Dembski's *The Design Inference*).

Like John Haught in his recent book *God After Darwin*, who argues that God "entices" matter and infuses it with "information," Kenneth Miller wants to have his Darwinian cake and eat it too. To avoid deism, in places he writes things like "God, the creator of space, time, chance and indeterminacy, would exercise exactly the degree of control He chooses." Yet at other times, to establish his Darwinian bona fides, he writes, "Evolution is not rigged, and religious belief does not require one to postulate a God who fixes the game, bribes the referees, or tricks natural selection." But which is it, God in control or not?

Putting the most consistent interpretation on this I can think of, perhaps Miller wanted to say that God was responsible for life taking the direction it did, but we can never prove it scientifically. That has a nice sound to it, seeming to make a role for faith, but in the end it too breaks down. "Evolution is not rigged" and "God exercises the degree of control he chooses" are compatible only if God chooses to exercise no control. To be consistent, either Miller has to give up Darwinism to allow for the active God of Christianity, or settle for a deistic God at best to allow for Darwinism.

10. Has Darwin Met His Match?

In 2002 the opinion magazine *Commentary* published an article questioning the canonical understanding of evolution, "Has Darwin Met His Match?,"[28] by the elegant writer and Darwin skeptic David Berlinski. It drew a blizzard of letters, pro and con. I contributed the following.

Letter to the editor, *Commentary* (March 2003): 18.

IALWAYS FIND DAVID BERLINSKI'S WRITING DELIGHTFUL, AND I agree with much that he says in "Has Darwin Met His Match?" Specific claims about how life arose in the murky past should always be examined skeptically, especially if accompanied by grand philosophizing.

On the other hand, the fact that much remains mysterious does not mean we cannot conclude anything at all with reasonable certainty.

On the general question of the sufficiency of unintelligent physical processes to produce the astonishing complexity of life, I think a negative answer is justified, for reasons I gave in my book *Darwin's Black Box* (1995). I quite agree with Mr. Berlinski that my argument against Darwinism does not add up to a deductive proof. No argument that rests on empirical observations can have such force. Yet, despite my sloppy prose in suggesting that "by definition... irreducibly complex systems cannot be approached gradually," I intended the argument to be a scientific one, not a logical/deductive one. In a scientific argument, conclusions are tentative, based on the preponderance of the physical evidence, and potentially falsifiable.

Here is my thumbnail sketch of the modern design argument as I see it: Either unintelligent processes can explain all of life or they cannot. Virtually everyone (including Darwinists) agrees that life appears to be intelligently designed. The only physical mechanism ever proposed that could plausibly mimic design is Darwinian natural selection. Yet, as I have argued, the irreducible complexity of biochemical systems is a barrier to direct evolutionary construction by natural selection, leaving Darwinists to hope for circuitous, indirect routes. No plausible indirect routes have been proposed, let alone experimentally demonstrated.

That leaves us with biological features that look designed, but only promissory notes for how they can be explained by unintelligent processes. What's more, we know why the features we see in biological systems look designed: they are at once functional and very unlikely—exactly what William Dembski, whose work Mr. Berlinski also discusses, means by his phrase "specified complexity." They look designed for the same reason that nonbiological artifacts like mousetraps look designed, and non-design explanations have turned out to be so much bluster.

It seems reasonable to me to conclude, while acknowledging our fallibility, that at least some features of life were really designed by an intelligent agent.

11. Hog's Tail or Bacon?

Jerry Coyne is a professor of evolutionary biology at the University of Chicago. Coyne wrote a best-selling book, *Why Evolution is True*, and maintains a blog of the same name. His views of the implications of Darwinism for life tend strongly toward the nihilistic. (For example, he argues there is no such thing as free will.) Like Richard Dawkins, Coyne is a "new atheist" and sees evolutionary implications reaching into virtually all areas of life.

"Hog's Tail or Bacon? Jerry Coyne in the *New Republic*,"
Uncommon Descent (website), February 25, 2009.

IN A LONG BOOK REVIEW[29] IN THE *NEW REPUBLIC*, UNIVERSITY OF Chicago evolutionary biologist Jerry Coyne calls Brown University cell biologist Ken Miller a creationist. No surprise there—"creationist" has a lot of negative emotional resonance in many intellectual circles, so it makes a fellow's rhetorical task a lot easier if he can tag his intellectual opponent with the label. (Kind of like calling someone a "communist" back in the 1950s.) No need for the hapless "creationist" to be a Biblical literalist, or to believe in a young earth, or to be politically or socially conservative, or have any other attribute the general public thinks of when they hear the "C" word. For Coyne, one just has to think that there may be a God who has somehow affected nature.

In fact Miller thinks that God set up the general laws of the universe in the knowledge that some intelligent species would emerge over time, to commune with Him. (To be convinced of the existence of God, the skeptical Coyne wants a nine-hundred-foot-tall Jesus to appear to the residents of New York City, or something equally dramatic. Some of us view the genetic code and the intricate molecular machinery of life as rather more spectacular than a supersized apparition.) Miller, however, is the poster boy for a religious scientist who thinks God would never abrogate the laws of nature (at least during the general development of the universe—special events in religious history are another matter). Calling Miller a creationist is stretching the word far out of shape, to make it simply a synonym for "theist," with the advantage to Coyne that the associated rhetorical baggage is dumped on Miller.

I would feel more sympathy for Miller except for the fact that he pulls the same trick when it suits his purpose. As Jerry Coyne notes, "One of Miller's keenest insights is that ID involves not just design but also supernatural creation." So even though proponents of ID such as (ahem) me explicitly deny the necessity of supernatural creation to ID, and go to great lengths to explain the difference between a scientific conclusion of design and a theological conclusion of creation, that's ignored, and both Coyne and Miller paint ID with the creationist brush, the better to dismiss it.

My purpose here, however, is not to either attack or defend Miller, who can fend for himself. Rather, it is to point out a rather large confusion in Jerry Coyne's thinking. He writes:

> What is surprising in all this is how close many creationists have come to Darwinism. Important advocates of ID such as Michael Behe, a professor at Lehigh University (and a witness for the defense in the Harrisburg case), accept that the earth is billions of years old, that evolution has occurred—some of it caused by natural selection—and that many species share common ancestors. In Behe's view, God's role in the development of life could merely have been as the Maker of Mutations, tweaking DNA sequences when necessary to fuel the appearance of new mutations and species. In effect, Behe has bought all but the tail of the Darwinian hog.

But if I have accepted everything but the teeny-weeny tail of Darwinism, why does Coyne get so upset with me (see his earlier review of *The Edge of Evolution*, also in the *New Republic*)? You'd think that if Jerry Coyne and I agreed on 99% of the important things, he'd be a bit friendlier. And why does he pummel Miller as a "creationist" for accepting everything but the very furthest tip of the hog's tail? In fact, why does Coyne use the same epithet for Miller as for someone who thinks the world was created in a puff of smoke six thousand years ago?

Because of course, contra Jerry Coyne, the question of purposeful design versus no design is hardly peripheral—it is central. The existence of intentional design is not the hog's tail; it is the bacon. A real "Maker of Mutations" or "Mover of Electrons" (as Coyne derisively designates Miller's view of God) cuts the heart out of Darwinism. As many people besides myself have pointed out over scores of years, Darwin's claim to fame was not to propose "evolution," teleological versions of which had been proposed by others before him. Rather, Darwin's contribution was to propose an apparently ateleological mechanism for evolution—random mutation and natural selection. Frankly, it is astounding that Jerry Coyne gets so confused about the significance of Darwin's theory.

12. Letter to the Journal of Chemical Education

Unlike many philosophy journals—or high school student newspapers for that matter—many science journals are remarkably unwilling to publish responses by people whose work has been attacked in their pages. You'd think a bit of controversy would be good for the ol' journal circulation, but I guess not. The following literary effort (and the next) never saw the light of day.

Letter to the editor, unpublished, *Journal of Chemical Education* (2004).

A S AN ACTIVE PROPONENT OF THE ARGUMENT FOR INTELLIGENT design[30] I enjoyed reading "Evidence from Biochemical Pathways in Favor of Unfinished Evolution rather than Intelligent Design."[31] However, educators who plan to use it for classroom discussions should be sure to alert students to the regrettable errors in the article's reasoning.

The most serious mistake is its repeated use of the invalid argument from ignorance.[32] The authors imply that since no reason is known why, say, DNA should be synthesized discontinuously on the lagging strand, then no good reason in fact exists. Yet not long ago the same sort of fallacious argument from ignorance was made concerning "junk DNA." In that case the reasoning went something like the following: no function was known for much genomic DNA; therefore none existed; no self-respecting designer would have made genomes containing long stretches of apparently purposeless sequences; therefore Darwinism is true.[33]

"Apparently," however, should never be confused with "actually," and the progress of science has laid bare the mistaken assumption about "functionless" DNA, as the *Scientific American* article "The Unseen Genome: Gems Among the Junk" recounts.[34] Similarly, it is entirely possible that good reasons exist for most or all of the systems cited by Behrman et al. of which they are unaware. Personally, I would bet on it.

Rather than merely opining that cells are not built to their taste ("We would have designed a system using two polymerase activities"), it would be much more persuasive to skeptics of Darwinian claims such as myself if the authors could demonstrate experimentally that random mutation and natural selection could indeed produce the astonishingly complex molecular machinery found in the cell.

13. Letter to Trends in Microbiology

Letter to the editor, unpublished, *Trends in Microbiology*. Later published at Uncommon Descent (website), May 21, 2009.

IN THEIR RECENT ARTICLE, "BACTERIAL FLAGELLAR DIVERSITY AND Evolution: Seek Simplicity and Distrust It?," Snyder et al.[35] attribute to me a view of irreducible complexity concerning the flagellum that I do not hold. They write: "One advocate of ID, Behe, has argued that the bacterial flagellum shows the property of 'irreducible complexity,' that is, that it cannot function if even a single one of its components is missing."

That isn't quite right. Rather, I argued that necessary structural and functional components cannot be missing. In *Darwin's Black Box* I wrote, "The bacterial flagellum uses a paddling mechanism. Therefore it must meet the same requirements as other such swimming systems. Because the bacterial flagellum is necessarily composed of at least three parts—a paddle, a rotor, and a motor—it is irreducibly complex."[36] A particular auxiliary component of the flagellar system, such as, say, a chaperone protein, may or may not be needed for the system to work under particular circumstances. However, if it is missing a necessary part, it simply cannot work.

That shouldn't be controversial. In fact Snyder et al. avail themselves of the same reasoning when they write about a hook-basal-body complex recently discovered in *Buchnera aphidicola*, which "lacks ... the gene for flagellin." They conclude it "must have some role other than motility." Well, why must it have some role other than motility? Because, of course, it is missing the paddling surface, and therefore can't work as a paddling propulsion system. In other words, in my sense of the term, it is irreducibly complex.

Snyder et al. think *Buchnera's* derived structure "illuminates flagellar evolution by providing an example of what a simpler precursor of today's flagellum might have looked like—a precursor dedicated solely to protein export rather than motility." That seems a bit too simple to me. The activity of a protein export system has no obvious connection to the activity of a rotary motor propulsion system. Thus the difficulty of accounting for the propulsive function of the flagellum and its irreducible complexity remains unaddressed. In regard to the flagellum's evolution, Snyder et al.'s advice to distrust simplicity is sound and should be followed consistently.

PART TWO: EVOLUTION'S IRREDUCIBLE CONUNDRUM

IRREDUCIBLE COMPLEXITY (IC)—THE CONCEPT THAT MANY FUNC-tioning systems need multiple parts in order to work at all—is the nemesis of Darwinism. That's because Darwin always envisioned evolution as acting very gradually, by "numerous, successive, slight modifications." Yet if some biological function requires multiple parts before it works, then gradualism becomes quite unlikely. What's more, IC is a double threat to Darwin. Since intelligent agents can plan ahead, envisioning distant goals that are achieved only after passing through unprofitable stages, then IC is also a potential marker for purposeful design. Because of its threat to Darwinian gradualism, the idea of irreducible complexity has been relentlessly attacked since I first introduced it in *Darwin's Black Box*. I think the attacks have been a real boon because, as the following essays show, they expose for all who are willing to see the flailing of Darwin's supporters in the teeth of what is an acute difficulty for the theory. Keep your eye on the ball as you read below of Darwinian counter-arguments and you'll see brazen attempts to redefine irreducible complexity to substitute a straw man, and hidden Darwinist hands trying to intelligently guide origins scenarios where they want them to go, where nature never would.

14. A MOUSETRAP DEFENDED

By the year 2000 the friends of Darwin had gotten their act together and published a number of rebuttals to *Darwin's Black Box* in books, essays, and on the web. Beginning with this essay, the next ten pieces are my formal replies to either the most common objections or to what I thought were the best rebuttals. In some cases what I covered relatively briefly in Part 1 will be covered below in a bit more depth, such as my defense of the blood clotting cascade as irreducibly complex.

"A Mousetrap Defended," Discovery Institute, July 31, 2000.

IN *DARWIN'S BLACK BOX: THE BIOCHEMICAL CHALLENGE TO EVOLUtion* I coined the term "irreducible complexity" in order to point out an apparent problem for the Darwinian evolution of some biochemical and cellular systems. In brief, an irreducibly complex system is one that needs several well-matched parts, all working together, to perform its function. The reason that such systems are headaches for Darwinism is that it is a gradualistic theory, wherein improvements can only be made step by tiny step,[1] with no thought for their future utility. I argued that a number of biochemical systems, such as the blood clotting cascade, intracellular transport system, and bacterial flagellum are irreducibly complex and therefore recalcitrant to gradual construction, and so they fit poorly within a Darwinian framework. Instead, I argued, they are best explained as the products of deliberate intelligent design.

In order to communicate the concept to a general audience, I used a mousetrap as an example of an irreducibly complex system in everyday life. The mousetrap I pictured in my book had a number of parts that all had to work together to catch mice. The usefulness of the mousetrap example was that it captured the essence of the problem I saw for gradualistic evolution at a level that could be understood by people who were unfamiliar with the fine points of protein structure and function—that is, nearly everyone. For that same reason, defenders of Darwinism have assailed it. Although it may seem silly to argue over a mousetrap, it is actually critical to allowing people who are not professional scientists to understand the issues involved. In this article I defend the mousetrap as an example of irreducible complexity that can't be put together by a series of small, undirected steps.

Mousetrap rebuttals have popped up in a variety of situations, including national television, but the most recent instance (June 2000) was at a conference I attended at Concordia University in Wisconsin where Kenneth Miller, professor of biology at Brown University, spent several minutes during his presentation attacking the mousetrap. In doing so he used images of mousetraps that were drawn by Professor John Mc-Donald of the University of Delaware and can be seen on his web site[2] (reproduced below with permission). In defense of the mousetrap I will make a number of points, including: 1) McDonald's reduced-component traps are not single-step intermediates in the building of the mousetrap I showed; 2) intelligence was intimately involved in constructing the series of traps; 3) if intelligence is necessary to make something as simple as a mousetrap, we have strong reason to think it is necessary to make the much more complicated machinery of the cell.

CONCEPTUAL PRECURSORS VS. PHYSICAL PRECURSORS

ON HIS web site Professor McDonald was careful to make a critical distinction. He clearly stated "the reduced-complexity mousetraps... are intended to point out the logical flaw in the intelligent design argument; they're not intended as an analogy of how evolution works." Nonetheless, Kenneth Miller discussed McDonald's examples in a way that would lead an audience to think that they were indeed relevant to Darwinian evolution. Only at the end of the presentation did he briefly mention the disanalogy. I believe such tactics are disingenuous at best, like tagging a brief warning onto the end of a cigarette commercial containing attractive images. The purpose of the images is to get you to buy the cigarettes, despite the warning. The purpose of citing McDonald's drawings is to get people to buy Darwinian evolution, despite the brief disclaimer.

The logical point Professor McDonald wished to make was that there are mousetraps that can work with fewer parts than the trap I pictured in my book. Let me say that I agree completely; in fact, I said so in my book (see below). For example, one can dig a steep hole in the ground for mice to fall into and starve to death. Arguably that has zero parts. One can catch mice with a glue trap, which has only one part. One can prop up a box with a stick, hoping a mouse will bump the stick and the box will fall on top of it. That has two parts. And so forth. There is no end to possible variation in mousetrap design. But, as I tried to emphasize in my book, the point that is relevant to Darwinian evolution is not whether one can make variant structures of varying degrees of complexity, but whether those structures lead, step-by-excruciatingly-tedious-Darwinian-step, to the structure I showed. I wrote:

> To feel the full force of the conclusion that a system is irreducibly complex and therefore has no functional precursors we need to distinguish between a physical precursor and a conceptual precursor. The trap described above is not the only system that can immobilize a mouse. On other occasions my family has used a glue trap. In theory at least, one can use a box propped open with a stick that could be tripped. Or one can simply shoot the mouse with a BB gun. However, these are not physical precursors to the standard mousetrap since they cannot be transformed, step-by-Darwinian-step, into a trap with a base, hammer, spring, catch, and holding bar.[3]

Since I agree with Professor McDonald that there could be mousetraps with fewer parts, the only relevant question is whether the mousetraps he drew are physical precursors, or merely conceptual precursors. Can they "be transformed, step-by-Darwinian-step" into the trap I pictured, essentially the same structure as the fifth trap shown below (in Figure 2.4), as some people have been led to believe? No, they can't.

From the First Trap to the Second

Professor McDonald started with a complete mousetrap and then showed ones with fewer parts. I will reverse that order, start with his simplest trap, and show the steps that would be necessary to convert it into the next more complex trap in his series. That, after all, is the way Darwinian evolution would have to work. If we are to picture this as a Darwinian process, then each separate adjustment must count as a "mutation." If several separate mutations have to occur before we go from one functional trap to the next, then a Darwinian process is effectively ruled out, because the probability of getting multiple unselected muta-

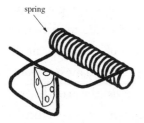
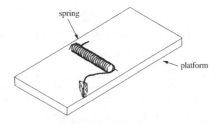

Figure 2.1. The first trap (left) and second trap (right).

tions that eventually lead to a specific complex structure is prohibitive. Shown above are the simplest and next-to-simplest traps.

The single-piece trap, consisting of just a spring with extended arms, is supposed to have one arm, under tension, propped up on the other arm. When a mouse jiggles it, the arm is released and comes down, pinning the mouse's paw against the other arm. Now, the first thing to notice is that the single-piece trap isn't a simple spring—it's got a very specific structure. If the lengths of the extended ends varied by much before their first bend, or if the angle of the bends differed somewhat, the trap wouldn't work. What's more, the strength of the material out of which the spring is made has to be consonant with the purpose of catching a mouse (for example, if it were made from an old Slinky it likely wouldn't work). It is not a simple starting point; it was intelligently selected.

Nonetheless, I realize that in coming up with an analogy we have to start somewhere. So I will not complain about an intelligently selected starting point. However, the involvement of intelligence at any other point along the way invalidates the entire exercise as an analogy to a Darwinian process. Because Darwinism wholly rejects intelligent direction, Darwinists must agree that the involvement of intelligence at any point in a scenario (after the agreed-on starting point) is fatal. That point occurs immediately for our mousetrap.

The second mousetrap (above) has a spring and a platform. One of the extended arms stands under tension at the very edge of the platform. The idea is that if a mouse in the vicinity jiggles the trap, the end of the arm slips over the edge and comes rushing down, and may pin the mouse's paw or tail against the platform. Now, the first thing to notice is that the arms of the spring are in a different relationship to each other than in the first trap. To get to the configuration of the spring in the second trap from the configuration in the first, it seems to me one would have to proceed through the following steps:[4] 1) twist the arm that has one bend through about 90° so that the end segment is perpendicular to the axis of the spring and points toward the platform; 2) twist the other arm through about 180° so the first segment is pointing opposite to where it originally pointed (the exact value of the rotations depend on the lengths of the arms); 3) shorten one arm so that its length is less than the distance from the top of the platform to the floor (so that the end doesn't first hit the floor before pinning the mouse) but still long enough to provide tension to the spring. While the arms were being rotated and adjusted, the original one-piece trap would have lost function, and the second trap would not yet be working.

At this point we bring in a new piece, the platform, which is a simple, flat piece of wood of appropriate shape and size. One now has a spring resting on top of a platform. However, the spring cannot be under tension in this configuration unless it is fixed in place. Notice that in the second mousetrap, not only has a platform been added, but two (barely visible) staples have been added as well. Thus we have gone not from a one-piece to a two-piece trap, but from a one- to a four-piece trap. Two staples are needed; if there were only one staple positioned as drawn, the tensed spring would be able to rotate out of position.

The staples have to be positioned carefully with respect to the platform. They have to be arranged within a very narrow tolerance so that one arm of the spring teeters perilously on the edge of the platform, or the trap doesn't work. If either of the staples is moved significantly from where they are drawn, the trap won't function.

I should add that I did not emphasize the staples in *Darwin's Black Box* because I was trying to make a simple point and didn't want to exhaust the readers with tedium. However, someone who wishes to seriously propose that the mousetrap I pictured is approachable in the tiny steps required by Darwinian processes would indeed have to deal with all the details, including the staples.

It is important to remember that the placement, size, shape, or any important feature (not just "piece") of a system can't just be chosen to fit the purposes of a person who wishes to simulate a Darwinian process. Rather, each significant feature has to be justified as being a small improvement. In the real world the occasional unselected feature might occur which serendipitously will be useful in the future, but invoking more than one unselected (neutral, nonadaptive) event in a Darwinian scenario seems to me impermissible because the improbability of the joint events starts to soar. In our current case the unselected event we are allowed was used up when we began with a special starting point.

I think the problems of rearranging the already-functioning first mousetrap shows the general difficulties one expects in trying to rearrange an already-functioning system into something else. The requirements ("selection pressures") that make a component suitable for one specialized system will generally make it unsuitable for another system without significant modification. Another problem we can note is that the second mousetrap is not an obvious improvement over the first; it is difficult to see how it would function any better than the one-piece trap. It's just that it's on the road to where we want to see the system end up—on the road to a distant target. That, of course, is intelligent direction.

The transition from the first to the second mousetrap is not analogous to a Darwinian process because: 1) a number of separate steps are required to make the transition; 2) each step has to fall within a narrow range of tolerance to get to the target trap; and 3) function is lost until the transition is completed. In fact, the situation of going from the first trap to the second trap is best viewed not as a transition, but as building a different kind of trap using some old materials from the first trap (with major modifications) and some new materials. Far from being an *analogy* to a Darwinian process, the construction of the second trap is an *example* of intelligent design.

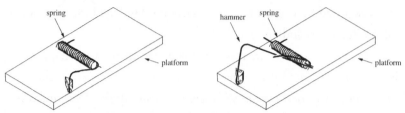

Figure 2.2. The second trap (left) and third trap (right).

From the Second Trap to the Third

The way the traps are drawn, the transition from the second to the third trap (see Figure 2.2) doesn't seem to be a big step. Both drawings are superficially similar. But when one thinks about the transition in detail, problems crop up. The first problem is that a new piece is added—the hammer. Unlike the platform that was added in the last transition (which I did not object to), the hammer is not a simple object. Rather it contains several bends. The angles of the bends have to be within relatively narrow tolerances for the end of the hammer to be positioned precisely at the edge of the platform, otherwise the system doesn't work. For the same reason, the length of the second segment of the hammer has to be within a narrow range of values. How does the hammer get into the third trap? It would seem that the extended arm of the second trap has to be held up while the newly fashioned hammer is inserted through the tunnel of the spring. Thus an intelligent agent has to actively push parts around to get to the configuration of the third trap. Again, there is no obvious improvement in function of the third trap compared to the second or first. Both the second and third traps appear to do the same thing as the first, but require more parts. Such an event is not expected in a Darwinian scenario. It seems the only reason they attract our attention is that they appear to be along the path we wish the process would go. That is intelligent design.

From the Third Trap to the Fourth

Going from the third trap to the fourth requires major rearrangements. The hammer is bent, lengthened, and an extra segment is added

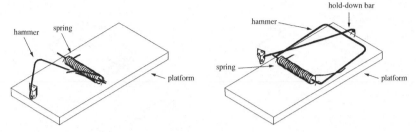

Figure 2.3. The third trap (left) and fourth trap (right).

to it. Two new pieces are added: the "hold-down bar" and a staple to hold down the hold-down bar. The end of the hold-down bar is endowed with a closed curl so that the staple has something to hang on to. The staple again has to be positioned in a specific region of the trap. Depending on details, this configuration may be an improvement over the first three traps because it appears that, depending on the tension of the spring, the trap could kill a mouse outright, rather than just pinning it (yet that feature could probably easily be built into the earlier versions). On the other hand, the arm of the spring is now being pushed through a much greater displacement in the fourth trap than previous versions. It seems unlikely a spring optimized for use in earlier traps would work well in the fourth trap (unless of course we are "looking ahead"). Rather than a "transition," this process is again better viewed as building a new trap using refashioned parts of the old trap plus new ones. This is intelligent design.

From the Fourth Trap to the Fifth
This is left as an exercise for the reader.

Discussion
I have to admit that even *I* find it tedious to discuss mousetraps in such excruciating detail. But the critical point is that that is exactly the level at which Darwinian evolution would have to work in the cell. *Every* relevant detail has to fit or the system fails. If an arm is too long or an angle not right or a staple placed incorrectly, the mouse dances free. If you

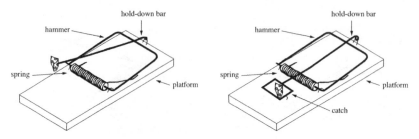

Figure 2.4. The fourth trap (left) and fifth trap (right).

want to get to a certain system, but the road there isn't a series of continual improvements, Darwinism won't take you there. It's important for those interested in these issues to realize that, when evaluating descriptive evolutionary scenarios (as opposed to experiments—see below), one has to attend to the tiniest details (as I did here) to see if intelligence is directing the show. On the other hand, if one doesn't pay the strictest attention, Darwinian scenarios look much more plausible because one sees only the possibilities, not the problems. It's easy for a speaker to persuade an audience that the McDonald mousetraps represent a series of Darwinian intermediates on the way to a standard trap—that they show irreducible complexity is no big deal. All one has to do is gloss over the difficulties. But although our minds can skip over details, nature can't.

In the real world of biology the staples, bends, and so forth would be features of molecules, of proteins in particular. If two proteins don't bind each other in the correct orientation (aren't stapled right), if they aren't placed in the right positions, if their new activity isn't regulated correctly, if many details aren't exactly correct, then the putative Darwinian pathway is blocked. Now, it's hard, almost impossible, for persons without the appropriate science background to tell where such difficulties would occur in Darwinian scenarios for blood clotting or ciliary function or other biological systems. When they read Darwinian stories in a book or hear them in lectures, they generally have no independent information to judge the scenario. In such a situation one should ask oneself, "If a simple mousetrap requires intelligent design, what is the likelihood that the much more complicated molecular machines of the cell could be built step-by-tiny-Darwinian-step?" Keeping that question in mind will foster a healthy skepticism toward optimistic scenarios.

Why do the McDonald mousetraps look persuasive to some people? Certainly one reason is the way they are drawn. Drawings of four of the five traps are dominated by the image of the large rectangular platform and prominent spring in the center. That makes them all look pretty much the same. The staples are barely visible and the various metal bars protruding here and there seem like insignificant details. In fact, they are critical. Another reason is that the scenario starts with the completed mousetrap. Any question about the placement of the parts, their size, stiffness, and so on doesn't easily arise because the parts were already placed where they needed to be for the ultimate goal in the original drawing (that is, the fifth mousetrap here, which is the first drawing in McDonald's series) and their properties could be inferred from the fact we started with a working trap. The universe of possibilities was tightly but implicitly circumscribed by the already-completed starting point. A third reason it seems persuasive is that the series is always presented as parts being removed from the complete mousetrap. Looking at it in such a backward manner—the reverse of what evolution would have to do—obscures the teleology of the building process. Going in a forward direction there is strong reason to think we would not end up at the fifth mousetrap when starting from the first, because the first works as well as the second and third, so greater complexity would be disfavored. In going backwards, however, lesser complexity is favored so it seems "natural" to move to simpler traps. Yet Darwinian evolution can't work like that.

A final reason for the persuasiveness of the example we can call the "Clever Hans effect." Clever Hans was the name of a horse who seemed to be pretty good at arithmetic. Its owner would give Hans a simple math problem such as 5 + 5, and the horse would stamp his hoof ten times, then stop. It eventually turned out that Clever Hans could pick up unconscious cues from its owner, who might raise his eyebrows or tilt his head when the horse's stamping reached the right value. The horse could even pick up unintentional cues from other people, not just the owner, who also apparently gave telltale reactions. In the case of Clever Hans, the human intelligence of the owner was inadvertently attributed to the horse. In my experience the same is invariably true of Darwinian scenarios—human intelligence is critical to guiding the scenario through difficulties toward the "proper" goal, but the intelligence is then attributed to natural selection. As with Clever Hans, the guidance is usually unconscious, but is intelligent nonetheless.

Clever Hans was exposed as mathematically clueless by carefully controlled experiments. To see whether natural selection can work wonders on its own—without the aid of human intelligence—we also have to do carefully controlled experiments. One way to do this is to ask bacteria in the laboratory if they can evolve irreducibly complex biochemical systems. (Kenneth Miller has called this the "acid test.") Bacteria are a good choice because they can be grown in huge numbers with short generation times—just what Darwinian evolution needs. However, when this was repeatedly tried over the course of twenty-five years for bacteria missing a comparatively simple biochemical system (called the "*lac* operon") natural selection came up empty (see "The Acid Test" later in this book). It could make the small changes typically termed "microevolution," but whenever it had to do a couple of things at once, such as would have to be done to make irreducibly complex systems, it got stuck.[5] Like Clever Hans on his own, natural selection seems to have much less intelligence than we had given it credit for. There is currently no experimental evidence to show that natural selection can get around irreducible complexity.

Darwinian scenarios, either for building mousetraps or biochemical systems, are very easy to believe if we aren't willing or able to scrutinize the smallest details, or to ask for experimental evidence. They invite us to admire the intelligence of natural selection. But the intelligence we are admiring is our own.

15. The Irreducibility of the Blood Clotting Cascade

"A system of this kind cannot just be allowed to free-wheel. The success of the coagulation process is due to the finely tuned modulation and regulation of all of the partial proteolytic digestions that occur. Too little or too much activity would be equally damaging for the organism. Regulation is a central issue in blood coagulation."
—Torben Halkier[6]

"In Defense of the Irreducibility of the Blood Clotting Cascade," Discovery Institute, July 31, 2000.

In *Darwin's Black Box: The Biochemical Challenge to Evolution*, I devoted a chapter to the mechanism of blood clotting, arguing that it is irreducibly complex and therefore a big problem for Darwinian evolution. Since my book came out, as far as I am aware there have been no papers published in the scientific literature giving a detailed scenario or experiments to show how natural selection could have built the system. However, three scientists publishing outside science journals have attempted to respond. The first is Russell Doolittle, a professor of biochemistry at the University of California at San Diego, member of the National Academy of Sciences, and expert on blood clotting. Second is Kenneth Miller, a professor of cell biology at Brown University and author of *Finding Darwin's God*.[7] The third scientist is Keith Robison, who at the time of his writing was a graduate student at Harvard University.

I will give their arguments below and my response. Here is a brief summary.

1) Professor Doolittle argued that new laboratory work showed two components of the blood clotting cascade could be eliminated ("knocked-out") from mice and the mice got along fine without them. However, Doolittle misread the laboratory work: the double-knock-out mice have severe problems and have no functioning blood clotting system. They are not models of evolutionary intermediates.

Although anyone can misread a paper, in my opinion the fact that an expert on blood clotting cited a recent and contradictory journal article, instead of a publication directly addressing the evolution of blood clotting, shows that there are indeed no detailed explanations for the evolution of blood clotting in the literature and that, despite Darwinian protestations, the irreducible complexity of the system is a significant problem for Darwinism.

2) Although embedded in a lengthy description of how blood clotting and other systems work, Professor Miller's actual explanation for how the vertebrate clotting cascade evolved consists of one paragraph. It is a just-so story that doesn't deal with any of the difficulties the evolution of such an intricate system would face. Even so, in the one paragraph Miller proposes what looks like a detrimental or fatal situation, akin to the knock-out mice (above) that lack critical components.

3) Keith Robison proposed that a cascade might begin with a single enzyme with three different properties. Upon duplication of the gene for the enzyme, the duplicate loses several of the properties, resulting in a two-component cascade. Repetition of the scenario builds cascades with more components. Although intriguing, the scenario starts with a complex, unjustified situation (the enzyme with multiple abilities) that already has all necessary activities. What's more, the proposed gene duplication and several steps needed to lose function are "neutral," unselected mutations. Stringing together several very specific neutral mutations to build a complex system is vastly improbable and amounts to intelligent design.

Russell Doolittle's Criticism

a. Mice lacking clotting factors have severe health problems.

In its issue of February/March 1997 *Boston Review* featured a symposium discussing *Darwin's Black Box* and Richard Dawkins's *Climbing Mount Improbable*. Among the dozen essays contributed by academics was one by University of California-San Diego biochemist Russell Doolittle. I had devoted a chapter of *Darwin's Black Box* to the blood clotting cascade, asserting that it is irreducibly complex and so does not fit well within a Darwinian framework. Doolittle, an expert on blood clotting, disagreed. Prefacing a discussion of globin homology, he remarked that "the genes for new proteins come from the genes for old ones by gene duplication," later adding, "this same kind of scenario can be reconstructed for a host of other physiological processes, including blood clotting."[8] Then, citing a paper by Bugge et al. entitled "Loss of Fibrinogen Rescues Mice from the Pleiotropic Effects of Plasminogen Deficiency,"[9] he commented:

> Recently the gene for plaminogen [*sic*] was knocked out of mice, and, predictably, those mice had thrombotic complications because fibrin clots could not be cleared away. Not long after that, the same workers knocked out the gene for fibrinogen in another line of mice. Again, predictably, these mice were ailing, although in this case hemorrhage was the problem. And what do you think happened when these two lines of mice were crossed? For all practical purposes, the mice lacking both genes were normal! Contrary to claims about irreducible complexity, the entire ensemble of proteins is not needed. Music and harmony can arise from a smaller orchestra.[10]

The implied argument appears to be that the cited work shows the clotting system is not irreducibly complex, so a simpler clotting cascade might be something like the one that lacked plasminogen and fibrinogen, which could be expanded into the modern clotting system by gene duplication. Perhaps there are other stable systems of lesser complexity, and the entire cascade could then be built up by small steps in what is thought to be the typical Darwinian pattern. However, that interpretation depends on a mistaken reading of Bugge et al.

Bugge et al. note that the lack of plasminogen in mice "results in high mortality, wasting, spontaneous gastrointestinal ulceration, rectal prolapse, and severe thrombosis. Furthermore, plasminogen-deficient mice display delayed wound healing following skin injury." On the other hand, if the gene for fibrinogen is knocked out, the result is failure to clot and frequent hemorrhage. Also, "pregnancy uniformly results in fatal uterine bleeding around the tenth day of gestation."[11] The point of Bugge et al. was that if one crosses the two knock-out strains, producing plasminogen-plus-fibrinogen deficiency in individual mice, the mice do not suffer the many problems that afflict mice lacking plasminogen alone.

Since the title of the paper emphasized that mice are "rescued" from some ill-effects, one might be misled into thinking that the double-knock-out mice were normal. They are not. As Bugge et al. state in their abstract, "Mice deficient in plasminogen and fibrinogen are phenotypically indistinguishable from fibrinogen-deficient mice." In other words, the double-knockouts have all the problems that mice lacking only fibrinogen have: they do not form clots, they hemorrhage, and the females die if they become pregnant. They are definitely not "for all practical purposes... normal."[12] (Figure 2.5)

The probable explanation is straightforward. The pathological symptoms of only-plasminogen-deficient mice apparently are caused by uncleared clots. But fibrinogen-deficient mice cannot form clots in the first place. So problems due to uncleared clots don't arise either in fibrinogen-deficient mice or in mice that lack both plasminogen and fibrinogen. Nonetheless, the severe problems that attend lack of clotting in fibrinogen-deficient mice continue in the double knockouts. Pregnant females still perish.

An important lesson exemplified by Bugge et al. is that it can be worse for the health of an organism to have an active-but-unregulated pathway (the one lacking just plasminogen) than no pathway at all (the one lacking fibrinogen, which exhibited fewer overt problems). This emphasizes that model scenarios for the evolution of novel biochemical systems have to deal with the issue of regulation from the inception of the system. Most important for the issue of irreducible complexity, however, is that the double-knock-out mice do not merely have a less sophisticated but still functional clotting system. They have no functional clotting system at all. They are not evidence for the Darwinian evolution of blood clotting. Therefore my argument, that the system is irreducibly complex, is unaffected by that example.

Lacking Plasminogen	Lacking Fibrinogen	Lacking Both
thrombosis	no clotting	no clotting
ulcers	hemorrhage	hemorrhage
high mortality	death in pregnancy	death in pregnancy

Figure 2.5. Symptoms of mice lacking clotting factors.

b. Gene duplication is not a Darwinian explanation.

I believe that the point about the knock-out mice is quite important because it helps illustrate the serious shortcomings of simple invocations of gene duplication as evolutionary explanations. Appeal to gene duplication has been quite common among scientists reviewing my book. Besides Russell Doolittle, it has been invoked by Allen Orr,[13] Douglas Futuyma,[14] Neil Blackstone,[15] Robert Dorit,[16] a committee of the National Academy of Sciences,[17] and others. The typical argument goes something like this: Modern biology has recognized that the sequences and structures of some proteins are quite similar to others (an example is hemoglobin vs. myoglobin), and this similarity is normally interpreted in terms of duplication and divergence of an ancestral gene. Many proteins in complicated pathways, such as the blood clotting cascade, are also similar to other proteins, consistent with the idea that they descended from a relatively few ancestor proteins. We can assume, then, (the argument continues) that although we don't know the details, most complicated pathways were built by natural selection using gene duplication.

A recent publication of the National Academy of Sciences nicely illustrates the argument in action:

> Modern-day intelligent design proponents argue that... molecular processes such as the many steps that blood goes through when it clots, are so irreducibly complex that they can function only if all the components are operative at once.... Complex biochemical systems can be built up from simpler systems through natural selection.... Jawless fish have a simpler hemoglobin than do jawed fish, which in turn have a simpler hemoglobin than mammals.... Genes can be duplicated, altered, and

then amplified through natural selection. The complex biochemical cascade resulting in blood clotting has been explained in this fashion.[18]

But the reaction to Bugge et al. is a paradox for the typical argument. On the one hand, structural and sequence evidence for gene duplication and domain swapping in the clotting cascade is very clear. So clear, in fact, that cascade proteins are used as textbook examples of those processes.[19] On the other hand, if there were indeed a robust Darwinian explanation for the origin of blood clotting by natural selection, or if sequence analyses had demonstrated how gene duplication might have produced the cascade, it would be difficult to understand why one would point to the knock-out mice as exemplifying Darwinian possibilities, when in reality they only underscore the serious problems facing the evolution of irreducibly complex systems. Detailed knowledge of the sequence, structure, and function of the proteins of the clotting cascade did not prevent a wholly unviable model from being proposed as a potential evolutionary intermediate. Why not?

The predicament is easily resolved when a critical point is recalled: *evidence of common descent is not evidence of natural selection.* Homologies among proteins (or organisms) are the evidence for descent with modification—that is, for evolution. Natural selection, however, is a proposed explanation for how evolution might take place—its mechanism—and so must be supported by other evidence if the question is not to be begged. This, of course, is a well-known distinction.[20] Yet, from reviewers' responses to my book, the distinction is often overlooked. Knowledge of homology is certainly very useful, can give us a good idea of the path of descent, and can constrain our hypotheses. Nonetheless, knowledge of the sequence, structure, and function of relevant proteins is by itself insufficient to justify a claim that evolution of a particular complex system occurred by natural selection. Gene duplication is not a Darwinian explanation because duplication points only to common descent, not to the mechanism of evolution.

c. What would an explanation look like?

If homology is not sufficient to justify a Darwinian conclusion, what is? The required amount of justification depends on the complexity of the system under consideration. For example, the task of getting from a simple oxygen-binding protein such as myoglobin, with one chain, to hemoglobin, with four chains that bind oxygen, does not appear to present substantial problems, as I discussed in *Darwin's Black Box*. In both cases the proteins simply bind oxygen, with more or less affinity, and neither globin has to interact critically with other proteins in a complex system. There seems to be a straightforward pathway of association leading from a simple myoglobin-like protein to a more complex hemoglobin-like one. In fact, its relative simplicity is probably the reason it is a favorite example in discussions of evolution by gene duplication.

Like hemoglobin/myoglobin, many proteins of the clotting cascade are similar to each other, and also similar to non-cascade proteins. So they too appear to have arisen by some process of gene duplication. I agree this is a good hypothesis. But does gene duplication lead straightforwardly to the blood clotting cascade? No. The important point to keep in mind is that a duplicated gene is simply a copy of the old one, with the same properties as the old one—it does not acquire sophisticated new properties simply by being duplicated. In order to understand how the present-day system got here, an investigator would have to explain how the duplicated genes acquired their new, sophisticated properties.

With clotting, however, the task of initiating and adding proteins to the cascade appears to be quite problematic. With one protein acting on the next, which acts on the next, and so forth, duplicating a given protein doesn't yield a new step in the cascade. Both copies of the duplicated protein would have the same target protein which they activate, and would themselves be activated by the same protein as before. In order to explain how the cascade arose, therefore, an investigator would have to propose a detailed route whereby a duplicated protein turns into a step in the cascade, with a new target, and a new activator. Furthermore, because clotting can easily go awry and cause severe problems when it is uncontrolled, a serious model for the evolution of blood clotting would have to include such things as: a quantitative description of the starting state, including tangentially interacting systems; a description of the initial regulatory mechanisms; a quantitatively justified proposal for a step-by-step route to the new state; a detailed plan for how regulatory mechanisms accommodated the changes; and more.

An alternative to presenting an exhaustively detailed model would be an experimental demonstration of the capability of natural selection to build a system whose complexity rivals that of the clotting cascade. In fact, experimental evidence is much preferred to mere model building, since it would be extremely difficult for models to predict whether proposed changes in complex systems might have unforeseen detrimental effects.

I pointed out in *Darwin's Black Box* that scenarios for the origin of biochemical systems lack essential detail. But since I am a proponent of an alternative explanation, some Darwinists have accused me of setting the evidentiary standard so high that it is impossible for them to meet it. The evidentiary standard, however, is set not by me, but by the complexity of the biochemical systems themselves. If malfunctioning of the blood clotting cascade or other complex system can cause a severe loss of fitness, then a Darwinian scheme for its evolution must show how this could be avoided. And if the system can malfunction when small details go awry, then the scheme has to be justified at least to the level of those details. Unless that is done, we remain at the level of speculation.

In noting that not much research has been done on the Darwinian evolution of irreducibly complex biochemical systems, I should emphasize that I do not prefer it that way. I would sincerely welcome more investigation of their supposed Darwinian origins. I fully expect that, as in the field of origin-of-life studies, the more we know, the more difficult the problem will be recognized to be.

Professor Doolittle continued to work on the blood clotting system until his recent retirement in his mid-eighties. In a book he wrote in this later period, *The Evolution of Vertebrate Blood Clotting*,[21] he investigated the clotting systems of ever-simpler vertebrates in the hope of drawing a complete picture and, I suspect, of answering my criticisms. As during his entire distinguished career, he did heroic work tracking down clotting components in obscure animals. A key to his later work was the availability of ever-improving techniques to sequence DNA and compare sequences between organisms. Large fractions of the genomes of lower vertebrates and even protochordates (creatures that in some ways are evolutionarily slightly simpler than vertebrates) have been sequenced. Doolittle notes that some simpler vertebrates have clotting systems with somewhat fewer components, and that some animals that don't have clotting systems do have proteins that resemble some clotting components. He concludes that "the story of how vertebrate clotting evolved is far from complete, but what we know of it so far is very satisfying."

He well deserves the satisfaction of his long career. However, the question of how Darwinian processes might have put together the clotting cascade is no further along now than when I wrote skeptically in *Darwin's Black Box* of Doolittle's work. Let's keep our eye on the ball here. As I explained in the section above, "gene duplication is not a Darwinian explanation," and "evidence of common descent is not evidence of natural selection." Darwin's theory makes the radical claim that random, unguided mutations can give rise to the molecular machinery of life. Simply pointing out that some proteins have sequences that are similar to others may be evidence for common descent, but it says nothing about what might have driven the construction of sophisticated systems. The best evidence of what undirected evolutionary processes can accomplish is provided by the experiments and observations on fibrinogen-deficient mice discussed earlier in this section and work discussed in *The Edge of Evolution*. They demonstrate that such processes overwhelmingly are restricted to degrading genomes, or slightly tweaking pre-existing machinery.

KENNETH MILLER'S CRITICISM

a. "From that point on..."

If an eminent scientist and expert on blood clotting such as Russell Doolittle does not know how blood clotting arose, nobody knows. Nonetheless, it is instructive to look at how several other scientists have addressed the issue. In this section I examine Kenneth Miller's writing.

In the chapter of *Finding Darwin's God*[22] that defends Darwinism from my criticisms, Professor Miller devotes the largest part—fully nine pages—to blood clotting. During the first five pages he gives an overview of how the blood clotting cascade works, as well as noticing that bleeding can be slowed by platelet aggregation, which is not a part of the clotting cascade. In the next two pages he writes of the sequence similarity of clotting factors and the phenomenon of gene duplication—facts well known to Russell Doolittle. In the final two pages he writes of a totally unrelated clotting system, that of lobsters. In other words, Miller spends almost all of the space writing about things other than how the vertebrate blood clotting cascade may have arisen step by Darwinian step.

His proposed model for the evolution of the vertebrate cascade is confined to *just one paragraph*. After postulating that, when a blood vessel breaks and they enter the new environment of a tissue, some blood proteins might be non-specifically cut by serine proteases and non-specifically aggregate, Miller writes:

> What happened next?... A series of ordinary gene duplications, many millions of years ago, copied some of these serine proteases. One of these duplicate genes was then mistargeted to the bloodstream, where its protein product would have remained inactive until exposed to an activating tissue protease—which would happen only when a blood vessel was broken. From that point on, each and every refinement of this mechanism would be favored by natural selection. Where does the many-layered complexity of the system come from? Again, the answer is gene duplication. Once an extra copy of one of the clotting protease genes becomes available, natural selection will favor slight changes that might make it more likely to activate the existing protease. An extra

level of control is thereby added, increasing the sensitivity of the cascade.[23]

Let's start with the last half of the paragraph ("From that point on..." and forward). The first thing to notice is that it's terminally fuzzy—too sketchy for much criticism. As I explained above, simply chanting "gene duplication" does not show how a complex system can be built, since duplication does not explain how new enzyme properties and targets arise. Russell Doolittle knew all about gene duplication, and yet postulated, as a model for an evolutionary intermediate, mice that turned out to be severely disabled. Professor Miller simply tries to use the term "gene duplication" as a magic wand to make the problem go away, but the problem does not go away. Miller's assertion that natural selection would favor each additional step is made quite problematic by the fact that each step in clotting has to be strictly regulated or else it is positively dangerous, as noted by Torben Halkier in the opening quotation of this document. In other words, what Halkier calls the "central issue" of regulation is ignored by Miller. Miller's statement does not even say what the newly duplicated proteases are envisioned to be acting on—whether the tissue protease, the original mistargeted circulating protease, plasma proteins, or everything at once.

Such a brief story is of no use at all in understanding how the irreducible complexity of the clotting cascade could be dealt with by natural selection. It strikes me that the main purpose of the paragraph is not to actually contribute to our understanding of how clotting actually may have arisen, but to persuade those who aren't familiar with biochemical complexity to believe Darwinism has the problem under control. It doesn't.

b. Problems from the get-go

Now let's look at the beginning of Miller's scenario. It turns out that as soon as he tries to get past the simple postulated beginning (that is, the nonspecific aggregation of proteins that have been nonspecifically degraded when a blood vessel is broken) his scenario runs into severe problems.

Miller's first step postulates a potentially deadly situation: a non-regulated zymogen circulating in the bloodstream with clottable proteins. Although we don't have access to Miller's imaginary organism to test the effects of this situation, to understand what it might mean we can look to several cases where regulatory proteins are missing from modern organisms: 1) Because the condition is very likely lethal in utero, no cases have been reported in the medical literature[24] of human patients missing antithrombin, a prominent regulator of the clotting cascade; 2) As described earlier, knock-out mice missing the gene for plasminogen to remove blood clots suffer severe thrombosis and increased mortality, as well as other debilitating symptoms.[25] It seems quite likely that Miller's hypothetical organism would experience the unregulated zymogen not as an improvement, but as a severe genetic defect.

Regulation is the key issue in clotting, but Miller does not address it at all. Miller's brief scenario does not even address potentially fatal difficulties—it ignores them. However, while Darwinists telling just-so stories can ignore difficulties, real organisms can't.

Here are several more problems with the brief scenario. First, it should be noted that the problem the scenario is trying to solve—hemostasis—can't initially be severe, because the starting point is a living organism, which must already be quite well adjusted to its environment. Second, the protein that Miller postulates to be mistargeted to the bloodstream would then no longer be doing its initial job; that would be expected to be detrimental to the organism. Third, Miller begins by postulating the mistargeting of a non-specific protease-precursor (a zymogen) to the bloodstream of some unfortunate organism. (Miller writes that if his scenario is correct, then "the clotting enzymes would have to be near-duplicates of a pancreatic enzyme...."[26] Pancreatic enzymes, which have to digest a wide variety of protein foodstuffs, are among the most nonspecific of enzymes). Now, that would pose a severe health threat to the mutant organism even greater than just an unregulated clotting cascade. For example, if the digestive enzyme precursor trypsinogen were mistargeted to the bloodstream, the potential for disaster would be very large. In the pancreas, misactivation of trypsinogen is prevented by the presence of trypsin inhibitor. In Miller's scenario one cannot plausibly suppose there to be a trypsin inhibitor fortuitously circulating in the plasma. If the mistargeted enzyme were accidentally activated, it would most likely cause generalized damage in the absence of a regulatory mechanism. It would not be a viable evolutionary intermediate.

Problems of regulation aside, it is difficult to see the advantage of the protease mistargeted to the bloodstream in the first place. While Miller's initial cellular or tissue protease would by necessity be localized to the site of a cut, a circulating zymogen would not. In modern organisms thrombinogen has a vitamin K-dependent gla-domain which allows it to localize to cell surfaces. In order to be effective before membrane-binding features had been acquired, it would seem that the postulated circulating protease would have to be present at rather high concentrations, exacerbating the regulatory difficulties discussed above.

As I wrote in *Darwin's Black Box*, the problem of blood clotting is not in just forming a clot—any precipitated protein might plug a hole.[27] Rather the problem is regulation. The regulatory problems of the clotting cascade are particularly severe since, as pointed out by Halkier,[28] error on either side—clotting too much or too little—is detrimental. As irreducible complexity would predict, Kenneth Miller's scenario has no problem postulating a simple clot (the initial nonspecific aggregation) but avoids the problem of regulation.

The take-home lesson is that Miller doesn't even try to deal with the problem of irreducible complexity and other obstacles that I pointed out in *Darwin's Black Box*—he just ignores them.

KEITH ROBISON'S PROPOSAL

SOON AFTER *Darwin's Black Box* was published Keith Robison posted some criticisms on talk.origins, one of which concerned the blood clotting cascade. I think his proposed scheme for adding steps to cascades, while it doesn't work, is the most serious and interesting one I have come across (grad students often come up with the best ideas). It spurred me to think more about the situation and has led me to formulate the concept of irreducible complexity in more explicitly evolutionary terms.

Robison's proposal was not focused on blood clotting per se, so he doesn't worry about forming a clot. Rather he concentrates on how new steps might be added to a cascade such as occurs in blood clotting, as well as other systems such as the complement cascade of the immune system. His starting point is a rather complex one, which I will grant for purposes of argument. He postulates a protein X which already has three pertinent properties: 1) it is activated by some external factor (perhaps by tissue trauma); 2) the activated protein X* then can activate more X by hydrolysis; and 3) activated X* cleaves some additional target. It's pretty much a cascade all by itself.

Fine, let's start there. Now begins the interesting scenario that I'll contest. To build a new step in the cascade, Robison then postulates several further steps. First is an initial gene duplication. Both genes make X, and the X from either gene when activated can activate the other. The second postulated event is a mutation in just one of the X genes that causes it to lose the ability to interact with the target. Nonetheless, it retains the ability to activate itself and the X coded by the original gene. The third step is loss of the yet-unmutated protein's ability to either respond to the external factor or activate itself and the other protein. At the end we have one of the proteins responsive to the external factor and able to activate both itself and the second protein, and just the second protein is able to cleave the target. Replication of the scenario yields more steps in the cascade, building irreducible complexity.

I argue that, while Robison's scenario does indeed build a new step in the cascade, it doesn't do it by Darwinian means. Rather, it does so by Robison's intelligent direction. Here are a few pertinent quotes from the several steps of his scenario: "This arrangement is neutral; the species has gained no advantage"; "Again, this genotype is neutral; it is neither beneficial nor detrimental"; "The initial steps are neutral, neither advantageous nor disadvantageous"; "The final step locks in the cascade. It is *potentially* advantageous."[29] (My emphasis—the "potentially" advantageous final step would require a further mutation to make it actually advantageous, so before that happens it is neutral.)

Thus his scenario postulates four successive, very specific steps: 1) gene duplication of the particular multi-talented enzyme; 2) the first loss-of-function step; 3) the second loss-of-function step; 4) a step to take advantage of the situation. As Robison emphasized, the first three steps are neutral; that is, they do the organism neither harm nor good. Only when the fourth step is completed is there a selective advantage. Now, it must be remembered that the Darwinian magic depends on natural selection. If a trait is advantageous, it has a good chance of taking over a population, thus providing a large base from which the next advantageous mutation might arise. However, if a trait is neutral, providing no advantage, it is far, far less likely to spread, so the odds of a second mutation appearing that depends on the first are not improved at all—they're pretty much the same as luckily getting the two specific mutations simultaneously. In the final analysis Robison's scenario is completely non-Darwinian. It postulates an already-functioning system that wasn't justified in Darwinian terms, and it then goes through three neutral, non-selected steps. Only at the very end is there a selectable property that wasn't postulated at the beginning.

To get a flavor of the difficulties Robison's scenario faces, note that standard population genetics says that the rate at which neutral mutations become fixed in the population is equal to the mutation rate. Although the neutral mutation rate is usually stated as about 10^{-6} per gene per generation, that is for any random mutation in the gene. When one is looking at particular mutations such as the duplication of a certain gene or the mutation of one certain amino acid residue in the duplicated gene, the mutation rate is likely about 10^{-10}. Thus the fixation of just one step in the population for the scenario would be expected to occur only once every ten billion generations. Yet Robison's scenario postulates multiple such events.

A Modest Conclusion

I WOULD like to pause here for a moment to point out that all three scientists who tried to meet the challenge to Darwinian evolution of blood clotting—Russell Doolittle, Kenneth Miller, and Keith Robison—foundered on exactly the same point, the point of irreducible complexity. Yet they foundered in three different ways. Doolittle mistakenly thought that even the current cascade might not be irreducibly complex, but experimental results showed him to be wrong. Miller either proposed unregulated steps or just waved his hands and shouted "gene duplication," avoiding the problem by obfuscation. Robison directly attacked a piece of the problem, but failed to see he was intelligently guiding events in a distinctly non-Darwinian scenario. Perhaps we may be allowed to conclude that when three scientists, highly intelligent and strongly motivated to discredit it, all come up empty, irreducible complexity is indeed a big hurdle for Darwinism.

An Evolutionary Perspective on Irreducible Complexity

In DARWIN'S *Black Box* I defined the concept of irreducible complexity (IC) in the following way: "By irreducibly complex I mean a single system which is composed of several well-matched, interacting parts that contribute to the basic function, and where the removal of any one of the parts causes the system to effectively cease functioning."[30]

While I think that's a reasonable definition of IC, and it gets across the idea to a general audience, it has some drawbacks. It focuses on already-completed systems, rather than on the process of trying to build a system, as natural selection would have to do. It emphasizes "parts," but says nothing about the properties of the parts, how complex they are, or how the parts get to be where they are. It speaks of "parts that contribute to the basic function," but that phrase can be, and has been, interpreted in ways other than what I had in mind (for example, talking about whole organs that contribute to complex functions such as "living"), muddying the waters in my view. What's more, the definition doesn't allow for "degrees" of irreducible complexity; a system either has it or it doesn't. Yet certainly some IC systems are more complex than others; some seem more forbidding than others.

While thinking of Keith Robison's scenario, I was struck by the idea that irreducible complexity could be better formulated in evolutionary terms by focusing on a proposed pathway, and on whether each step that would be necessary to build a certain system using that pathway was selected or unselected. If a system has to pass through one unselected step on the way to a particular improvement, then in a real evolutionary sense it is encountering irreducibility: two things have to happen (the mutation passing through the unselected step and the mutation that gives a selectable system) before natural selection can kick in again. If it has to pass through three or four unselected steps (like Robison's scenario), then in an evolutionary sense it is even more irreducibly complex. The focus is off of the "parts" (whose number may stay the same even while the nature of the parts is changing) and re-directed toward "steps."

Envisioning IC in terms of selected or unselected steps thus puts the focus on the process of trying to build the system. A big advantage, I think, is that it encourages people to pay attention to details; hopefully it would encourage really detailed scenarios by proponents of Darwinism (ones that might be checked experimentally) and discourage just-so stories that leap over many steps without comment. The original definition of IC was accurate and worked well for its purposes, but definitions can always be refined. So with those thoughts in mind, I offer the following tentative "evolutionary" definition of irreducible complexity: *An irreducibly complex evolutionary pathway is one that contains one or more unselected steps (that is, one or more necessary-but-unselected mutations). The degree of irreducible complexity is the number of unselected steps in the pathway.*

That definition has the advantage of promoting research: to state clear, detailed evolutionary pathways; to measure probabilistic resources; to estimate mutation rates; to determine if a given step is selected or not. It allows for the proposal of any evolutionary scenario a Darwinist (or others) may wish to submit, asking only that it be detailed enough so that relevant parameters might be estimated. If the improbability of the pathway exceeds the available probabilistic resources (roughly the number of organisms over the relevant time in the relevant phylogenetic branch) then Darwinism is deemed an unlikely explanation and intelligent design a likely one.

16. Reply to Michael Ruse

This article appeared in a discussion of intelligent design published in *Research News and Opportunity in Science and Theology* (now discontinued). Other contributors included Karl Giberson, Michael Ruse, Eugenie Scott, William Dembski, Robert Pennock, and Jonathan Wells. Philosopher Ruse unwisely stepped out of his area of expertise to weigh in on a scientific dispute between Russell Doolittle and myself.

"A Reply to Michael Ruse," *Research News & Opportunity in Science and Theology* (July/August 2002).

IF NOTHING ELSE, MICHAEL RUSE HAS CHUTZPAH.

Let me tell a little story about blood clotting, Russell Doolittle, and Michael Ruse. In 1996, in *Darwin's Black Box*, I argued ("notoriously") that the blood clotting cascade is irreducibly complex (that is, if a part is removed the cascade doesn't work), and so is a problem for Darwinian evolution and is better explained by intelligent design. However, Russell Doolittle—professor of biochemistry at the University of California-San Diego, member of the National Academy of Sciences, and lifelong student of the blood clotting system—disagreed. Writing in 1997 in *Boston Review*, a publication of MIT, Doolittle pointed to a then-recent report which, he claimed, showed that several parts of the clotting system—plasminogen and fibrinogen—could be "knocked out" of mice without ill effect. (Fibrinogen is the fabric of the clot. Plasminogen removes clots once healing is complete.) According to Professor Doolittle, if one component is removed, the mice are in bad shape, but if two components are removed, the mice were normal.[31] While that would be an interesting result, it's incorrect. Doolittle misread the report.

The authors of the paper wrote in their abstract that "Mice deficient in plasminogen and fibrinogen are phenotypically indistinguishable from fibrinogen-deficient mice."[32] In other words, mice lacking both components have all the problems that mice lacking just fibrinogen have. Those problems include failure to clot, hemorrhage, and death of females during pregnancy. The mice are very far from "normal." They are decidedly not promising evolutionary intermediates.

A year later, apparently unaware of Doolittle's mistake, Ruse instructed the readers of *Free Inquiry* on why intelligent design proponents are scorned.

> For example, Behe is a real scientist, but this case for the impossibility of a small-step natural origin of biological complexity has been trampled upon contemptuously by the scientists working in the field. They think his grasp of the pertinent science is weak and his knowledge of the literature curiously (although conveniently) outdated. For example, far from the evolution of clotting being a mystery, the past three decades of work by Russell Doolittle and others has thrown significant light on the ways in which clotting came into being. More than this, it can be shown that the clotting mechanism does not have to be a one-step phenomenon with everything already in place and functioning. One step in the cascade involves fibrinogen, required for clotting, and another, plaminogen [*sic*], required for clearing clots away.[33]

And Ruse went on to quote the passage from Doolittle I quoted above (see essay 15 in this volume, "The Irreducibility of the Blood Clotting Cascade"). Ruse was so impressed with Doolittle's work that he even *copied his typo/misspelling*, "plaminogen." Let me state clearly what this means. Ruse is a prominent academic Darwinian philosopher. Yet he apparently didn't even bother to look up and understand the original paper on the hemorrhaging mice before deciding Doolittle was right and I was contemptibly wrong. To this day he takes sides in a scientific dispute he shows no signs of understanding.

But perchance Ruse is so confident because "the rest of the scientific community agrees" with Doolittle (how does Ruse know this?) that I'm "simply not up to date." Well, maybe many scientists do agree with Doolittle. But those who do are as wrong as he was. In my travels I've had quite a few scientists sneeringly throw his erroneous *Boston Review* argument at me. Just recently Neil S. Greenspan, a professor of pathology at Case Western Reserve University, wrote in the *Scientist*, "The Design advocates also ignore the accumulating examples of the reducibility of biological systems. As Russell Doolittle has noted in commenting on the writings of one ID advocate..."[34] and Greenspan goes on to approvingly cite Doolittle's mistaken argument in *Boston Review*.

Then with innocent irony Greenspan continues, "These results cast doubt on the claim by proponents of ID that they know which systems exhibit irreducible complexity and which do not." But since the results of the hemorrhaging-mice study were precisely the opposite of what Doolittle along with Ruse, Greenspan, and other copycats thought, the shoe is on the other foot. The Doolittle incident shows that Darwinists in fact don't know how natural selection could assemble complex biochemical systems. Worse, it shows that they either cannot or will not recognize problems for their theory.

I'll bet a philosopher like Ruse could think of some other reasons why a lot of the scientific community is up in arms over intelligent design besides spurious claims that we "fail to understand the workings of evolution."

17. MILLER VERSUS LUSKIN

"Miller vs. Luskin, Part 1," Uncommon Descent (website), January 12, 2009;
"Miller vs. Luskin, Part 2," Uncommon Descent (website), January 15, 2009.

BROWN UNIVERSITY PROFESSOR KENNETH MILLER HAS GOTTEN into a little tiff with Discovery Institute's Casey Luskin over what I said/meant about the blood clotting cascade in *Darwin's Black Box*. Let me make a few comments.

In Chapter 4 of *Darwin's Black Box* I first described the clotting cascade and then, in a section called "Similarities and Differences," analyzed it in terms of irreducible complexity. Near the beginning of that part I had written: "Leaving aside the system before the fork in the pathway, where details are less well known, the blood clotting system fits the definition of irreducible complexity... The components of the system (beyond the fork in the pathway) are fibrinogen, prothrombin, Stuart factor, and proaccelerin."

Casey Luskin concludes that from that point on I was focusing my argument on the system beyond the fork in the pathway, containing those components I named.[35] That is a reasonable conclusion because, well, because that's what I said I was doing, and Mr. Luskin can comprehend the English language.

Apparently Professor Miller can't. He breathlessly reports that one page after I had qualified my argument I wrote, "Since each step necessarily requires several parts, not only is the entire blood-clotting system irreducibly complex, but so is each step in the pathway," and he asserts that this meant I had inexplicably switched back to considering the whole cascade, including the initial steps. It seems not to have occurred to Miller that that sentence should be read in the context of the previous page, so he focuses on the components before the fork, the better to construct a straw man to knock down. In fact, in that section containing the second quote ("Since each step...") I was arguing about the difficulty of inserting a new step into the middle of a generic, pre-existing cascade ("One could imagine a blood clotting system that was somewhat simpler than the real one—where, say, Stuart factor, after activation by the rest of the cascade, directly cuts fibrinogen to form fibrin, bypassing thrombin"), and likened it to inserting a lock in a ship canal. It could be done if an intelligent agent were directing it, but it would be really difficult to do by chance/selection. All that seems to have passed Miller by.

In philosophy there is something called the "principle of charitable reading." In a nutshell, it means that one should construe an author's argument in the best way possible, so that the argument is engaged in its strongest form. Unfortunately, in my experience Miller does the opposite—call it the "principle of malicious reading." He ignores (or doesn't comprehend) context, ignores (or doesn't comprehend) the distinctions an author makes, and construes the argument in the worst way possible.

Good salesmanship. Bad scholarship.

At the end of the post squabbling with Luskin, Miller refers to some great new work by UC San Diego professor and member of the National Academy of Science Russell Doolittle. Doolittle, of course, has worked on the blood clotting cascade for about fifty years! (I discussed some of his work in Chapter 4 of *Darwin's Black Box*.) In a new paper Doolittle and coworkers analyze DNA sequence data from a primitive vertebrate, the lamprey, thinking that it might have a simpler clotting cascade than higher vertebrates.[36] It is difficult work, because the sequences of lamprey proteins—even ones that are indeed homologous to the proteins of other vertebrates—are significantly diverged from, say, mammalian proteins.

They argue that most of the core clotting cascade proteins are present, but two seem to be absent: the lamprey has single proteins that act as Factor V/VIII (proaccelerin/anti-hemophilic factor) and Factor IX/X (Christmas factor/Stuart factor). The authors then infer that either gene or genome duplication led to separation of the factors. Although it's interesting work, Doolittle's conclusions are only suggestive (and the authors clearly say that the data are only suggestive). They found four copies of genes that are similar to Factors V/VIII, as well as to the nonclotting proteins ceruloplasmin and hephaestin (Figure 2 in their paper). They argue that only one is a real blood clotting factor and the other three aren't, but the arguments are pretty tentative. The same for Factors IX/X. The authors identify two "Factor X" genes. Might one of those be acting as a Factor IX gene? At the conclusion of the paper the authors say they may try to support their arguments with biochemical experiments. I'm looking forward to reading the results of those.

Whether or not their conclusion is correct, however, as far as the argument for intelligent design is concerned the only relevant part of

Doolittle's paper is Figure 10, which purports to show the clotting pathway in lampreys vs. other vertebrates. (Intelligent design is wholly compatible with common descent—including descent by gene duplication/rearrangement. Rather, ID argues against the Darwinian claim that complex, functional molecular systems could be built by a random, unguided process.) Yet to get from one arrangement to the other one would take multiple steps, not just one: whole genome duplication, re-targeting of Factor IX, re-targeting of Factor VIII, and so on. (The problems are essentially the same, as I pointed out in "In Defense of the Irreducibility of the Blood Clotting Cascade," included above in this collection.) So even if the suggested events occurred, they were extremely unlikely to have occurred by a Darwinian mechanism of random mutation/natural selection (the authors make no argument for a Darwinian mechanism). Guided, yes. Random, no.

It's pertinent to remember here the central point of my second book, *The Edge of Evolution*. We now have data in hand that show what Darwinian processes can accomplish, and it ain't much. We no longer have to rely on speculative scenarios that overlook barriers and problems that nature would encounter. Random mutation/natural selection works great in folks' imaginations, but it's a bust in the real world.

18. "A True Acid Test"

"'A True Acid Test,'" Discovery Institute, July 31, 2000.

IN THIS ESSAY I REPLY TO WHAT I CONSIDER TO BE THE MOST IMPORtant claim made by any critic of intelligent design: that direct experimental evidence has shown that evolution can indeed generate irreducibly complex biochemical systems. As I will show below, the claim is false.

Briefly, in his book *Finding Darwin's God*[37] Kenneth Miller quite rightly says that a "true acid test" of Darwinism is to see if it could regenerate an irreducibly complex system that was knocked out using the tools of molecular biology. He then discusses work from the laboratory of Barry Hall of the University of Rochester on the *lac* operon of the bacterium *E. coli*. Miller strongly implies that natural selection pieced together the whole pathway in Hall's experiments, but in fact it only replaced one component (and even then it could only replace the component with a spare near-copy of the original component). When two or more components were deleted, or when the bacterium was cultured in the absence of an artificial chemical (called IPTG), no viable bacteria could be recovered. Just as irreducible complexity would predict, when several steps must be taken at once, natural selection is a poor way to proceed.

Since Miller calls this work the "acid test," that of course means that other examples he discusses in his book are not "acid tests"; they are at best indirect arguments. The more indirect the argument, the easier for Darwinists to overlook or conceal difficulties.

"A TRUE ACID TEST"

BROWN UNIVERSITY cell biologist Kenneth Miller has written a book recently defending Darwinism from a variety of critics, including me. In a chapter devoted to rebutting *Darwin's Black Box*, he marshals an array of examples which, he asserts, tell against claims of irreducible complexity. However, for all of his counterexamples, I either disagree that he is dealing with irreducibly complex systems, disagree that he is focusing on the irreducibly complex aspects of a system, or disagree that his brief scenarios successfully answer the challenge of irreducible complexity (for an example, see my critique above of his blood clotting scenario). In this section I focus on his most serious claim—that an experiment has shown natural selection can construct an irreducibly complex system.

Professor Miller correctly states that "a true acid test" of the ability of Darwinism to deal with irreducible complexity would be to use "the tools of molecular genetics to wipe out an existing multipart system and then see if evolution can come to the rescue with a system to replace it."[38] Therefore the most important and novel part of Miller's rebuttal is his claim that experimental work in a bacterial system has actually succeeded in producing an irreducibly complex system by natural selection. In a section entitled "Parts Is Parts," in which he discusses the careful work over the past quarter-century by Barry Hall of the University of Rochester on the experimental evolution of a lactose-utilizing system in *E. coli*, Miller excitedly remarks:

> Think for a moment—if we were to happen upon the interlocking biochemical complexity of the reevolved lactose system, wouldn't we be impressed by the intelligence of its design? Lactose triggers a regulatory sequence that switches on the synthesis of an enzyme that then metabolizes lactose itself. The products of that successful lactose metabolism then activate the gene for the *lac* permease, which ensures a steady supply of lactose entering the cell. Irreducible complexity. What good would the permease be without the galactosidase?... No good, of course. By the very same logic applied by Michael Behe to other systems, therefore, we could conclude that the system had been designed. Except we know that it was not designed. We know it evolved because we watched it happen right in the laboratory![39]

I will show this picture is grossly exaggerated.

Here is a brief description of how the *lac* operon functions. The *lac* operon of *E. coli* contains genes coding for several proteins which are involved in metabolism of the disaccharide lactose. One protein of the *lac* operon, called a permease, imports lactose through the otherwise impermeable cell membrane. Another protein is an enzyme called β-galactosidase, which can hydrolyze the disaccharide to its two constituent monosaccharides, galactose and glucose, which the cell can then process further. Because lactose is rarely available in the environment, the bacterial cell switches off synthesis of the permease and β-galactosidase to conserve energy until lactose is available. The switch is controlled by another protein called a repressor, whose gene is located next to the operon. Ordinarily the repressor binds to the *lac* operon, shutting it off by physically interfering with expression of the operon. In the presence of the natural "inducer" allolactose (a by-product of lac β-galactosidase activity) or the artificial chemical inducer isopropylthiogalactoside

(IPTG), however, the repressor binds to the inducer and releases the operon, allowing the lac operon enzymes to be synthesized by the cell.

When I first read this section of Miller's book, I was quite impressed by the prospect that actual experiments—not theoretical, "just-so" stories—had produced a genuine, non-trivial counterexample to irreducible complexity. After going back to read Professor Hall's publications, however, I found that the situation was considerably different. Not only were Hall's results not what I expected based on Miller's description, in fact they fit most naturally within a framework of irreducible complexity and intelligent design. The same work that Miller points to as an example of Darwinian prowess I would cite as showing the limits of Darwinism and the need for design.

ADAPTIVE MUTATION

So WHAT did Barry Hall actually do? To study bacterial evolution in the laboratory, in the mid-1970s Hall produced a strain of E. coli in which the gene for just the β-galactosidase of the lac operon was deleted. He later wrote, "All of the other functions for lactose metabolism, including lactose permease and the pathways for metabolism of glucose and galactose, the products of lactose hydrolysis, remain intact, thus re-acquisition of lactose utilization requires only the evolution of a new β-galactosidase function."[40]

Thus, contrary to Miller's own criterion for "a true acid test," a multipart system was not "wiped out"—only one component of a multipart system was deleted.

Without β-galactosidase, Hall's cells could not grow when cultured on a medium containing only lactose as a carbon source. However, when grown on a plate that also included alternative, useable nutrients, bacterial colonies could be established. When the other nutrients were exhausted the colonies stopped growing. However, Hall noticed that after several days to several weeks, hyphae grew on some of the colonies. Upon isolating cells from the hyphae, Hall saw that they frequently had two mutations, one of which was in a gene for a protein he called "evolved β-galactosidase" ("*ebg*"), which allowed it to metabolize lactose efficiently. (Despite considerable efforts by Hall to determine it, the natural function of ebg remains unknown.)[41] The *ebg* gene is located in another operon, distant from the *lac* operon, and is under the control of its own repressor protein. The second mutation Hall found was always in the gene for the ebg repressor protein, which caused the repressor to bind lactose with sufficient strength to de-repress the ebg operon.

The fact that there were two separate mutations in different genes—neither of which by itself allowed cell growth[42]—startled Hall, who knew that the odds against the mutations appearing randomly and independently were prohibitive.[43] Hall's results and similar results from other laboratories led to research in the area dubbed "adaptive mutations."[44] As Hall later wrote,

> Adaptive mutations are mutations that occur in nondividing or slowly dividing cells during prolonged nonlethal selection, and that appear to be specific to the challenge of the selection in the sense that the only mutations that arise are those that provide a growth advantage to the cell. The issue of the specificity has been controversial because it violates our most basic assumptions about the randomness of mutations with respect to their effect on the cell.[45]

The mechanism(s) of adaptive mutation are currently unknown. While they are being sorted out, it is misleading to cite results of processes which "violate our most basic assumptions about the randomness of mutations" to argue for Darwinian evolution, as Miller does.

A Nearly Identical Active Site

THE NATURE of adaptive mutation aside, a strong reason to consider the lac/ebg results as quite modest is that the ebg proteins—both the repressor and β-galactosidase—are homologous to the *E. coli* lac proteins and overlap the proteins in activity. Both of the unmutated ebg proteins already bind lactose. Binding of lactose even to the unmutated ebg repressor induces a one-hundred fold increase in synthesis of the ebg operon.[46] Even the unmutated ebg β-galactosidase can hydrolyze lactose at a level of about 10% that of a "Class II" mutant β-galactosidase that supports cell growth.[47] These activities are not sufficient to permit growth of E. coli on lactose, but they already are present. The mutations reported by Hall simply enhance pre-existing activities of the proteins. In a recent paper[48] Professor Hall pointed out that both the lac and ebg β-galactosidase enzymes are part of a family of highly-conserved β-galactosidases, identical at thirteen of fifteen active site amino acid residues, which apparently diverged by gene duplication more than two billion years ago. The two mutations in ebg β-galactosidase that increase its ability to hydrolyze lactose change two nonidentical residues back to those of other β-galactosidases in ebg's phylogenetic class, so that their active sites are identical. Thus—before any experiments were done—the ebg active site was already a near-duplicate of other β-galactosidases, and only became more active by becoming a complete duplicate. Significantly, by phylogenetic analysis Hall concluded that those two mutations are the only ones in E. coli that confer the ability to hydrolyze lactose:

> The phylogenetic evidence indicates that either Asp-92 and Cys/Trp-977 are the only acceptable amino acids at those positions, or that all of the single base substitutions that might be on the pathway to other amino acid replacements at those sites are so deleterious that they constitute a deep selective valley that has not been traversed in the 2 billion years since those proteins diverged from a common ancestor.[49]

Such results hardly support extravagant claims for the creativeness of Darwinian processes.

CAVEATS UNMENTIONED

A CRITICAL caveat not mentioned by Kenneth Miller is that the mutants that were initially isolated would be unable to use lactose in the wild—they required the artificial inducer IPTG to be present in the growth medium. The reason is that a permease is required to bring lactose into the cell. However, ebg only has a β-galactosidase activity, not a permease activity, so the experimental system had to rely on the pre-existing lac permease. Since the lac operon is repressed in the absence of either allolactose or IPTG, Hall decided to include the artificial inducer in all media up to this point so that the cells could grow. Thus the system was being artificially supported by intelligent intervention.

"At this point it is important to discuss the use of IPTG in these studies," Hall wrote. "Unless otherwise indicated, IPTG is always included in media containing lactose or other β-galactoside sugars. The sole function of the IPTG is to induce synthesis of the lactose permease, and thus to deliver lactose to the inside of the cell. Neither the constitutive nor the inducible evolved strains grew on lactose in the absence of IPTG."[50]

With further growth and selection, Hall isolated secondary mutants with improved β-galactosidase activity. These mutants all had the same two changes (mentioned above) at positions 92 and 977 of ebg β-galactosidase. Hall discovered that, in addition to hydrolyzing lactose, the double mutants could also synthesize some allolactose, just as the homologous lac β-galactosidase can do, allowing them to induce expression of the *lac* operon without further need of IPTG. Critically again, however, the lac permease induced by the action of the double mutant ebg is a pre-existing protein, part of the original lac operon, and was not produced in the experiment by the selection procedures. In the absence of that required component, the bacteria cannot use lactose.

Miller's prose ("Irreducible complexity. What good would the permease be without the galactosidase?"[51]) obscures the facts that most of the system was already in place when the experiments began, that the system was carried through nonviable states by inclusion of IPTG, and that the system will not function without pre-existing components. In contrast to Miller, Hall himself is cautious and clear about the implications of his results:

> The mutations described above have been deliberately selected in the laboratory as a model for the way biochemical pathways might evolve so that they are appropriately organized with respect to both the cell and its environment. It is reasonable to ask whether this model might have any relationship to the real world outside the laboratory. If it is assumed that the selection is strictly for lactose utilization, then a growth

advantage exists only when all three mutations are present simultaneously.[52]

Hall is nonetheless optimistic: "Any one of the mutations alone could well be neutral (it is unlikely that any would be disadvantageous); but neutral mutations do enter populations by random chance events, and are fixed by a chance process termed genetic drift."[53]

However, if a mutation is not selected, the probability of its being fixed in a population is independent of the probability of the next mutation. Such a system is irreducibly complex, requiring several steps to be taken independently of each other before having selective value. If three mutations are required before there is any selective value, then the cumulative probability starts to become very small indeed, even considering the size of bacterial populations. In the present case Hall argued that a small selective value might accrue after the second mutation (in the ebg repressor).[54] However, I find his rationale unconvincing and having little experimental support. Furthermore, Professor Hall does not discuss the implications of the requirement for the preexisting lac permease gene.

CONCLUSION

MILLER ENDS the section in his typical emphatic style:[55] "No doubt about it—the evolution of biochemical systems, even complex multipart ones, is explicable in terms of evolution. Behe is wrong."[56]

I disagree. Leaving aside the still-murky area of adaptive mutation, the admirably careful work of Hall involved a series of micromutations stitched together by intelligent intervention. He showed that the activity of a deleted enzyme could be replaced only by mutations to a second, homologous protein with a nearly identical active site; and only if the second repressor already bound lactose; and only if the system were also artificially supported by inclusion of IPTG; and only if the system were also allowed to use a preexisting permease. Such results are exactly what one expects of irreducible complexity requiring intelligent intervention, and of the limited capabilities for Darwinian processes.

19. COMMENTS ON MILLER'S REPLY

"Comments on Ken Miller's Reply to My Essays,"
Discovery Institute, January 8, 2001.

KENNETH MILLER, BROWN UNIVERSITY PROFESSOR OF BIOLOGY
and author of *Finding Darwin's God*, has posted a response to my essays.[57] Overall I'm satisfied with his reply because, although he continues
to defend his position, from the substance of his writing I think it should
be plain to most open-minded readers that he is struggling to fend off
examples that weigh heavily against Darwinism. So, for the most part,
I am content to let the exchange end here. I simply urge all who are interested to read my essays as well as his response and come to their own
conclusions. I do, however, want to make a few additional comments, in
just two areas, to keep the issues in focus.

ON THE "ACID TEST"

PROFESSOR MILLER claims that the careful work of University of Rochester biologist Professor Barry Hall is an experimental demonstration
of the ability of Darwinian evolution to produce an irreducibly complex
biochemical system. (Barry Hall himself never made such a claim.) I
disagree with Miller here. The fact that the artificial chemical inducer
IPTG was added to the lactose-utilizing system effectively mitigated its
irreducibility, turning the system into one that could be improved a step
at a time.

"Does Barry Hall's ebg system fit the definition of irreducible complexity?" Miller asks. "Absolutely. The three parts of the evolved system
are: (1) A lactose-sensitive ebg repressor protein that controls expression
of the galactosidase enzyme; (2) The ebg galactosidase enzyme; (3) The
enzyme reaction that induces the lac permease. Unless all three are in
place, the system does not function, which is, of course, the key element
of an irreducibly complex system." Miller's claim is incorrect because in
the presence of IPTG the three features he lists are not all needed.

In the presence of IPTG, the "enzyme reaction that induces the lac
permease" is not required because IPTG itself induces the lac permease.

Thus, in the presence of IPTG the system is not irreducibly complex. And, as I wrote in my original essay, Barry Hall clearly noted that in the absence of IPTG—when the system actually is irreducibly complex—no viable mutants have been found in his twenty-five years of investigation.

The inclusion of IPTG was the result of the decision of an intelligent agent (Barry Hall) to deliberately alleviate the irreducibility of the system. In the absence of that intelligent action, Darwinian processes alone were ineffective. That is exactly what intelligent design theorists would expect.

Miller also writes, "The ebg gene is actually only 34% homologous to the gene whose activity it replaces (meaning that about 2/3 of the protein is quite different from the galactosidase gene whose function it replaces)." Yet he knows as well as I do that 34% general sequence homology makes it virtually certain that the three-dimensional structures of the two enzymes are essentially identical. And since the active sites (the business end) of the enzymes are much more similar (they are identical in 13 of 15 residues), the ebg enzyme is pretty much a spare copy of the lac enzyme. Thus it seems to me that the taking over of lac galactosidase function by ebg hardly even rises to the level of microevolution.

What is actually surprising—even to a design theorist such as myself—is Barry Hall's finding that no enzyme other than ebg could fill in for the missing lac galactosidase. I would have expected otherwise. Perhaps even changes we would consider to be "microevolution" are oftentimes beyond the reach of Darwinian processes. Perhaps even I give natural selection too much credit.

On the Mousetrap Example

Professor Miller writes, "MacDonald's drawings address Behe's contention that 'all components have to be in place before any mice are caught.' They don't, of course, because there is more than one way to construct a mousetrap from mechanical parts."

But they do—all of the parts of my mousetrap do indeed have to be in place for it to function. I noted clearly in my book that I am very much aware "there is more than one way to construct a mousetrap from mechanical parts." Nonetheless, if you just take away pieces from the trap I pictured—and don't manipulate it further—the trap doesn't work. What Miller actually means is that if you take away some components and then go on to, for instance, twist a couple of metal pieces in just the right way and add a few staples in the correct positions, you can construct a new kind of working trap, which may superficially resemble the starting trap. That, however, is intelligent design. Neither Miller nor anyone else has shown that the mousetrap I pictured in my book can be constructed by a series of small changes, one at a time, as Darwinian evolution would have to do. The important take-home lesson is that even things that look superficially similar, such as the series of traps Miller showed, may not be able to be transformed into each other through a Darwinian process.

Towards the end of his essay Miller writes, "If simpler versions of this mechanical device [the mousetrap] can be shown to work, then simpler versions of biochemical machines could work as well... and this means that complex biochemical machines could indeed have had functional precursors."

However, consider the general statement, "System A looks like system B but cannot be transformed into system B by a series of small, Darwinian-like steps." Miller explicitly agrees that is true for the mousetrap series.[58] Yet in the statement above he seems to think that the very fact it is true for mousetraps somehow shows it is not true for biochemical systems.

That reasoning isn't completely backwards, but it's pretty close. If the simpler mousetraps can't be transformed into the similar-looking-but-more-complex ones by something analogous to a Darwinian process—which, again, Miller freely admits in his essay—then the mousetrap series gives us no reason to think that postulated simpler biochemical systems could have been transformed by natural selection into the complex systems we see today. On the contrary, the mousetrap series gives us a prima facie case to think they couldn't.

20. IRREDUCIBLE COMPLEXITY AND EVOLUTIONARY LITERATURE

This essay pertains to the paltry state of the evolutionary literature concerning molecular machinery around the turn of the millennium, relatively soon after the appearance of *Darwin's Black Box*. Since then, however, there has developed a veritable cottage industry publishing papers on the topic, devoted to countering the intelligent design threat. Emphasizing in one's paper how the results falsify irreducible complexity or easily explain some astounding molecular system virtually guarantees that it will be accepted by *Nature*, *Science*, or some other top-tier journal (usually accompanied by a story in the *New York Times*). To a person with the least skepticism, however, all such papers so far (many discussed later in this book) either confuse common descent with Darwinism, pass over profound difficulties, actually demonstrate greater problems for unintelligent evolution than had previously been appreciated, or otherwise are built on one or more other fatal defects. At this juncture—some twenty years after the publication of *Darwin's Black Box*, and in spite of the dedicated efforts of many extremely intelligent scientists who utterly despise its conclusion— the case for the intelligent design of life is even more compelling, and the case against Darwinism even more damning, than even I had thought in those early years.

"Irreducible Complexity and the Evolutionary Literature:
Response to Critics," Discovery Institute, July 31, 2000.

ALTHOUGH SEVERAL PERSONS HAVE CITED NUMEROUS REFERENCES from the scientific literature purporting to show that the problem of irreducible complexity I pointed out in *Darwin's Black Box* is being seriously addressed, the references show no such thing. Invariably the cited papers or books either deal with non-irreducibly complex biochemical systems, or do not deal with actual irreducibly complex systems in enough detail for critical evaluation. I strongly emphasize, however, that I do not prefer it that way. I would sincerely welcome much more serious, sustained research in the area of irreducible complexity. I fully expect such research would heighten awareness of the difficulties of Darwinian evolution.

WEB SPINNERS

THE NECESSARY starting point of *Darwin's Black Box* was the contention that, despite the common assumption that natural selection accounts for adaptive complexity, the origins of many intricate cellular systems have not yet been explained in Darwinian terms. After all, if the systems have already been explained, then there's no need to write. While most scientist-reviewers disagreed (often emphatically) with my proposal of intelligent design, most also admitted to a lack of Darwinian explanations. For example, microbiologist James Shapiro of the University of Chicago declared in *National Review*, "There are no detailed Darwinian accounts for the evolution of any fundamental biochemical or cellular system, only a variety of wishful speculations."[59] In *Nature* University of Chicago evolutionary biologist Jerry Coyne conceded that "there is no doubt that the pathways described by Behe are dauntingly complex, and their evolution will be hard to unravel," and that "we may forever be unable to envisage the first proto-pathways."[60]

In a particularly scathing review in *Trends in Ecology and Evolution* Tom Cavalier-Smith, an evolutionary biologist at the University of British Columbia, nonetheless wrote, "For none of the cases mentioned by Behe is there yet a comprehensive and detailed explanation of the probable steps in the evolution of the observed complexity. The problems have indeed been sorely neglected—though Behe repeatedly exaggerates this neglect with such hyperboles as 'an eerie and complete silence.'"[61] Evolutionary biologist Andrew Pomiankowski agreed in *New Scientist*: "Pick up any biochemistry textbook, and you will find perhaps two or three references to evolution. Turn to one of these and you will be lucky to find anything better than 'evolution selects the fittest molecules for their biological function.'"[62] In *American Scientist* Yale molecular biologist Robert Dorit averred, "In a narrow sense, Behe is correct when he argues that we do not yet fully understand the evolution of the flagellar motor or the blood clotting cascade."[63]

A prominent claim I made in *Darwin's Black Box* is that not only are irreducibly complex biochemical systems unexplained, there have been very few published attempts even to try to explain them. This contention has been vigorously disputed not so much by scientists in the relevant fields as by Darwinian enthusiasts on the internet. Several web-savvy fans of natural selection have set up extensive, sophisticated sites that appear to receive a significant amount of notice. They influence college students, reporters and, sometimes, academic reviewers of my book, such as Cal State-Fullerton biochemist Bruce Weber, who listed the addresses of the web sites in his review in *Biology and Philosophy* as "summaries of the current research that Behe either missed or misrepresented,"[64] and Oxford physical chemist Peter Atkins, who wrote:

> Dr. Behe claims that science is largely silent on the details of molecular evolution, the emergence of complex biochemical pathways and processes that underlie the more traditional manifestations of evolution at the level of organisms. Tosh! There are hundreds, possibly thousands, of scientific papers that deal with this very subject. For an entry into this important and flourishing field, and an idea of the intense scientific effort that it represents (see the first link above) [*sic*].[65]

The link Atkins referred to is a website[66] set up by a man named John Catalano, an admirer of Oxford biologist Richard Dawkins. (Catalano's larger site is devoted to Dawkins's work, schedule, etc.) It contains citations to a large number of papers and books which Catalano believes belie my claim that "there has never been a meeting, or a book, or a paper on details of the evolution of complex biochemical systems."[67] The citations were solicited on the web from anyone who had a suggestion, and then compiled by Catalano.

Something, however, seems to be amiss. The assertion here that very many papers have been published clashes with statements of the reviews I quoted earlier which say, for example, that "the problems have indeed been sorely neglected."[68] Would reviewers such as Jerry Coyne and Tom Cavalier-Smith—both antagonistic to my proposal of intelligent design—be unaware of the "hundreds, possibly thousands, of scientific papers that deal with this very subject"? Both claims—that the problems have been neglected and that the problems are being actively investigated—cannot be correct. Either one set of reviewers is wrong, or there is some confusion about which publications to count. Which is it?

In the context of my book it is easy to realize that I meant there has been little work on the details of the evolution of irreducibly complex biochemical systems by Darwinian means. I had clearly noted that of course a large amount of work in many books and journals was done under the general topic of "molecular evolution," but that, overwhelmingly, it was either limited to comparing sequences (which, again, does not concern the mechanism of evolution) or did not propose sufficiently detailed routes to justify a Darwinian conclusion. Yet the Catalano site lists virtually any work on evolution, whether it pertains to irreducible complexity or not.

For example, it lists semi-popular books such as *Patterns in Evolution: The New Molecular View* by Roger Lewin, and general textbooks on molecular evolution such as *Molecular Evolution* by Wen-Hsiung Li. Such books simply don't address the problems I raise. *Molecular Evolution* by Wen-Hsiung Li[69] is a fine textbook which does an admirable job of explicating current knowledge of how genes change with time. That knowledge, however, does not include how specific, irreducibly complex biochemical systems were built. The text contains chapters on the molecular clock, molecular phylogenetics, and other topics which essentially are studies in comparing gene sequences. As I explained in *Darwin's Black Box*, comparing sequences is interesting but cannot explain how molecular machines arose. Li's book also contains chapters on the mechanisms (such as gene duplication, domain shuffling, and concerted evolution of multigene families) that are thought to be involved in evolution at the molecular level. Again, however, no specific system is justified in Darwinian terms.

Here is an illustration of the problem. Li spends several pages discussing domain shuffling in the proteins of the blood clotting cascade.[70] However, Li himself has not done work on understanding how the obstacles to the evolution of the clotting cascade may have been circumvented. Since those investigators who do work in that area have not yet published a detailed Darwinian pathway in the primary literature,[71] we can conclude that the answer will not be found in a more general text. We can further assume that the processes that text describes (gene duplication, etc.), although very significant, are not by themselves sufficient to understand how clotting, or by extension any complex biochemical system, may have arisen by Darwinian means.

Catalano's site lists other books that I specifically discussed in *Darwin's Black Box*, where I noted that, while they present mathematical models or brief general descriptions, they do not present detailed biochemical studies of specific irreducibly complex systems.[72] There is no explanation on Catalano's web site of why he thinks they address the questions I raised. The site also points to papers with intriguing titles, but which are studies in sequence analysis, such as "Molecular Evolution of the Vertebrate Immune System"[73] and "Evolution of Chordate Actin Genes: Evidence from Genomic Organization and Amino Acid Sequences."[74] As I explained in *Darwin's Black Box*, sequence studies by themselves can't answer the question of what the mechanism of evolution is. Catalano's compendium also contains citations to papers concerning the evolution of non-irreducibly complex systems, such as hemoglobin and metabolic pathways, which I specifically said may have evolved by natural selection.[75]

EQUIVOCAL TERMS

ANOTHER WEBSITE that has drawn attention (as evidenced from the inquiries I receive soliciting my reaction to it) is authored by David Ussery,[76] associate research professor of biotechnology at The Technical University of Denmark. One of his main goals is to refute my claim concerning the dearth of literature investigating the evolution of irreducibly complex systems. For example, in a section on intracellular vesicular transport he notes that I stated in *Darwin's Black Box* that a search of a computer database "to see what titles have both evolution and vesicle in them comes up completely empty."[77] My search criterion, of having both words in the title, was meant to be a rough way to show that nothing much has been published on the subject. Ussery, however, writes that, on the contrary, a search of the PubMed database using the words evolution and vesicle identifies well over a hundred papers. Confident of his position, he urges his audience, "But, please, don't just take my word for it—have a look for yourself!"[78]

The problem is that, as I stated in the book, I had restricted my search to the titles of papers, where occurrence of both words would probably mean they concerned the same subject. Ussery's search used the default PubMed setting, which also looks in abstracts.[79] By doing so he picked up papers such as "Outbreak of nosocomial diarrhea by *Clostridium difficile* in a department of internal medicine."[80] This paper discusses the "clinical evolution" (i.e., course of development) of diarrhea in hospitalized patients, who also had "vesicle catherization." Not only do the words evolution and vesicle in this paper not refer to each other, the paper does not even use the words evolution and vesicle in the same sense as I did. Since the word evolution has many meanings, and since the word vesicle can mean just a container (like the word "box"), Ussery picked up equivocal meanings.

The paper cited above shows Ussery's misstep in an obvious way. However, there are other papers resulting from an Ussery-style search where, although they do not address the question I raised, the unrelatedness is not so obvious to someone outside the field. An example of a paper that is harder for someone outside the field to evaluate is "Evolution of the Trappin Multigene Family in the *Suidae*."[81] The authors examine the protein and gene sequences for a group of secretory proteins (the trappin family) which "have undergone rapid evolution" and are similar to "seminal vesicle clotting proteins." The results may be interesting, but the seminal vesicle is a pouch in the male reproductive tract for storing semen—not at all the same thing as the vesicle in which intracellular transport occurs. And trappins are not involved in intracellular transport.

A second example is "Syntaxin-16, a putative Golgi t-SNARE."[82] This paper actually does concern a protein involved in intracellular vesicular transport. However, as the abstract states, "Database searches identified putative yeast, plant and nematode homologues of syntaxin-16, indicating that this protein is conserved through evolution." The database searches are sequence comparisons. Once again I reiterate, sequence comparisons by themselves cannot tell us how a complex system might have arisen by Darwinian means.

Instead of listing further examples let me just say that I have not seen a paper using Ussery's search criteria that addresses the Darwinian evolution of intracellular vesicular transport in a detailed manner, in keeping with what I had originally asserted in my book.

It is impossible for me to individually address the "hundreds, possibly thousands" of papers listed in these web sites. But perhaps I don't have to. If competent scientists who are not friendly to the idea of intelligent design nonetheless say that "there are no detailed Darwinian accounts for the evolution of any fundamental biochemical or cellular system, only a variety of wishful speculations,"[83] and that "we may forever be unable to envisage the first proto-pathways,"[84] then it is unlikely that much literature exists on these problems. So after considering the contents of the web sites, we can reconcile the review of Peter Atkins with those of other reviewers. Yes, there are a lot of papers published on "molecular evolution," as I had clearly acknowledged in *Darwin's Black Box*. But a grand total of zero of them concern Darwinian details of irreducibly complex systems, which is exactly the point I was making.

KENNETH MILLER

IN *FINDING Darwin's God*[85] Kenneth Miller is also anxious to show my claims about the literature are not true (or at least are not true now, since the handful of papers he cites in his section "The Sound of Silence" were published after my book appeared). Yet none of the papers he cites deals with irreducibly complex systems.

The first paper Miller discusses concerns two structurally similar enzymes, both called isocitrate dehydrogenase. The main difference between the two is simply that one uses the organic cofactor NAD while the other uses NADP. The two cofactors are very similar, differing only in the presence or absence of a phosphate group. The authors of the study show that by mutating several residues in either enzyme, they can change the specificity for NAD or NADP.[86] Although the study is very interesting, at the very best it is microevolution of a single protein, not an irreducibly complex system.

The next paper Miller cites concerns "antifreeze" proteins.[87] Again, these are single proteins that do not interact with other components; they are not irreducibly complex. In fact, they are great examples of what I agree evolution can indeed do—start with a protein that accidentally binds something (ice nuclei in this case, maybe antibiotics in another case) and select for mutations that improve that property. But they don't shed light on irreducibly complex systems.

Another paper Miller cites concerns the cytochrome c oxidase proton pump,[88] which is involved in electron transfer. In humans six proteins take part in the function; in some bacteria fewer proteins are involved. While quite interesting, the mechanism of the system is not known in enough detail to understand what's going on; it remains in large part a black box. Further, the function of electron transfer does not necessarily require multiple protein components, so it is not necessarily irreducibly complex. Finally, the study is not detailed enough to criticize, saying things such as "It makes evolutionary sense that the cytochrome bc1 and cytochrome c oxidase complexes arose from a primitive quinol terminal oxidase complex via a series of beneficial mutations." In order to judge whether natural selection could do the job, we have to know what the "series of beneficial mutations" is. Otherwise it's like saying that a five-part mousetrap arose from a one-part mousetrap by a series of beneficial mutations.[89]

Finally Miller discusses a paper which works out a scheme for how the organic-chemical components of the tricarboxylic acid (TCA) cycle, a central metabolic pathway, may have arisen gradually.[90] There are several points to make about it. First, the paper deals with the chemical interconversion of organic molecules, not the enzymes of the pathway or their regulation. As an analogy, suppose someone described how petroleum is refined step by step, beginning with crude oil, passing through intermediate grades, and ending with, say, gasoline. He shows that the chemistry of the processes is smooth and continuous, yet says nothing about the actual machinery of the refinery or its regulation, nothing about valves or switches. Clearly that is inadequate to show refining of petroleum developed step by step. Analogously, someone who is seriously interested in showing that a metabolic pathway could evolve by Darwinian means has to deal with the enzymic machinery and its regulation.

The second and more important point is that, while the paper is very interesting, it doesn't address irreducible complexity. Either Miller hasn't read what I said in my book about metabolic pathways, or he is deliberately ignoring it. I clearly stated in *Darwin's Black Box* that metabolic pathways are not irreducibly complex,[91] because components can be gradually added to a previous pathway. Thus metabolic pathways simply aren't in the same category as the blood clotting cascade or the bacterial flagellum. Although Miller somehow misses the distinction, other scientists do not. In a recent paper Thornhill and Ussery write that something they call serial-direct-Darwinian-evolution "cannot generate irreducibly complex structures." But they think it may be able to generate a reducible structure, "such as the TCA cycle (Behe, 1996 a, b)."[92] (In other words Thornhill and Ussery acknowledge the TCA cycle is not irreducibly complex, as I wrote in my book. Miller seems unable or unwilling to grasp that point.

A PLEA FOR MORE RESEARCH

IN POINTING out that not much research has been done on the Darwinian evolution of irreducibly complex biochemical systems I should emphasize that I do not prefer it that way. I would sincerely welcome more research (especially experimental research, such as done by Barry Hall—see my discussion in an essay earlier in this volume, "The Acid Test," about Hall's work) into the supposed Darwinian origins of the complex systems I described in my book. I fully expect that, as in the field of origin-of-life studies, the more we know, the more difficult the problem will be recognized to be.

21. PHILOSOPHICAL OBJECTIONS TO INTELLIGENT DESIGN

"Philosophical Objections to Intelligent Design,"
Discovery Institute, July 31, 2000.

SOME REVIEWERS OF *DARWIN'S BLACK BOX*[93] HAVE RAISED PHILO-sophical objections to intelligent design. I will discuss several of these over the next few sections, beginning with the question of falsifiability.

IS INTELLIGENT DESIGN FALSIFIABLE?

To DECIDE whether, or by what evidence, it is falsifiable, one first has to be sure what is meant by "intelligent design." By that phrase someone might mean that the laws of nature themselves are designed to produce life and the complex systems that undergird it. In fact, something like that position has been taken by the physicist Paul Davies and the geneticist Michael Denton in their recent books, respectively, *The Fifth Miracle: The Search for the Origin and Meaning of Life*[94] and *Nature's Destiny: How the Laws of Biology Reveal Purpose in the Universe.*[95] That stance also seems to pass muster with the National Academy of Sciences:

> Many religious persons, including many scientists, hold that God created the universe and the various processes driving physical and biological evolution and that these processes then resulted in the creation of galaxies, our solar system, and life on Earth. This belief, which sometimes is termed 'theistic evolution,' is not in disagreement with scientific explanations of evolution. Indeed, it reflects the remarkable and inspiring character of the physical universe revealed by [science].[96]

In such a view, even if we observe new complex systems being produced by selection pressure in the wild or in the laboratory, design would not be falsified because it is considered to be built into natural laws. Without commenting on the merits of the position, let me just say that that is not the meaning I assign to the phrase. By "intelligent design" I mean to imply design beyond the laws of nature. That is, taking the laws of nature as given, are their other reasons for concluding that life and its component systems have been intentionally arranged? In my book, and in this essay, whenever I refer to intelligent design (ID), I mean this stronger sense of design-beyond-laws. Virtually all academic critics of my book have taken the phrase in the strong sense I meant it.

In the strong sense, ID is no longer approved by the National Academy, for a specific reason. They say that it isn't science because it's "not testable by the methods of science."[97] In his review of *Darwin's Black Box* for the journal *Nature*, Jerry Coyne, professor of evolutionary biology at the University of Chicago, explains why he also thinks intelligent design is unfalsifiable:

> If one accepts Behe's idea that both evolution and creation can operate together, and that the Designer's goals are unfathomable, then one confronts an airtight theory that can't be proved wrong. I can imagine evidence that would falsify evolution (a hominid fossil in the Precambrian would do nicely), but none that could falsify Behe's composite theory. Even if, after immense effort, we are able to understand the evolution of a complex biochemical pathway, Behe could simply claim that evidence for design resides in the other unexplained pathways. Because we will never explain everything, there will always be evidence for design. This regressive ad hoc creationism may seem clever, but it is certainly not science.[98]

Coyne's conclusion that design is unfalsifiable, however, seems to be at odds with the arguments of other reviewers of my book. Clearly, Russell Doolittle,[99] Kenneth Miller,[100] and others have advanced scientific arguments aimed at falsifying ID. (See my articles on blood clotting and the "acid test" on this web site.) If the results with knock-out mice[101] had been as Doolittle first thought, or if Barry Hall's work[102] had indeed shown what Miller implied, then they correctly believed that my claims about irreducible complexity would have suffered quite a blow. And since my claim for intelligent design requires that no unintelligent process be sufficient to produce such irreducibly complex systems, then the plausibility of ID would suffer enormously. Other scientists, including those on the National Academy of Science's Steering Committee on Science and Creationism, in commenting on my book have also pointed to physical evidence (such as the similar structures of hemoglobin and myoglobin) which they think shows that irreducibly complex biochemical systems can be produced by natural selection: "However, structures and processes that are claimed to be 'irreducibly' complex typically are not on closer inspection."[103]

Now, one can't have it both ways. One can't say both that ID is unfalsifiable (or untestable) and that there is evidence against it. Either it is unfalsifiable and floats serenely beyond experimental reproach, or it can be criticized on the basis of our observations and is therefore testable. The fact that critical reviewers advance scientific arguments against ID (whether successfully or not) shows that intelligent design is indeed falsifiable.

In fact, my argument for intelligent design is open to direct experimental rebuttal. Here is a thought experiment that makes the point clear. In *Darwin's Black Box* I claimed that the bacterial flagellum was irreducibly complex and so required deliberate intelligent design. The flip side of this claim is that the flagellum can't be produced by natural selection acting on random mutation, or any other unintelligent process. To falsify such a claim, a scientist could go into the laboratory, place a bacterial species lacking a flagellum under some selective pressure (for mobility, say), grow it for ten thousand generations, and see if a flagellum—or any equally complex system—was produced. If that happened, my claims would be neatly disproven.[104]

How about Professor Coyne's concern that, if one system were shown to be the result of natural selection, proponents of ID could just claim that some other system was designed? I think the objection has little force. If natural selection were shown to be capable of producing a system of a certain degree of complexity, then the assumption would be that it could produce any other system of an equal or lesser degree of complexity. If Coyne demonstrated that the flagellum (which requires approximately forty gene products) could be produced by selection, I would be rather foolish to then assert that the blood clotting system (which consists of about twenty proteins) required intelligent design.

Let's turn the tables and ask, how could one falsify the claim that, say, the bacterial flagellum was produced by Darwinian processes? (Professor Coyne's remarks about a Precambrian fossil hominid are irrelevant since I dispute the mechanism of natural selection, not common descent. I would no more expect to find a fossil hominid out of sequence than he would.) If a scientist went into the laboratory and grew a flagellum-less bacterial species under selective pressure for many generations and nothing much happened, would Darwinists be convinced that natural selection is incapable of producing a flagellum? I doubt it. It could always be claimed that the selective pressure wasn't the right one, or that we started with the wrong bacterial species, and so on. Even if the experiment were repeated many times under different conditions and always gave a negative result, I suspect many Darwinists would not conclude that Darwinian evolution was falsified. Of complex biochemical systems Coyne himself writes, "We may forever be unable to envisage the first proto-pathways. It is not valid, however, to assume that, because one man cannot imagine such pathways, they could not have existed."[105] If a person accepts Darwinian paths which are not only unseen, but which we may be forever unable to envisage, then it is effectively impossible to make him think he is wrong.

Kenneth Miller announced an "acid test" for the ability of natural selection to produce irreducible complexity. He then decided that the test was passed, and unhesitatingly proclaimed intelligent design falsified ("Behe is wrong"[106]). But if, as it certainly seems to me, E. coli actually fails the lactose-system "acid test," would Miller consider Darwinism to be falsified? Almost certainly not. He would surely say that the experiment started with the wrong bacterial species, used the wrong selective pressure, and so on. So it turns out that his "acid test" was not a test of Darwinism; it tested only intelligent design. The same one-way testing was employed by Russell Doolittle. He pointed to the results of Bugge et al.[107] to argue against intelligent design. But when the results turned out to be the opposite of what he had originally thought, Professor Doolittle did not abandon Darwinism.

It seems then, perhaps counterintuitively to some, that intelligent design is quite susceptible to falsification, at least on the points under discussion. Darwinism, on the other hand, seems quite impervious to falsification. The reason for that can be seen when we examine the basic claims of the two ideas with regard to a particular biochemical system like, say, the bacterial flagellum. The claim of intelligent design is that "No unintelligent process could produce this system." The claim of Darwinism is that "Some unintelligent process (involving natural selection and random mutation) could produce this system." To falsify the first claim, one need only show that at least one unintelligent process could produce the system. To falsify the second claim, one would have to show the system could not have been formed by any of a potentially infinite number of possible unintelligent processes, which is effectively impossible to do.

I think Professor Coyne and the National Academy of Sciences have it exactly backwards. A strong point of intelligent design is its vulnerability to falsification. (Indeed, some of my religious critics dislike intelligent design theory precisely because they worry that it will be falsified, and thus theology will appear to suffer another blow from science. See, for example, Flietstra.[108]) A weak point of Darwinian theory is its resistance to falsification. What experimental evidence could possibly be found that would falsify the contention that complex molecular machines evolved by a Darwinian mechanism?

WHAT IS "IRREDUCIBLE COMPLEXITY" AND WHAT DOES IT SIGNIFY?

SOME REVIEWERS have criticized the concept of irreducible complexity. In *Boston Review* University of Rochester evolutionary biologist H. Allen Orr agrees that many biological systems are "irreducibly complex," but argues that Darwinian evolution can, at least in theory, directly account for them. However, as I will show, his argument depends on changing the definition of irreducible complexity, which obscures the difficulty.

In his review Orr initially seems to clearly understand what I meant by "irreducible complexity" (quoted earlier). Of the example I used in *Darwin's Black Box* he writes: "A mousetrap has a clear function (crushing mice) and is made of several parts (a platform, a spring, a bar that does the crushing). If any of these parts is removed, the trap doesn't work. Hence it's irreducibly complex." So far, so good. Nonetheless, later in the review he seems to lose hold of the concept:

> An irreducibly complex system can be built gradually by adding parts that, while initially just advantageous, become—because of later changes—essential. The logic is very simple. Some part (A) initially does some job (and not very well, perhaps). Another part (B) later gets added because it helps A. This new part isn't essential, it merely improves things. But later on, A (or something else) may change in such a way that B now becomes indispensable. This process continues as further parts get folded into the system. And at the end of the day, many parts may all be required.[109]

Now, how can we square this paragraph with his initial agreement that if any part of a mousetrap is removed, it doesn't work? Thinking of the mousetrap example, what would correspond to "Some part (A)" that "initially does some job"? In fact, the whole point of the mousetrap example was to show that there is no "part (A)" that will initially do the job. There is no "part (B)" that helps gradually improve "part (A)." A gradual addition of parts is not possible for the mousetrap example (or at least it is very far from obvious that it is possible). Orr later gives a biological example of what he has in mind.

The transformation of air bladders into lungs that allowed animals to breathe atmospheric oxygen was initially just advantageous: such beasts could explore open niches—like dry land—that were unavailable to their lung-less peers. But as evolution built on this adaptation (modifying limbs for walking, for instance), we grew thoroughly terrestrial and lungs, consequently, are no longer luxuries—they are essential. The punch-line is, I think, obvious: although this process is thoroughly Darwinian, we are often left with a system that is irreducibly complex.[110]

In Orr's example, however, what is the irreducibly complex system? Is it the swim bladder? The lung? The whole organism? What is the function of the system? Is it "swimming," "breathing," "living," or something else? If we assume he meant that the irreducibly system is, say, the lung, can the lung be considered "a single system," as my definition requires?[111] What are the parts of the lung without which it will stop working, like a mousetrap without a spring? What is "part (A)" and what is "part (B)"? None of these things is clear at all—certainly not as clear as the parts and function of a mousetrap.

Let me preface my remaining remarks on this subject by acknowledging that it is often notoriously difficult to rigorously define a concept, as exemplified by the problems encountered in trying to define "science," "life," or "species." Furthermore, I am no philosopher; my end purpose is not to come up with a string of words that completely and exhaustively defines the phrase "irreducible complexity." Rather, my purpose is to focus attention on a class of biochemical systems that pose a particular challenge to Darwinian evolution. The examples I gave in my book—a mousetrap, cilium, clotting cascade, and so on—clearly show the necessity for some systems of having a number of discrete parts working together on a single function. The examples, I think, better get across the concept of irreducible complexity than does the definition I offered,[112] although I think the definition I gave does an adequate job.

With those comments in mind, it can be seen that Orr simply switched concepts in mid-review, as shown by his conflicting remarks quoted above. He jumped from my idea of irreducible complexity to a hazy concept that can perhaps be paraphrased as, "If you remove this part, the organism will eventually die." I'm happy to agree for purposes of discussion that a class of biological phenomena exists which are required for life and which can be changed gradually by natural selection, perhaps even including the swim bladder/lungs Orr mentions (although it is not nearly so obvious as he assumes it to be). It's just that they are not the same types of things as, nor do they somehow obviate the problem of, irreducibly complex systems like mousetraps and cilia. If they were, then Orr could have explained them away as easily as he does swim bladders and lungs. (After all, lung tissue contains cilia plus many, many other components; Orr should thus find it easier to explain cilia alone, rather than cilia-plus-other-components.) Implicitly changing the definition of irreducible complexity, as Orr did, does not tell us how the blood clotting cascade or the bacterial flagellum could have been produced. On the contrary, it distracts our attention from those features of the systems that make them recalcitrant to Darwinian explanation.

Other scientific reviewers have made arguments similar to Orr's which depend on implied definitions of irreducible complexity different from what I used. Writing in the *Wall Street Journal*, Paul Gross compares biochemical systems to cities, where features can be added over time.[113] But the analogy is poorly chosen because no city completely stops working when a part is removed, as does a mousetrap or cilium. In *Boston Review* Douglas Futuyma writes, "In mammals, successive duplications of the beta gene gave rise to the gamma and epsilon chains, which characterize the hemoglobin of the fetus and early embryo respectively, and enhance uptake of oxygen from the mother. Thus a succession of gene duplications, widely spaced through evolutionary time, has led to the "irreducibly complex" system of respiratory proteins in mammals."[114]

But the several hemoglobins that Futuyma calls the "'irreducibly complex' system of respiratory proteins" in fact do not constitute an irreducibly complex system in my sense of the term. They do not interact with each other, as do the parts of a mousetrap or clotting cascade. They go their separate ways, and for the most part aren't even present at the same time in the organism. Like Allen Orr, Futuyma implicitly switches the meaning of "irreducibly complex." Unfortunately, that does not solve the problem I pointed out, but only obscures it. (As an aside, it is difficult to understand what Futuyma intends by the quotation marks around the phrase *irreducibly complex*. He can't be quoting me; I never used the term in connection with hemoglobin—quite the opposite. He may intend them to be taken as "scare quotes," to warn the reader to take the phrase with a grain of salt. But since he is the one who decided to use the term in conjunction with hemoglobins and then to argue against it, the effect is that of setting up a straw man.)

A different question about irreducible complexity is asked by David Ussery on his web site. He notes that, whereas a bacterial flagellum in *E. coli* requires about forty different proteins, in *H. pylori* only thirty-three are required. Since fewer proteins are required, how can the flagellum be irreducibly complex? Two responses can be made. First, some systems may have parts that are necessary for a function, plus other parts that, while useful, are not absolutely required. Although one can remove the radio from a car and the car will still work, one can't remove the battery or some other parts and have a working car. Ussery himself seems to recognize this when he writes, "I would readily admit that there is STILL the problem of the evolution of the 'minimal flagellum,'"[115] but he hopes gene duplication will explain that. Second, one must be careful not to identify one protein with one "part" of a biochemical machine. For example, genes coding for two proteins in one organism may be joined into a single gene in another. A single protein in one organism may be doing the jobs of several polypeptides in another. Or two proteins may combine to do one job (an example is the α- and β-subunits of tubulin, which together make microtubules, a "part" of the eukaryotic cilium).

In his review Ussery mistakenly attributes to me the belief that 240 separate proteins are required for the bacterial flagellum. The confusion apparently arose because at the end of a chapter on the eukaryotic cilium and bacterial flagellum, I stated that a typical cilium contains over two hundred different kinds of proteins. In the next paragraph I wrote,

"The bacterial flagellum, in addition to the proteins already discussed, requires about forty other proteins for function."[116] Although I meant in addition to the flagellar proteins I had discussed a few pages earlier in the chapter, Ussery interpreted the statement to include the several hundred ciliary proteins as well. Ordinarily I would simply overlook such a mistaken attribution, since it should be obvious to informed readers that I wouldn't be lumping the proteins of cilia and flagella together— after all, they are completely different structures that occur in separate kinds of organisms. In his review in *Biology and Philosophy*, however, Bruce Weber writes, "Behe cannot imagine how anything short of the full 240 components of the flagellum could propel a bacterium. But only 33 proteins are needed to produce a functional flagellum for *Helicobacter pylori*."[117] And Weber then cites Ussery's web site as his source. Since Ussery's misreading of my book seems to be spreading, and since naive readers might be more impressed by a drop from 240 to 33 than by a change from 40 to 33, I have to state for the record that I did not mean the bacterial flagellum requires the proteins of the eukaryotic cilium, which is precisely why I treated them as separate and distinct systems in my book *Darwin's Black Box*.

Several reviewers have questioned whether irreducible complexity is necessarily a hallmark of intelligent design. James Shapiro, who has worked on adaptive mutations, writes in *Boston Review*[118] of "some developments in contemporary life science that suggest shortcomings in orthodox evolutionary theory" while arguing for "a growing convergence between biology and information science which offers the potential for scientific investigation of possible intelligent cellular action in evolution." Thus, Shapiro appears to think that irreducibly complex biochemical structures might be explained in a non-Darwinian fashion without invoking intelligence beyond the cells themselves. In *Biology and Philosophy* Bruce Weber[119] writes that the work of Stuart Kauffman and others on self-organizing phenomena "disrupts the dichotomy Behe has set up of selection or design." Most explicitly, Shanks and Joplin argue in *Philosophy of Science* that self-organizing phenomena such as the Belousov-

Zhabotinsky reaction demonstrate that irreducible complexity is not necessarily a pointer to intelligent design.[120] I have responded to Shanks and Joplin's argument in a separate paper.[121] Briefly, complexity is a quantitative feature; systems can be more or less complex. Although it produces some complexity, the self-organizing behavior so far observed in the physical world has not produced complexity and specificity comparable to irreducibly complex biochemical systems. There is currently little reason to think that self-organizing behavior can explain biochemical systems such as the bacterial flagellum or blood clotting cascade.

The underlying point of all these criticisms that needs to be addressed, I think, is that it is possible future work might show irreducible complexity to be explainable by some unintelligent process (although not necessarily a Darwinian one). And on that point I agree that the critics are entirely correct. I acknowledge that I cannot rule out the possibility that future work might explain irreducibly complex biochemical systems without the need to invoke intelligent design, a point I made in *Darwin's Black Box*.[122] I agree I cannot prove that studies of self-organization will not eventually show it to be capable of much more than we know now. Nor can I definitively say that Professor Shapiro's ideas about self-designing cells might not eventually prove true, or that currently unknown theories won't prevail. But the inability to guarantee the future course of science is common to everyone, not just those who are supportive of intelligent design. For example, no one can warrant that the shortcomings of self-organization will not be exacerbated by future research, rather than overcome, or that even more difficulties for natural selection will not become apparent.

I agree with the commonsense point that no one can predict the future of science. I strongly disagree with the contention that, because we can't guarantee the success of intelligent design theory, it can be dismissed, or should not be pursued. If science operated in such a manner, no theory would ever be investigated, because no theory is guaranteed success forever. Indeed, if one ignores a hypothesis because it may one day be demonstrated to be incorrect, then one paradoxically takes unfalsifiability to be a necessary trait of a scientific theory. Although philosophers of science have debated whether falsifiability is a requirement of a scientific theory, no one to my knowledge has argued that unfalsifiability is a necessary mark.

Because no one can see the future, science has to navigate by the data it has in hand. Currently there is only one phenomenon that has demonstrated the ability to produce irreducible complexity, and that is the action of an intelligent agent. It seems to me that that alone justifies pursuing a hypothesis of intelligent design in biochemistry.

In his recent book *Tower of Babel: The Evidence against the New Creationism*, however, philosopher of science Robert Pennock argues that science should avoid a theory of intelligent design because science must of necessity embrace "methodological naturalism."[123] I have responded to Pennock elsewhere.[124] Briefly, science should follow the data wherever it appears to lead, without preconditions. Further, the question of the identity of the designer remains open (see below)—just as the cause of the Big Bang has been open for decades. Thus, science can pursue theories with extra-scientific implications (such as the Big Bang[125] or intelligent design) as far as it can, using its own proper methods.

CAN WE—MAY WE—DETECT DESIGN IN THE CELL?

SEVERAL REVIEWERS have argued against the legitimacy of reasoning to a conclusion of intelligent design based on biochemical evidence. In the same review discussed above, Allen Orr raises an intriguing question of how we apprehend design. He writes:

> We know that there are people who make things like mousetraps. (I'm not being facetious here—I'm utterly serious.) When choosing between the design and Darwinian hypotheses, we find design plausible for mousetraps only because we have independent knowledge that there are creatures called humans who construct all variety of mechanical contraptions; if we didn't, the existence of mousetraps would pose a legitimate scientific problem.[126]

So, Orr says, we know mousetraps are designed because we have seen them (or other contraptions) being designed by humans, but we have not seen irreducibly complex biochemical systems being designed, so we can't conclude they were.

Although he makes an interesting point, I think his reasoning is incorrect. Consider the SETI project (Search for Extraterrestrial Intelligence), in which scientists scan space for radio waves that might have been sent by aliens. Those scientists believe that they can distinguish a designed radio wave (one carrying a message) from the background radio noise of space. However, we have never observed space aliens sending radio messages; we have never observed aliens at all. Nonetheless, SETI workers, funded for years by the federal government, are confident that they can detect intelligently designed phenomena, even if they don't know who produced them.

The relevance to intelligent design in biochemistry is plain. Design is evident in the designed system itself, rather than in pre-knowledge of who the designer is. Even if the designer is an entity quite unlike ourselves, we can still reach a conclusion of design if the designed system has distinguishing traits (such as irreducible complexity) that we know require intelligent arrangement. (One formal analysis of how we come to a conclusion of design is presented by William Dembski in his Cambridge University Press monograph *The Design Inference*.[127])

We can probe Orr's reasoning further by asking how we know that something was intelligently designed even if it indeed resulted from human activity. After all, humans engage in all sorts of activities which we would not ascribe to intelligence. For example, in walking through the woods a person might crush plants by his footsteps, accidentally break tree branches and so on. Why do we not ascribe those marks to purposeful activity? On the other hand, when we see a small snare (made of sticks and vines) in the woods, obviously designed to catch a rabbit, why do we unhesitatingly conclude the parts of the snare were purposely arranged by an intelligent agent? Why do we apprehend purpose in the snare but not in the tracks? As Thomas Reid argued in response to the skepticism of David Hume, intelligence is apprehended only by its effects; we cannot directly observe intelligence.[128] We know humans are intelligent by their outward actions. And we discriminate intelligent from non-intelligent human actions by external evidence. Intelligence, human or not, is evident only in its effects.

Michael Ruse in *Boston Review* raises another objection, saying that scientists qua scientists simply can't appeal to design: "Design is not something you add to science as an equal—miracles or molecules, take your pick. Design is an interpretation which makes some kind of overall metaphysical or theological sense of experience."[129]

Contrary to Ruse's argument, however, many scientists already appeal to design. I mentioned the SETI program above; clearly those scientists think they can detect design (and nonhuman design at that.) Forensic scientists routinely make decisions of whether a death was designed (murder) or an accident. Archaeologists decide whether a stone is a designed artifact or just a chance shape. Cryptologists try to distinguish a coded message from random noise. It seems unlikely that any of those scientists view their work as trying to make "metaphysical or theological sense of experience." They are doing ordinary science.

Ruse probably meant that scientists can't specifically appeal to God or the supernatural. Evolutionary biologist Douglas Futuyma echoes Ruse's sentiment with rousing rhetoric: "When scientists invoke miracles, they cease to practice science.... Behe, claiming a miracle in every molecule, would urge us to admit the defeat of reason, to despair of understanding, to rest content in ignorance. Even as biology daily grows in knowledge and insight, Behe counsels us to just give up."[130]

In speaking of "miracles"—relying for rhetorical effect on that word's pejorative connotations when used in a scientific context—Ruse and Futuyma are ascribing to me a position I was scrupulous in my book to avoid. Although I acknowledged that most people (including myself) will attribute the design to God—based in part on other, non-scientific judgments they have made—I did not claim that the biochemical evidence leads ineluctably to a conclusion about who the designer is. In fact, I directly said that, from a scientific point of view, the question remains open.[131] In doing so I was not being coy, but only limiting my claims to what I think the scientific evidence will support. To illustrate, Francis Crick has famously suggested that life on earth may have been deliberately seeded by space aliens.[132] If Crick said he thought that the clotting cascade was designed by aliens, I could not point to a biochemical feature of that system to show he was wrong. The biochemical evidence strongly indicates design, but does not show who the designer was.

I should add that, even if one does think the designer is God, subscribing to a theory of intelligent design does not necessarily commit one to "miracles." At least no more than thinking that the laws of nature were designed by God—a view, as we've seen, condoned by the National Academy of Sciences. In either case one could hold that the information for the subsequent unfolding of life was present at the very start of the universe, with no subsequent "intervention" required from outside of nature. In one case, the information is present just in general laws. In the other case, in addition to general laws, information is present in other factors too. The difference might boil down simply to the question of whether there was more or less explicit design information present at the beginning—hardly a point of principle.

While we're on the subject of God, another point should be made: A number of prominent scientists, some of whom fault me for suggesting design, have themselves argued for atheistic conclusions based on biological data. For example, Professor Futuyma has written, "Some shrink from the conclusion that the human species was not designed, has no purpose, and is the product of mere mechanical mechanisms—but this seems to be the message of evolution."[133] And Russell Doolittle remarks concerning the blood clotting cascade that "no Creator would have designed such a circuitous and contrived system."[134] It is rather disingenuous for those who use biological data to argue that life shows no evidence of design, to complain when others use biological evidence to argue the opposing view.

"Giving Up" in "Ignorance"

SOME SCIENTIFIC reviewers have dismissed the conclusion of design as an "argument from ignorance," or a "God of the gaps" argument. This can take several forms. One form of the objection is presented by University of London evolutionary biologist Andrew Pomiankowski, who writes: "Most biochemists have only a meagre understanding of, or interest in, evolution. As Behe points out, for the thousand-plus scholarly articles on the biochemistry of cilia, he could find only a handful that seriously addressed evolution. This indifference is universal."[135]

So, Pomiankowski argues, we do not have answers because nobody has looked, and biochemists haven't looked because they have little interest in the subject.

Although initially plausible, this interpretation suffers from the fact that there is demonstrable interest in evolution among molecular bioscientists. (One doesn't have to officially call oneself a "biochemist" to address such problems. Molecular biologists, geneticists, immunologists, embryologists—investigators in all of these disciplines are in a position to work on them.) The authors of the large number of books and papers listed on John Catalano's and David Ussery's web sites are clearly interested in evolution (see my early discussion of the evolutionary literature on that web site), as are the authors of numerous other studies that involve sequence comparisons. Since many papers are published in the general area of molecular evolution, we have to ask why there are so few in the particular area of the Darwinian evolution of irreducibly complex systems. Pomiankowski proposes that it is because the problem is so difficult;[136] I suggest it is difficult because irreducibly complex systems fit poorly within a gradualistic theory such as Darwinism.

A less reasonable form, I think, of the "ignorance" accusation is presented by Neil Blackstone. An evolutionary biologist at Northern Illinois University, Blackstone levels a formal charge at me of an error in logic—the "argumentum ad ignorantiam," as his review is titled.[137] He even cites a philosophy textbook by Irving Copi to give the charge authority. Those who chop logic to rule out a hypothesis, however, should make sure they are on very firm logical ground. Blackstone is not.

Copi defines the fallacy as follows: "The *argumentum ad ignorantiam* is committed whenever it is argued that a proposition is true simply on the basis that it has not been proved false, or that it is false because it has not been proved true."[138] But I certainly did not argue that the Darwinian evolution of biochemical complexity is false "simply on the basis" that it has not been proved true. Nor did I say that intelligent design is true "simply on the basis" that it has not been proved false. To lay the groundwork for a proposal of intelligent design I did argue extensively that the blood clotting cascade and other systems have not been explained by Darwinism. That, of course, was necessary because many people have the impression that Darwinian theory has already given a satisfactory account for virtually all aspects of life. My first task was to show the readership that that impression is not correct.

But my argument did not stop there. I spent many pages throughout the book showing that there is a structural reason—irreducible complexity—for thinking that Darwinian explanations are unlikely to succeed. Furthermore, I argued that irreducible complexity is a hallmark of intelligent design, took several chapters to explicate how we apprehend design, showed why some biochemical systems meet the criteria, and addressed objections to the design argument. Truncating my case for intelligent design and then saying I commit the fallacy of argumentum ad ignorantiam is not, in my opinion, fair play.

Let's explore the intricacies of formal logic a little further. Although Blackstone didn't mention it, Copi has more to say on the argument from ignorance: "A qualification should be made at this point. In some circumstances it can be safely assumed that if a certain event had occurred, evidence of it could be discovered by qualified investigators. In such circumstances it is perfectly reasonable to take the absence of proof of its occurrence as positive proof of its non-occurrence."[139]

Although I did not limit my argument to the lack of evidence for the Darwinian evolution of irreducibly complex biochemical systems, when qualified investigators (such as, say, those investigating blood clotting) come up empty, it is "perfectly reasonable" to weigh that against Darwinism. (By itself, of course, it is not positive evidence for design.) Although lack of progress is not "proof" of the failure of Darwinism, it certainly is a significant factor to consider.

In a milder variation of the "argument from ignorance" complaint, other scientific reviewers have objected that an appeal to intelligent design is tantamount to "giving up." For example, in the *Forward* Emory University evolutionary biologist Marc Lipsitch remarks: "[Behe] correctly suggests that a complete theory of evolution would include an account of how the intricate chemical systems inside our bodies arose (or might have arisen) from inanimate molecules, one step at a time. Mr. Behe's question is a fair one, but instead of suggesting a series of experiments that could address the question, he throws up his hands."[140]

Unfortunately, the point is made with circular logic: it depends on the presupposition that life is not designed, which is the point at issue. If life is not designed then, yes, a theory of intelligent design is ultimately a blind alley (if not quite "giving up"). However, if aspects of life are indeed designed, then the search for the putative unintelligent mechanisms that built them is the blind alley. But how do we decide ahead of time which is correct?

We can't decide the correct answer ahead of time. Science can only follow the data where they lead, as they become available.

22. A RESPONSE TO PAUL DRAPER

"Philosophical-ish Objections to Intelligent Design: A Response to Paul Draper," Discovery Institute, February 20, 2020.

RECENTLY I WAS ASKED BY SEVERAL PEOPLE WHETHER I HAD EVER responded to an old review of *Darwin's Black Box* by Purdue University philosopher of religion Paul Draper. I had not done so, but will use the occasion to respond now and clear up a couple of philosophical-ish objections that have been raised against intelligent design over the years.

In 2002 Draper—then on the faculty of Florida International University—published a paper in the journal *Faith and Philosophy: Journal of the Society of Christian Philosophers*, entitled "Irreducible Complexity and Darwinian Gradualism: A Reply to Michael J. Behe."[141] Draper wrote, "My goal in this paper will be to show that, while this challenge is both more original and, with a few modifications, more powerful than many of Behe's critics realize, it is incomplete and for that reason does not refute Darwinism."

I appreciate his recognition of the argument as a potent one, and I'm grateful for Draper's defense of it against some other objections. However, his own main line of criticism echoes one from the evolutionary biologist H. Allen Orr[142] (Draper cites Orr) that I had already addressed,[143] and Draper's charge of incompleteness rests on neglecting differences between a philosophical argument and a scientific one.

I Say Mousetrap; He Responds... "A Then B"

Let's start with Allen Orr. In 1996 H. Allen Orr reviewed *Darwin's Black Box* for *Boston Review*. Orr correctly described the problem of irreducible complexity I had pointed out: "You cannot, in other words, gradually improve a mousetrap by adding one part and then the next. A trap having half its parts doesn't function half as well as a real trap; it doesn't function at all. So Darwinism's problem is obvious: it requires that each step in the evolution of a system be functional and adaptive."[144]

Orr also agreed that such systems can't be developed by recruiting parts from other systems. "We might think that some of the parts of an irreducibly complex system evolved step by step for some other purpose and were then recruited wholesale to a new function," he wrote. "But this is also unlikely. You may as well hope that half your car's transmission will suddenly help out in the airbag department. Such things might happen very, very rarely, but they surely do not offer a general solution to irreducible complexity."[145]

Orr then proposed a solution:

Behe's colossal mistake is that, in rejecting these possibilities, he concludes that no Darwinian solution remains. But one does. It is this: An irreducibly complex system can be built gradually by adding parts that, while initially just advantageous, become—because of later changes—essential. The logic is very simple. Some part (A) initially does some job (and not very well, perhaps). Another part (B) later gets added because it helps A. This new part isn't essential, it merely improves things. But later on, A (or something else) may change in such a way that B now becomes indispensable. This process continues as further parts get folded into the system. And at the end of the day, many parts may all be required.[146]

I pointed out in my response to his review that his proposed solution was quite vague and that he had dodged the critical question: "*Some* part initially does *some* job? Which part of the mousetrap is he talking about? A mouse has nothing to fear from a 'trap' that consists of just an unattached holding bar, or spring, or platform, with no other parts."[147] Yes, new ideas are often hazy, but in order to be considered scientific they must progress beyond hand-waving. Orr gives credit for the original "A then B" scenario to the early twentieth-century geneticist Hermann Muller. Yet in 1918—when, according to Orr, Muller first proposed it—no one knew what genes even were! The structures of proteins—the machinery of the cell—would not begin to be elucidated for another four decades. Since the 1950s, however, biology has made tremendous progress in understanding the molecular basis of life. In the twenty-first century—more than a hundred years after Muller's proposal—it is well beyond time to stop hiding behind letters. Exactly what are those magical parts "A" and "B"? And exactly what are they supposed to be doing?

Orr went on to suggest that the transformation of air bladders of fish into lungs of terrestrial vertebrates illustrated his point: "The transformation of air bladders into lungs that allowed animals to breathe atmospheric oxygen was initially just advantageous: such beasts could explore open niches—like dry land—that were unavailable to their lungless peers. But as evolution built on this adaptation (modifying limbs for walking, for instance), we grew thoroughly terrestrial and lungs, consequently, are no longer luxuries—they are essential."[148]

So apparently he hadn't grasped the concept of irreducible complexity after all. I had stressed in *Darwin's Black Box* that it is the *molecular level* of life that must be examined to decide about irreducibility. Complex tissues such as air bladders and lungs are not "single [i.e., molecular] systems." When one writes glibly about the evolution of whole organs (as Darwin himself did with the eye[149]), whose molecular components are many and mostly unknown, one quickly descends into fantasy. Notice that Orr doesn't tell us in sufficient detail—or *any* detail, for that matter—how a Darwinian process could transform even an air bladder into a lung. Like most other evolutionary biologists, he simply assumes it can.

Six Years Later

Professor Draper's line of reasoning from 2002, which is also an imaginary exercise of evolution-by-letter, misses the same point as does Orr's:

> An irreducibly complex two-part system AB that performs function F could evolve directly as follows. Originally, Z performs F, though perhaps not very well. (This is possible because, from the fact that AB cannot perform F without A or B, it doesn't follow that Z cannot perform F by itself.) Then A is added to Z because it improves the function, though it is not necessary. B is also added for this reason. One now has a *reducibly* complex system composed of three parts, Z, A and B. Then Z drops out, leaving only A and B. And without Z, both A and B are required for the system to function. One might object that in this scenario AB is not really produced directly because, although the route from Z to AB involves no change in function, it must involve a change in mechanism since interaction between two distinct parts is essential to AB's mechanism and so Z cannot function by the same mechanism as AB. If this is right, then no complex system of interacting parts, whether irreducibly complex or not, could evolve both gradually and directly from scratch. An indirect route would always be required in order to get started. But that's no problem for this model. For I can simply stipulate that Z itself is composed of two parts (Z1 and Z2),

which do perform AB's function by the same mechanism, and then add that $Z1$ drops out first, followed by $Z2$.[150]

In addition to being terminally fuzzy (like Orr's), Draper's proposal overlooks key problems of his own scenario. For example:

1) The fact that the necessary parts of Z ("$Z1$ and $Z2$") may be attached to each other (that's how I understand "composed"), as he stipulates, instead of physically separated, does not obviate the problem, as he seems to think; rather, it makes the problem *worse*. The parts must be connected to each other in the correct way—*more* irreducible requirements. For example, all the parts of a mousetrap are attached in specific orientations (if, say, the spring were facing the wrong way, the trap would fail). Yet the unguided production of even that comparatively simple machine has resisted the efforts of Darwin defenders on the internet for twenty-five years.[151]

2) If, as Draper concedes, Z is composed of two necessary parts, then of course one is *starting* with irreducible complexity, not *explaining* it.

3) Even if Z is already working (somehow) and the subsequent addition of A and then B improves it, as Draper posits, why in the world should we think A and B can later take over Z's role? *Improving* the function of Z is a different task from *performing* the function of Z. Placing a drop of oil on the spring and a piece of cheese on the platform might improve a mousetrap's function, but certainly neither alone nor both together would make the other parts dispensable. If A and B did take over Z's role later, it would seem to be a gratuitous development, no different in kind or difficulty from the postulated irreducibly complex starting point.

Another putative route to irreducible complexity that Draper explores is also an imaginative exercise in evolution-by-letter.

The sort of route I have in mind occurs when an irreducibly complex and irreducibly specific system S that serves function F evolves from a precursor S* that shares many of S's parts but serves a different function F*. Notice that parts that S and S* share and that are required for S to perform F need not be required for S* to perform F* even if they contribute to F*, and parts that are irreducibly specific relative to F may be only reducibly specific relative to F*. Thus, both S* and the specificity of its parts may have been gradually produced via a direct evolutionary path. Then one or more additional parts are added to S*, resulting in a change of function from F* to F. And relative to F, the parts and their specificity, which had not been essential for F*, are now essential.[152]

I'm afraid I find that scenario very unclear. Wouldn't it be great if Professor Draper offered a concrete illustration of what he meant here? But he doesn't, and I can only conclude that he can't. Rather, the argument strikes me as little more than wordplay—just a riff on Hermann Muller's hundred-year-old idea from a time when no one knew much about the molecular foundation of life. The only physical illustration Draper cites is a standard mousetrap that has been degraded and purposely manipulated. In an endnote Draper explains:

> These claims are easy to demonstrate. Simply remove the catch from a trap and bend the holding bar so that (roughly) the middle of it can be placed just barely under the tip of the slightly curved end of the spring that extends under the hammer when it is armed. This will make it possible to arm the hammer without a catch, and the closer to the tip of the spring the holding bar is placed, the more insecurely the hammer will be armed. Additionally, one can make the trap even more sensitive to pressure on the platform by allowing the end of the holding bar to extend below the platform when the hammer is armed (so that the platform will not lie flat on the floor when the trap is set).[153]

With his proposed very careful rearranging of the functioning standard mousetrap, the intelligent agent Paul Draper runs squarely into a problem I flagged in *Darwin's Black Box*:

> To feel the full force of the conclusion that a system is irreducibly complex and therefore has no functional precursors we need to distinguish between a *physical* precursor and a *conceptual* precursor. The trap described above is not the only system that can immobilize a mouse. On other occasions my family has used a glue trap. In theory at least, one can use a box propped open with a stick that could be tripped. Or one

can simply shoot the mouse with a BB gun. However, these are not physical precursors to the standard mousetrap since they cannot be transformed, step-by-Darwinian-step, into a trap with a base, hammer, spring, catch, and holding bar.[154]

Perhaps because of professional inclinations, philosopher Draper does not even *try* to argue that his imagined mousetrap is a physical (rather than just a conceptual) precursor to the standard mousetrap I discussed. I responded to a similar intelligently guided mousetrap scenario by Professor John McDonald of the University of Delaware years before Draper's paper was published.[155] Mousetraps can of course be built in many different ways. But McDonald's and Draper's examples are not Darwinian, single-random-step precursors to a standard mousetrap. Rather, they are intelligently re-engineered systems that resemble the trap I pictured only because that is the goal they are intended to reach. As I wrote in my response to McDonald, those may be conceptual precursors to a standard mousetrap, but they are not physical precursors and therefore not Darwinian precursors either.[156]

It's a lot easier to imagine that, through a long series of unexplained steps, "A" could morph into "B" than that, say, a glue trap could morph into a standard mechanical mousetrap. I think that is one crucial difference between philosophical approaches and scientific ones. If intermediate steps are not enumerated and tested for a physical pathway, then a theorist is in serious danger of getting stuck in a fantasy world. Recall that evolutionary biologist Allen Orr agreed that borrowing parts used for something else is quite unlikely to work for irreducibly complex systems: "You may as well hope that half your car's transmission will suddenly help out in the airbag department." Draper's schemes seem to be built entirely on such hopes. Yet if Draper and others can't account for even a humble mousetrap (even if we supposed, for the sake of argument, that mousetraps reproduced and were subject to small mutations), why should we think Darwin's theory can account for the sophisticated machinery of the cell?

In *Darwin's Black Box* I argued that for all practical purposes irreducibly complex biochemical systems met Darwin's challenge that "if it could be demonstrated that any complex organ existed which could not possibly have been formed by numerous, successive, slight modifications, my theory would absolutely break down."[157] Twenty-five years later there is still no reason to think otherwise.

THE STATE OF THE LITERATURE CIRCA 2020

NEAR THE end of his review from 2002, Draper writes:

> A defender of Behe might respond that the improbability of all Darwinian paths is established by the silence of the scientific literature on the issue of how Behe's systems evolved.... Should we conclude, then, that this silence provides substantial indirect evidence for Behe's position that Darwinism cannot explain the development of these systems? In other words, is it likely that the literature would be much noisier if Behe's position were false? I am inclined to give negative answers to these questions for two reasons. First, the discipline of biochemistry is very young.[158]

The discipline of biochemistry is now nearly twenty-five years older than when *Darwin's Black Box* was published. Yet, as I document in my recent book, *Darwin Devolves*, the state of the literature regarding irreducibly complex systems is unchanged—even as problems for Darwin have multiplied.[159] Although the science of life at the molecular level has advanced by leaps and bounds, that new work conspicuously does not include explanations for how Darwinian processes could produce irreducibly complex systems. Despite the intervening years, despite the immense progress of science, and despite the intense dislike of intelligent design to motivate many very smart scientists, even the examples I highlighted in 1996 have gone completely unexplained. That in itself is strong evidence that Darwinists have been barking up the wrong tree.

Draper adds:

> Second, any indirect evidence that the silence of the literature provides in support of Behe's position is offset by indirect evidence against Behe's position. For in assessing the probability that some system evolved gradually, one cannot just examine the specifics of that system. One

must also consider what we know about how other systems evolved. And Behe admits, as well he should, that much evolutionary change is both gradual and driven by Darwinian mechanisms. These Darwinian success stories raise the probability that Darwinian evolution produced Behe's biochemical systems, even if we cannot yet specify step by step exactly how that happened.[160]

Time has not been kind to Draper's position. As I discuss in *Darwin Devolves*, we are now in a much better position than we were in 1996 to see how Darwinian processes work in nature, rather than in people's imaginations. It turns out there is indeed much evolutionary change that is both gradual and driven by Darwinian mechanisms, as Draper noted. However, the surprise—made possible by recent advances in DNA sequencing techniques—is that the large majority of even *helpful, beneficial* mutations swept through a population by natural selection are *degradative*. That is, they *break* or *damage* pre-existing genes. (More on this later in the present volume.) One can call examples of evolutionary degradation "success stories" if one wishes. Degradative processes, however, are prohibitively unlikely to have initially constructed the elegant biochemical machinery modern science has discovered in the cell.

THE CONFLICT LIES ELSEWHERE

IN HIS 2011 book, *Where the Conflict Really Lies: Science, Religion, and Naturalism*, the Christian über-philosopher Alvin Plantinga discusses my *Darwin's Black Box* in connection with Paul Draper's objections and says explicitly that he thinks Draper is right. Yet, in support of his agreement, Professor Plantinga quotes the same fuzzy paragraph of Draper's about S, F, S*, and F* that I quoted here. Plantinga then emphasizes a critical difference between philosophical and scientific arguments:

> It's important to note that the possibilities Draper suggests are merely abstract possibilities. Draper doesn't argue or even venture the opinion that in fact there are routes of these kinds that are not prohibitively improbable; he simply points out that Behe has not eliminated them. And of course this is quite proper, inasmuch as Draper is doing no more than evaluating Behe's argument. All he is trying to show is that Behe's conclusion doesn't deductively follow from his premises.[161]

But *no* scientific argument simply follows deductively from its premises. An argument about nature necessarily concerns detailed empirical facts—evidence—so it *cannot* be a completely deductive one. Although scientific thinking of course involves logic and deduction, *all* scientific arguments—including the argument for the irreducible complexity of the molecular machinery of life—rest most heavily on empirical data. No scientific explanation has ever ruled out all possible rival explanations by dint of deductive logic alone. Draper used a wholly inappropriate standard to evaluate my argument, one that is not applied to any other scientific theory.

Plantinga goes on to discuss my argument from *The Edge of Evolution* that empirical results—especially data on the development of resistance to the antimalarial drug chloroquine—show there is a real limit to what Darwinian processes can explain in life. He then wonders whether this makes it likely that life was designed:

> How does the argument go? One possibility: the main alternative to intelligent design is unguided evolution, but the probability that unguided evolution should produce these protein machines is so low that we must conclude that it is false. Is this right? Not clearly. First, exceedingly improbable things do happen, and happen all the time. Consider a deal in a hand of bridge. If you distribute the fifty-two cards into four groups of thirteen cards each, there are some 10^{28} possible combinations. Therefore the probability that the cards should be dealt just as they are dealt is in the neighborhood of 10^{28}. So consider a rubber of bridge that takes four deals: the probability that the cards should be dealt precisely as they are, for those four deals, is about 10^{-112}.[162]

Since improbable-yet-fair bridge hands are dealt all the time, he implies, low probability alone can't be the criterion for design.

True enough, as far as it goes. But that doesn't mean we can't confidently recognize design in low-probability events. This line of thought was explored by William Dembski in his 1998 *The Design Inference*. He explained that an inference to design required not only an event of small probability, but also one that was *specified*—that is, one that hit a predetermined or retrospectively surprising target. For example, there are of course very large numbers of possible bridge hands, but if a person at a table kept getting hands with 13 cards of the same suit, eyebrows would be raised. The "lots of things are improbable" tack to explain away the elegant structure of the machinery of life is simply not a serious response.

Professor Plantinga then raises an objection first put forward by the philosopher of science Elliott Sober, that we need to have a positive reason to think a designer would *want* to make cellular machinery to conclude that a designer did.[163] Plantinga seems to agree — a designer who didn't want to design machinery probably wouldn't do so.

> Well, should we instead compare P(protein machines/unguided evolution), the probability of the existence of these protein machines given unguided evolution, with P(protein machines/intelligent designer), the probability of the existence of these protein machines given the existence of an intelligent designer? Here, again, the problem is that we don't have a very good grasp of either of those probabilities.... Suppose there is an intelligent designer: how likely is it that he or she (or it) would design and cause to come to be just *these* protein machines?... An intelligent designer that really hated life, or proteins, or protein machines, would be very unlikely to design protein machines.[164]

Try as I might, and as admiring as I am of other work of both Plantinga and Sober, I see absolutely no force to this argument, and it is quite susceptible to *reductio ad absurdum*: Sure, a designer who didn't want to design something probably didn't design it. But let's not stop there. After all, a designer who really *did* want to design machinery, but got distracted and forgot, wouldn't have designed it either. Nor would a designer who wanted to design but went on vacation instead. Or one that came down with the flu. Or was stopped by someone else. Or got its head stuck in a vase. Do we really have to individually eliminate all those and many other possibilities before coming to a conclusion of design?

I don't think so. Rather, the rejection of all those hypotheticals is baked into the *positive, affirmative* design argument. The problem with the Sober approach, I think, is that it seems (perhaps unconsciously) focused on a particular candidate for designer—maybe God, maybe a space alien, whatever—and imagines that the proper question to ask is, how do we know that *this particular candidate* has the motivation to design something? Wrong question. The intelligent design argument rightly starts with the *physical system itself* and asks, does this system exhibit a *strongly purposeful arrangement of parts?*

(As an aside, the major focus of my books on demonstrating the stark inability of Darwinian processes to achieve what proponents claim for them is of course necessary in order to dispel that common misconception. But a conclusion of design is not, as the caricature frames it, simply a default judgment rendered in the absence of some other plausible explanation. Rather, as I have explained in a number of places,[165] design—the work of a mind—is *positively* perceived in a purposeful arrangement of parts.)

It would be silly to examine, say, Mount Rushmore, or the monolith found on the Moon in the movie *2001: A Space Odyssey*, and say no inference about their possible design could be made because we didn't know whether a designer wanted to make either of them. Rather, we conclude *from observation of a system itself* that it was designed. Indeed, interesting secondary questions (such as *who* designed it, *when, how, why,* and so on) can't even be posed unless we already suspect that a system was designed.

23. Self-Organization and Irreducible Complexity

This essay is one of several gratifying interactions I've had with philosophy journals, which don't shy away from controversy and have the admirable policy of allowing a fellow whose work is attacked in their pages to reply.

"Self-Organization and Irreducibly Complex Systems: A Reply to Shanks and Joplin," *Philosophy of Science* 67, no. 1 (March 2000): 155–162.)

SOME BIOCHEMICAL SYSTEMS REQUIRE MULTIPLE, WELL-MATCHED parts in order to function, and the removal of any of the parts eliminates the function. I have previously labeled such systems "irreducibly complex," and argued that they are stumbling blocks for Darwinian theory. I have proposed that they are, instead, best explained as the result of deliberate intelligent design. In a recent article Shanks and Joplin analyze and find wanting the use of irreducible complexity as a marker for intelligent design. Their primary counter-example is the Belousov-Zhabotinsky reaction, a self-organizing system in which competing reaction pathways result in a chemical oscillator. In place of irreducible complexity they offer the idea of "redundant complexity," meaning that biochemical pathways overlap so that a loss of one or even several components can be accommodated without complete loss of function. Here I note that complexity is a quantitative property, so that conclusions we draw will be affected by how well matched the components of a system are. I also show that not all biochemical systems are redundant. The origin of non-redundant systems requires a different explanation than that for redundant ones.

INTRODUCTION

IN THE past half-century biology has made astonishing progress in understanding the molecular and cellular basis of life. In light of this progress it is fair to ask whether Darwin's mechanism of natural selection acting on random variation appears to be a good explanation for the origin of all, or just some, of the molecular systems science has discovered. In *Darwin's Black Box: The Biochemical Challenge to Evolution* I argued that some biochemical systems, such as the blood clotting cascade or bacterial flagellum, are resistant to Darwinian explanation because they are irreducibly complex. I defined an irreducibly complex system as a system which is composed of several well-matched, interacting parts that contribute to the basic function, and where the removal of any one of the parts causes the system to effectively cease functioning. And I said, "The difficulty for Darwinian theory is that an irreducibly complex system cannot be produced directly (that is, by continuously improving the initial function, which continues to work by the same mechanism) by slight, successive modifications of a precursor system, because any precursor to an irreducibly complex system that is missing a part is by definition nonfunctional."[166]

To illustrate the concept with a familiar example for a general readership, I pointed to a simple mechanical mousetrap, composed of several parts such as the base, hammer, spring and so on, and noted that the absence of any of the parts destroys the mouse-catching ability of the trap. Darwin's vision of natural selection gradually improving function in "numerous, successive, slight modifications"[167] appears not to fit well with such systems. I went on to argue that, since intelligent agents are the only entities known to be able to construct irreducibly complex systems, the biochemical systems are better explained as the result of deliberate intelligent design.

But are gradual Darwinian natural selection and intelligent design the only potential explanations? Shanks and Joplin[168] direct our attention to complexity theory, which concerns the ability of systems to self-organize abruptly, sometimes in surprising ways. They suggest that irreducibly complex biochemical systems might in principle be explained by self-organization, eliminating the need to invoke intelligence. They then go on to argue that biochemical systems are "redundantly complex"— that is, contain components that can be removed without entirely eliminating function.

After briefly describing the Belousov-Zhabotinsky reaction—Shanks and Joplin's main counterexample—I will first argue that the reaction does not meet the definition of irreducibly complex, because the interacting components are not "well-matched." I will then agree that redundant complexity exists, but show that not all of biochemistry is redundant.

A Closer Look at Chemical Self-Organization

THE DISSIPATION of energy in nature can organize matter and produce reaction pathways. A simple example is the clumping of matter into stars under the influence of gravity. More complex examples are tornados and the stellar nuclear pathways that lead to the production of the heavy elements. These examples, however, have no direct relevance to the origin of biochemical systems. Shanks and Joplin[169] offer what they think is a more pertinent example—the Belousov-Zhabotinsky reaction, a self-organizing chemical system discovered in the 1950s by B. P. Belousov in an attempt to model the Krebs cycle. The term "BZ reaction" is applied to a group of chemical reactions in which an organic substrate is oxidized by bromate ions in the presence of a transition metal ion and acid. Instead of proceeding monotonically to equilibrium, the reaction oscillates between two pathways because of a competition between bromide ion and bromous acid for reaction with bromate ion. Bromate oxidizes the metal ions, which in turn are re-reduced by reaction with organic substrate. When the reaction is well stirred, the visible result is a solution that switches from one color to another at constant time intervals until the reaction materials are consumed. When the same reaction is set up as a thin, unstirred layer, waves of color change propagate through the layer. For details of the reaction pathways see Gray and Scott[170] and references therein.

Shanks and Joplin write that the BZ reaction "satisfies Behe's criteria for irreducible chemical complexity" because if any of the chemical components is removed "the characteristic behavior of the system is disrupted." Thus "irreducible complexity in a self-organizing system" can be generated "without the aid of a designing *deus ex machina*."[171]

I disagree that the BZ reaction "satisfies Behe's criteria" for an irreducibly complex system. Although it does have interacting parts that are required for the reaction, the system lacks a crucial feature—the components are not "well-matched." The appearance of the modifier "well-matched" in the definition I constructed (above) reflects the fact that complexity is a quantitative property. A system can be more or less complex, so the likelihood of coming up with any particular interactive system by chance can be more or less probable. As an illustration, contrast the greater complexity of a mechanical mousetrap (mentioned above) with the much lesser complexity of a lever and fulcrum. Together a lever and fulcrum form an interactive system which can be used to move weights. Nonetheless, the parts of the system can have a wide variety of shapes and sizes and still function. Because the system is not well-matched, it could easily be formed by chance.

Systems requiring several parts to function that need not be well-matched, we can call "simple interactive" systems (designated 'SI'). Ones that require well-matched components are irreducibly complex ('IC'). The line dividing SI and IC systems is not sharp, because assignment to one or the other category is based on probabilistic factors which often are hard to calculate and generally have to be intuitively estimated based on always-incomplete background knowledge. Moreover, no law of physics automatically rules out the chance origin of even the most intricate IC system. As complexity increases, however, the odds become so abysmally low that we reject chance as an explanation.[172]

Just as I think that a gradual origin by natural selection is a good explanation for some things, I agree that a discontinuous origin by self-organization explains some things too. Nonetheless, I do not think either explains irreducible complexity. I argue that Shanks and Joplin's counterexample—the BZ reaction—is not IC; it is SI, because the components are not well-matched.

To justify my position, let me first illustrate a well-matched system using the blood clotting cascade.[173] The active form of one protein of the cascade is called thrombin, which cleaves the soluble protein fibrinogen to produce fibrin, the insoluble meshwork of a blood clot. The chemistry catalyzed by thrombin is simply the hydrolysis of a certain fibrinogen peptide bond. However, all proteins are made of amino acid residues joined by peptide bonds. A typical protein contains several hundred peptide bonds. There is nothing remarkable about the bond in fibrinogen that is cleaved by thrombin. Yet thrombin selects that particular bond for cleavage out of literally hundreds of thousands of peptide bonds in its environment and ignores almost all others. It can do this because the shape of thrombin is well-matched to the shape of fibrinogen around the bond it cleaves. It "recognizes" not only the bond it cuts, but also a number of other features of its target. The other proteins of the clotting cascade (Stuart factor, proaccelerin, tissue factor, and so on) have similar powers of discrimination. So do virtually all of the components of the molecular machines I discussed in *Darwin's Black Box*.

Let's contrast this biochemical specificity with a comparable chemical reaction lacking such specificity. The peptide bonds of proteins can also be cleaved by simple chemicals. A typical procedure calls for incubating the protein in 6N hydrochloric acid at 110°C for twenty-four hours. If fibrinogen were incubated under those conditions, the peptide bond that thrombin cleaves would be broken, but so would every other peptide bond in the protein. It would be completely reduced to amino acids. If thrombin were in the mix, it too would be completely destroyed. If the other proteins of the clotting cascade were there, no clotting would take place, even though the peptide bonds that are cleaved in the cascade would be cleaved, because all other peptide bonds would be hydrolyzed too. There is virtually no specificity to the chemical hydrolysis beyond the type of bond that is cleaved.

Similarly, the reactants of the BZ reaction are small organic or inorganic chemicals that show little specificity for each other. One ingredient, sodium bromate, is a general-purpose oxidizing reagent and is capable of degrading a very large spectrum of chemicals besides the ones used in BZ reactions (thus its transport aboard airlines is forbidden). Another requirement of the reaction is simply for a transition metal that can change its oxidation state, and a number of such metals are known, including iron, cerium, and manganese ions.[174] A third requirement is for an organic molecule that can be oxidized. Many candidates could fulfill this role (ones that have been used include malonic, citric, maleic, and malic acids), and organic molecules can be oxidized by many reagents other than bromate. The last ingredient is simply a high concentration of sulfuric acid. As Field noted, setting up BZ reactions "is an exceedingly easy task as they will occur over a wide range of concentrations and conditions."[175]

The BZ class of self-organizing reactions—chemical oscillations—is surprising and interesting. Nonetheless, its complexity can be likened to that of other self-organizing systems found outside of biology, such as, say, tornados, which, although they command our attention, do not approach the specificity of well-matched, irreducibly complex biochemical systems.

BIOCHEMICAL SELF-ORGANIZATION: BEHAVIOR VS. ORIGIN

THE DYNAMICAL behavior of the BZ reaction has been modeled by a set of two ordinary differential equations.[176] Because some biological systems can be modeled by similar mathematics, Shanks and Joplin conclude that self-organization can explain the behavior of the biological systems. There are several reasons to question the relevance of their point. First, they also note that "the substrates and products in these systems are very different from those in the BZ reaction."[177] In other words, we have traveled far from cerium, sodium bromate, and the other constituents of the chemical system. Second, and more importantly, the behavior of a system must be distinguished from its origin. As an illustration, consider highway traffic flow. A number of mathematical models have been used to describe traffic flow, some drawing on theories of self-organization.[178] The mathematics, however, have not called the automobiles into being. The mathematics simply try to describe the typical behavior of traffic when a certain density is reached under conditions of restricted movement on a highway.

Examples of biological processes that show BZ dynamical behavior include glycolysis and aggregation of dispersed cells of the slime mold *Dictyostelium discoideum* into a slug. But consider the sophisticated components of the aggregation-signaling system of *D. discoideum*, which include: a cyclic AMP membrane receptor protein that can exist in an active and inactive form; an adenylate cyclase that binds to the active form of the receptor and itself becomes activated; a protein to export cyclic AMP into the extracellular medium; and more.[179] All of that complicated machinery is ignored in BZ models—treated as a black box. Oscillations in the cellular concentration of glycolytic intermediates are due in large part to the multi-talented phosphofructokinase (PFK), a tetrameric enzyme that can exist in two conformational states (an active form and a less-active one) and which has binding sites for a dozen different activators and inhibitors.[180] Mathematical models of BZ behavior don't explain the origin of the impressive abilities of PFK any more than models of traffic flow explain the origin of brakes or gas pedals. Thus, even if a biological system displays self-organizing behavior, the question of its origin remains.

NOT ALL BIOCHEMICAL SYSTEMS ARE REDUNDANT

IN CONTRAST to claims about irreducible complexity, Shanks and Joplin write, "Real biochemical systems, we argue, manifest redundant complexity—a characteristic result of evolutionary processes."[181] By this they mean that biochemical pathways overlap and are interconnected, so that removal of one or even several components does not completely destroy the function. In support of their position they cite a diverse array of biochemical examples: the synthesis of an alternate pine tree lignin with increased content of dihydroconiferyl alcohol; viable mice in which the gene for the tumor suppressor p53 was knocked out; and more.

Their initial illustration is the metabolic pathways for the synthesis of glucose-6-phosphate. They point out that the molecule can be made by "several different isoforms or variants of hexokinase, and all are present, as a result of gene duplication, in varying proportions in different tissues." What's more, "Knock out one enzyme isoform and the other isoforms in the tissue can take over its function."

True enough. The observation that some biochemical systems are redundant, however, does not entail that all are. And, in fact, some are not redundant. Consider the following examples of nonredundant metabolic pathways. Primates, including humans, cannot synthesize ascorbic acid (vitamin C) because they lack a functional gene for L-gulono-gamma-lactone oxidase, although a pseudogene is present (Nishikimi and Yagi 1991).[182] Vitamin C is made by no other pathway. Hexosaminidase A is required to catabolize ganglioside GM2; its loss results in Tay-Sachs disease.[183] These enzymes are parts of "real biochemical systems," but they do not "manifest redundant complexity." (For many, many additional examples, see *The Metabolic and Molecular Bases of Inherited Disease*[184] or other texts on inborn errors of metabolism.) Therefore, arguments developed about the origin of redundant systems do not necessarily apply to all biochemical systems.

Shanks and Joplin's argument for redundant complexity has the same strengths and weaknesses when the subject moves from metabolic pathways to other biochemical systems. That is, they are right to notice that some systems or components are redundant, but wrong to extrapolate the conclusion to all systems. For example, they point to mice in which the gene for the protein p53 has been knocked out. Protein p53 is "involved in a number of fundamental cell processes, such as affecting gene transcription, acting as control points in the cell cycle, initiating programmed cell death," and more. Shanks and Joplin write that "looking at this case from the standpoint of a 'genetic mousetrap model,' one would naturally predict that the removal of this gene... would lead to catastrophic collapse of the developmental process.... Such is not the case."[185] Yet contrast this case with that of mice in which the gene for either fibrinogen,[186] tissue factor,[187] or prothrombin[188] has been knocked out. Those proteins are all components of the blood clotting cascade, which I discussed prominently in *Darwin's Black Box*,[189] claiming it is irreducibly complex. The loss of any one of those proteins prevents clot formation—the clotting cascade is broken. Thus Shanks and Joplin's concept of redundant complexity does not apply to all biochemical systems.

24. Wolfram's A New Kind of Science

"Review of *A New Kind of Science* by Stephen Wolfram," *First Things* (November 2002).

ONE SHOULD HESITATE TO REVIEW A BOOK THAT THREATENS A hernia from lifting it. Apart from the danger to health, it likely will turn out to have been written by: 1) the Unabomber or some other crank; 2) a fellow who unwisely declined the services of an editor; or 3) a genius who ranges over subjects way beyond one's (that is, my) ability to comment intelligently. Stephen Wolfram's *A New Kind of Science*[190] fits firmly into categories 2 and 3. I'm undecided about 1.

Wolfram fills twelve hundred pages (not counting the index) with discussions of subjects as disparate as free will, the fundamental laws of physics, evolution, the free market, extraterrestrials, and much, much more. In fact he not only discusses them, but places them in relationship to each other. For example, in the index there are entries for "Economics, and extraterrestrial trade" and "Economics, and free will." The sheer breadth of the book elicited several waves of reaction in this reader. At first a sense of awe that any person could know so much about so much. Next, excitement at the prospect that Wolfram might really be on to something fundamental. But, finally, déjà vu, as the subject eventually turned to topics I know something about.

Wolfram was a prodigy even among geniuses, being the youngest recipient of a MacArthur Foundation award. He published his first physics paper at age fifteen, earned a PhD from Caltech in a single year, and joined the faculty there at the age of twenty. A few years later he headed to the Institute for Advanced Studies at Princeton (yes, the one where Einstein spent much of his career). By the age of thirty he had written a computer program called Mathematica, which helps engineers and scientists deal more easily with complicated mathematics. Marketing the software has made him independently wealthy, requiring neither an academic position nor the largesse of government research grants to pursue his interests.

With a start like that it's understandable that Wolfram is enormously self-confident, thinking he can entirely change the way science views the world—pretty much by himself, thank you very much. For the past decade he closeted himself in his study, working on his book and neither publishing his interim results nor attending scientific meetings. Yet somehow the word leaked out that he was writing a putative scientific blockbuster and, given Wolfram's reputation, in some circles the book has been very anxiously anticipated.

As I write this, it is the twenty-fifth anniversary of the death of Elvis Presley and media hype fills the air. Though not nearly of the same order of magnitude, the Wolfram phenomenon has gotten quite a ride from the media, too. A LexisNexis search on the word "wolfram" shows 193 articles in the past six months, many mentioning Wolfram and his book, including stories in *Newsweek*, *Business Week*, and the *Guardian*, *Telegraph*, and *Times* of London. The *New York Times* has run three articles in the past few months, one from the science desk, another from the arts and ideas desk, plus an official book review. The leading science journal *Nature* ran an article plus a book review.

Certainly the bulk of the attention is due to Wolfram's great biography and mysterious ways. But at least some of the anticipation is because many people think that, despite John Horgan's claims in *The End of Science*, something big is missing from our understanding of the world. Too many problems that should have been solved by now remain intractable—problems like forecasting the weather or the course of epidemics, understanding the origin of life or the behavior of economic markets, or even predicting things as trivial as how a rising column of smoke will curl. There must be some simple trick or basic insight, the thinking goes, that is being overlooked. Once we figure out the trick, everything will fall into place.

In the past few decades there have been a number of candidates for the role of New Insight—catastrophe theory, chaos theory, complexity theory. All claimed to have wide-ranging implications, from physics to economics and social interactions. All have been trumpeted for a while. None have lived up to the hype.

Although Wolfram draws connections to virtually every facet of the universe, *A New Kind of Science* is basically a book about computing. In particular, Wolfram investigates a type of computer program called a cellular automaton (CA). To get the gist of what a CA does, imagine a very large piece of graph paper, with a grid of squares. In the middle of the top line color one square black. Now move down a line and choose a simple rule to decide if the square underneath should be blackened or not. For example, one rule might be to color a square black on the next line if the square directly above was itself black, its left-hand neighbor was also black, and its right-hand neighbor was white. Now move down another line and apply the same rule again. Repeat forever. There are exactly 256 simple rules like that (Wolfram pictures over half of them), where the outcome depends just on nearest-neighbors and which involve only two colors. Most of the rules give pretty simple patterns—a solid black triangle, a triangle with alternating white and black squares, and so on.

But some rules give complex patterns. Some show nested patterns of triangles within triangles, others yield triangles where the left side consists of straight lines, but the right side swirls with larger and smaller triangles in no discernible pattern. With other rules a regular pattern is suddenly cut off by a random jagged line of dots apparently coming out of left field. (About half of the book consists of pictures of such results.) Wolfram stresses that, unlike previous ideas such as chaos theory, where randomness and uncertainty derived from the physical environment, the randomness of CAs is generated by a simple, regular, discrete, numerical process. We start with the simplicity of pure integers, yet somehow end up with capriciousness and unpredictability.

From this basic result, Wolfram draws a number of large conclusions. He says that, since the colors of squares in some CAs appear random, there's no way to predict ahead of time what color a given square will be when the computer program is run. This illustrates a principle he calls computational irreducibility. In short, the outcome of many types of processes can't be calculated from an equation or set of equations. There is no mathematical shortcut. One simply has to watch the actual process unfold to see what will happen. This pretty much spells doom to long-range weather forecasting.

Another result is that some CAs can act as "universal computers," performing any calculation that your desktop computer can perform. For example, he shows that the output of a particular CA contains lines spaced according to the prime numbers. Another big conclusion is called the Principle of Computational Equivalence (Wolfram's capitalization). This is a little harder to grasp. He summarizes it thus: "Whenever one sees behavior that is not obviously simple—in essentially any system—it can be thought of as corresponding to a computation of equivalent sophistication." One consequence, he thinks, is that our minds can't grasp most complex processes because those processes have the same computational sophistication as our brains. Our thinking can't outrace the processes.

Much of this is interesting and, for all I know, may have important implications for mathematics and computational theory. But Wolfram wants his work to go beyond just computation to have it explain all of nature. Here he is unconvincing. The problem is that there are no obvious physical rules in nature that correspond even to Wolfram's simplest computer CAs, let alone to the more complex ones. The fact that a program running on a computer lights up pixels in a pattern reminiscent of, say, curling cigarette smoke does not at all mean that the same causes underlie both the computer image and the real smoke. When Wolfram turns to biological evolution he, like proponents of complexity theory before him, rudely disses Darwinian theory as overblown. But, again like those before him, as support for his own ideas Wolfram points to a few simple features of life that may possibly fit with his math, but passes over in silence the more complex underlying features that don't. For example, he writes of the geometric patterns of coloration of butterflies and sea shells and muses that the patterns may arise from simple physical constraints rather than specific coding in DNA. Well, maybe so. But the pigments that cause the coloration are the products of enzymes which are made by ribosomes that contain scores of complex macromolecular components. Do CAs have anything to do with that? Wolfram give us no reason to think so.

All that, however, is simply overreaching, which, even if disappointing, is at least understandable. A more serious problem is that in one very important area Stephen Wolfram, boy genius, goes off the deep end—although admittedly he has plenty of company among modern scientists. He emphasizes that underwriting much of his thinking is a peculiar presupposition: "All processes, whether they are produced by human effort or occur spontaneously in nature, can be viewed as computations." And he does mean all processes. In his view rocks rolling down a hill are computers, taking input at each step and updating the system according to a set of rules, just like a PC does. Indeed, "fluid turbulence in the gas around a star" has done "more computation than has by most measures ever been done throughout the whole course of human intellectual history." And, in the same physical, unknowing sense, the human brain is only a computer. The main reason Wolfram thinks his ideas can be applied to all of nature is because by this definition the universe itself is a computer.

Stripped of the computer talk, this is just good old-fashioned materialism: in either a desktop computer, the universe at large, or a human brain, there's nothing but particles bouncing around. That's fine with Wolfram, who repeatedly states that one of his goals is to remove all notions of purpose from science. But when a definition of computation explicitly repudiates the mind that comprehends the calculations, we quickly descend into absurdity. When prodded in a *New York Times* interview, Wolfram agreed that a bucket of nails rusting quietly in a corner is a universal computer, comparable in pertinent features to the human mind. So, by his own premise, in the end Wolfram's own genius is reduced to the equivalent of a bucket of rusty nails.

How ironic that Wolfram wrote his book to overcome faulty premises in science.

25. OBSTACLE TO DARWINIAN EVOLUTION

Whenever professors get together to talk, somebody eventually
says, "Hey, let's write a book on this!" (We get to add it to our CVs.)
For *Debating Design*, the philosopher of biology Michael Ruse and
design theorist William Dembski gathered contributions in 2004
from some true academic luminaries: geneticist Francisco Ayala,
philosopher of science Elliott Sober, complexity theorist Stuart
Kauffman, physicist Paul Davies, theologians John Polkinghorne
and Richard Swinburne, and more. Also included were Brown
University biologist Kenneth Miller and myself, taking shots at
each other.

> Excerpt from "Irreducible Complexity: Obstacle to Darwinian Evolution,"
> in *Debating Design: From Darwin to DNA*, eds. William Dembski and
> Michael Ruse (New York: Cambridge University Press, 2004).

FINALLY, RATHER THAN SHOWING HOW THEIR THEORY COULD HAN-
dle the obstacle, some Darwinists are hoping to get around irreducible complexity by verbal tap dancing. At a debate between proponents
and opponents of intelligent design sponsored by the American Museum
of Natural History in April 2002, Kenneth Miller actually claimed (the
transcript is available at the website of the National Center for Science
Education[191]) that a mousetrap isn't irreducibly complex because subsets
of a mousetrap, and even each individual part, could still "function" on
their own. The holding bar of a mousetrap, Miller observed, could be
used as *a toothpick*, so it still had a "function" outside the mousetrap. Any
of the parts of the trap could be used as a paperweight, he continued, so
they all had "functions." And since any object that has mass can be a paperweight, then any part of anything has a function of its own. Presto!—
there is no such thing as irreducible complexity! Thus the acute problem

for gradualism that any child can see in systems like the mousetrap is smoothly explained away.

Of course the facile explanation rests on a transparent fallacy, a brazen equivocation. Miller uses the word "function" in two different senses. Recall that the definition of irreducible complexity notes that removal of a part "causes the *system* to effectively cease functioning." Without saying so, in his exposition Miller shifts the focus from the separate function of the intact *system* itself to the question of whether we can find a different use (or "function") for some of the *parts*. However, if one removes a part from the mousetrap I pictured, it can no longer catch mice. The *system* has indeed effectively ceased functioning, so the *system* is irreducibly complex, just as I had written. What's more, the functions that Miller glibly assigns to the parts—paperweight, toothpick, key chain, etc.—have little or nothing to do with the function of the system of catching mice (unlike the mousetrap series proposed by John McDonald discussed earlier), so they give us no clue as to how the system's function could arise gradually. Miller explained precisely nothing.

With the problem of the mousetrap behind him, Miller moved on to the bacterial flagellum—and again resorted to the same fallacy. If nothing else, one has to admire the breathtaking audacity of verbally trying to turn another severe problem for Darwinism into an advantage. In recent years it has been shown that the bacterial flagellum is an even more sophisticated system than had been thought. Not only does it act as a rotary propulsion device; it also contains within itself an elegant mechanism to transport the proteins that make up the outer portion of the machine, from the inside of the cell to the outside.[192] Without blinking, Miller asserted that the flagellum is not irreducibly complex because some proteins of the flagellum could be missing and the remainder could still transport proteins, perhaps independently. (Proteins similar—but not identical—to some found in the flagellum occur in the type III secretory system of some bacteria.[193]) Again he was equivocating, switching the focus from the function of the system to act as a rotary propulsion machine to the ability of a subset of the system to transport proteins across a membrane. However, taking away the parts of the flagellum certainly destroys the ability of the system to act as a rotary propulsion machine, as I have argued. Thus, contra Miller, the flagellum is indeed irreducibly complex. What's more, the function of transporting proteins has as little directly to do with the function of rotary propulsion as a toothpick has to do with a mousetrap. So discovering the supportive function of transporting proteins tells us precisely nothing about how Darwinian processes might have put together a rotary propulsion machine.

26. EVEN IF PARTS HAVE OTHER FUNCTIONS

"Irreducible Complexity Is an Obstacle to Darwinism Even If Parts of a System Have Other Functions," Discovery Institute, February 18, 2004.

IN A RECENT COLUMN IN THE *WALL STREET JOURNAL*,[194] SCIENCE writer Sharon Begley repeated some false claims about the concept of irreducible complexity (IC) that have been made by Darwinists, in particular by Kenneth Miller, a professor of biology at Brown University. After giving a serviceable description in her column of why I argue that a mousetrap is IC, Begley added the Darwinist poison pill to the concept. The key misleading assertion in the article is the following: "Moreover, the individual parts of complex structures supposedly serve no function." In other words, opponents of design want to assert that if the individual parts of a putatively IC structure can be used for anything at all other than their role in the system under consideration, then the system itself is not IC. So, for example, Kenneth Miller has seriously argued that a part of a mousetrap could be used as a paperweight, so not even a mousetrap is IC. Now, anything that has mass could be used as a paperweight. Thus, by Miller's tendentious reasoning, any part of any system at all has a separate "function." Presto! There is no such thing as irreducible complexity.

That's what often happens when people who are adamantly opposed to an idea publicize their own definitions of its key terms—the terms are manipulated to wage a PR battle. The evident purpose of Miller and others is to make the concept of IC so brittle that it easily crumbles. Put another way, they are building a straw man. I never wrote that individual parts of an IC system couldn't be used for any other purpose. (That would be silly—who would ever claim that a part of a mousetrap couldn't be used as a paperweight, or a decoration, or a blunt weapon?) Quite the opposite, I clearly wrote in *Darwin's Black Box* that even if the individual parts had their own functions, that still does not account for the irreducible complexity of the system. In fact, it would most likely exacerbate the problem, as I stated when considering whether parts lying around a garage could be used to make a mousetrap without intelligent intervention.

> In order to catch a mouse, a mousetrap needs a platform, spring, hammer, holding bar, and catch. Now, suppose you wanted to make a mousetrap. In your garage you might have a piece of wood from an old Popsicle stick (for the platform), a spring from an old wind-up clock, a piece of metal (for the hammer) in the form of a crowbar, a darning needle for the holding bar, and a bottle cap that you fancy to use as a catch. But these pieces, even though they have some vague similarity to the pieces of a working mousetrap, in fact are not matched to each other and couldn't form a functioning mousetrap without extensive modification. All the while the modification was going on, they would be unable to work as a mousetrap. The fact that they were used in other roles (as a crowbar, in a clock, etc.) does not help them to be part of a mousetrap. As a matter of fact, their previous functions make them ill-suited for virtually any new role as part of a complex system.[195]

The reason why a separate function for the individual parts does not solve the problem of IC is that IC is concerned with the function of the system: "By irreducibly complex I mean a single system which is composed of several well-matched, interacting parts that contribute to the basic function, and where the removal of any one of the parts causes the system to effectively cease functioning."[196]

The system can have its own function, different from any of the parts. Any individual function of a part does not explain the separate function of the system.

Miller applies his crackerjack reasoning not only to the mousetrap, but also to the bacterial flagellum—the extremely sophisticated, ultra-complex biological outboard motor that bacteria use to swim—pictured on the cover of the present volume. I discussed it in *Darwin's Black Box* and since then it has become something of a poster child for intelligent design. No wonder. Anyone looking at a drawing of the flagellum immediately apprehends the design. Since the flagellum is such an embarrassment to the Darwinian project, Miller tries to distract attention from its manifest design by pointing out that parts of the structure can have functions other than propulsion. In particular, some parts of the flagellum act as a protein pump, allowing the flagellum to aid in its own construction—a level of complexity that was unsuspected until relatively recently.

Miller's argument is that since a subset of the proteins of the flagellum can have a function of their own, then the flagellum is not IC and Darwinian evolution could produce it. That's it! He doesn't show how natural selection could do so; he doesn't cite experiments showing that such a thing is possible; he doesn't give a theoretical model. He just points to the greater-than-expected complexity of the flagellum (which Darwinists did not predict or expect) and declares that Darwinian processes could produce it. This is clearly not a fellow who wants to look into the topic too closely.

In fact, the function of a pump has essentially nothing to do with the function of the system to act as a rotary propulsion device, any more than the ability of parts of a mousetrap to act as paperweights has to do with the trap function. And the existence of the ability to pump proteins tells us nil about how the rotary propulsion function might come to be in a Darwinian fashion. For example, suppose that the same parts of the flagellum that were unexpectedly discovered to act as a protein pump were instead unexpectedly discovered to be, say, a chemical factory for synthesizing membrane lipids. Would that alternative discovery affect Kenneth Miller's reasoning at all? Not in the least. His reasoning would still be simply that a part of the flagellum had a separate function. But how would a lipid-making factory explain rotary propulsion? In the same way that protein pumping explains it—it doesn't explain it at all.

The irreducible complexity of the flagellum remains unaltered and unexplained by any unintelligent process, despite Darwinian smoke-blowing and obscurantism.

I have pointed all this out to Ken Miller on several occasions, most recently at a debate in 2002 at the American Museum of Natural History. But he has not modified his story at all.

As much as some Darwinists might wish, there is no quick fix solution to the problem of irreducible complexity. If they want to show their theory can account for it (good luck!), then they'll have to do so by relevant experiments and detailed model-building—not by wordplay and sleight-of-hand.

27. DARWINISM GONE WILD

"Darwinism Gone Wild: Neither Sequence Similarity nor
Common Descent Address a Claim of Intelligent Design," Evolution
News and Views, Discovery Institute, April 19, 2007.

OKAY, SO ONE DAY A GUY WALKS UP TO YOU AND SAYS IRREDUCIBLE complexity is no problem for a random, Darwinian-like evolutionary process. In fact, he can explain how a mousetrap could be made step by step. That's great, you reply, tell me. Easy, says he. He has just finished a detailed analysis of the standard mechanical mousetrap and discovered that, except for the wooden base, all the parts are made of metal! What's more, he's even looked at non-standard mechanical traps, and their pieces are all made of metal, too! Also, after much sleuthing he's noticed that the mousetrap spring has a lot in common with the spring inside his ballpoint pen—both are made of metal, and both are curled into spirals.

Fascinating, you reply, please go on.

Go on? What, are you blind? Don't you see? asks he. The mousetrap spring must have arisen from something like the pen's spring, to make the beginning of the mousetrap. Then the spring duplicated to form the other metal parts, which were added one by one to make the trap we see today. What more could a reasonable person ask for?

You point out that it isn't quite obvious to you how that helps, that the function of the mousetrap would seem to be missing from all those parts, and that while all the parts were being added, the system still wouldn't work like a trap. In fact, you note that the scenario says nothing at all about how the mouse-trapping function arose.

IDiot!, he mutters.

Common Descent Versus Random Mutation/ Natural Selection

That's pretty much the scenario being played out after the recent on-line publication of a paper by Liu and Ochman in the *Proceedings of the National Academy of Sciences*.[197] The gist of the paper is that the workers compared the sequences of the dozens of proteins of the flagellum of the gut bacterium *E. coli* to each other, to other *E. coli* proteins, and to the flagellar proteins of other kinds of bacteria. They noted plausible sequence similarities among the flagellar proteins to each other, but not so much to other bacterial proteins. So Liu and Ochman concluded that all twenty-four proteins of the flagellum core must have descended from a single gene for a single protein!

As I'll mention below, other people find that claim very dubious, but let's leave that aside for now. Let's concentrate on the fact that this is being touted as an answer to claims of intelligent design. As I've pointed out many times, beginning with *Darwin's Black Box* in 1996, the argument for intelligent design in biology has little to do with protein-sequence similarity or common ancestry, for the same reason that knowing all the parts are made of metal doesn't explain the mousetrap. Even if all those parts are made of metal, and even if they derived serially from each other or from some primordial piece of metal, that doesn't even begin to explain how a mousetrap could be built step by step by a random process. In the same way, even if all the proteins of the flagellum derived serially one from the other, or from some magical precursor protein, that doesn't even try to explain how a flagellum could be built step by step by a Darwinian process.

Let me emphasize the point: Common descent is one thing. Random mutation and natural selection is something completely different. Evidence for common descent is NOT evidence for evolution by random mutation/natural selection. At the very best, protein sequence comparisons may say something about common descent, but they aren't support for Darwin's crucial claim that the startlingly elegant, functional complexity of life arose by random mutation culled by natural selection. The PNAS paper is quite irrelevant to that. The bottom line is that, the paper does not even try to address the irreducible complexity of the flagellum or its need for intelligent design.

CURIOSITY-CHALLENGED

THE PNAS paper reaches conclusions that other workers find very questionable. Nicholas Matzke of the pro-Darwinian National Center for Science Education and Panda's Thumb blog declares the work to be of "canine quality," that is, "a dog."[198] (Although a geographer by training, Matzke has acquired some skills in the area and earlier published his own sequence comparisons of flagellar proteins in Nature Reviews Microbiology.) The bottom line is that Matzke is quite skeptical that the two dozen kinds of proteins in the flagellum core could be derived from a single protein. His point is well taken. Yet neither of the scientists that Science magazine journalist Jennifer Cutraro called for comments expressed any curiosity concerning that startling claim.[199]

Nor were they curious about some other pretty obvious challenges facing PNAS paper: 1) What kind of amazing protein would it take to actually be able to give rise to the disparate physical parts of the flagellum? 2) The authors of the paper find few homologies between flagellar proteins and other proteins; yet if that primordial protein were indeed so plastic, why hasn't it been co-opted to perform many other functions in the bacterial cell? 3) Their prodigy protein supposedly gave rise to all the core parts of the flagellum billions of years ago, before the common ancestor of major classes of bacteria. Yet since that time it has not been heard from. A single protein which blossoms to give one coherent, astoundingly complex structure and then, its work complete, is never heard from again—that hardly seems like what one should expect on Darwinian grounds.

GRANDIOSER AND GRANDIOSER

IT SEEMS that the grandiosity of Darwinian claims against ID is rapidly accelerating. Just one year ago the supposed big breakthrough was a paper by Thornton[200] showing that, if he himself personally changed a couple amino acids of a receptor protein in his lab, he could slightly alter the ligand it bound. So just last year, one worker strained to account for a couple amino acid changes to a single protein affecting one property. Yet twelve months later, the *PNAS* paper blithely claims to account for dozens of whole proteins with many different functions.

All in all, this paper is a marvelous example of Darwinism-gone-wild, where imagination does almost all the work, experiment none of it. I'm hopeful that my book, *The Edge of Evolution*, will provide a sorely needed reality check when it comes out in June. It will demonstrate the enormous difficulty of putting together by random mutation and selection even two coherent amino acid changes, let alone a multi-protein complex.

28. REDUCIBLE VERSUS IRREDUCIBLE SYSTEMS

"Reducible Versus Irreducible Systems and Darwinian
Versus Non-Darwinian Processes," Evolution News and
Views, Discovery Institute, September 14, 2009.

RECENTLY A PAPER APPEARED ONLINE IN THE JOURNAL *PROCEED-ings of the National Academy of Sciences*, entitled "The Reducible Complexity of a Mitochondrial Molecular Machine."[201] As you might expect, I was very interested in reading what the authors had to say. Unfortunately, as is all too common on this topic, the claims made in the paper far surpassed the data, and distinctions between such basic ideas as "reducible" versus "irreducible" and "Darwinian" versus "non-Darwinian" were pretty much ignored.

Since *PNAS* publishes letters to the editor on its website, I wrote in. Alas, it seems that polite comments by a person whose work is the clear target of the paper are not as welcome as one might suppose from reading the journal's letters-policy announcement. ("We wish to provide readers with an opportunity to constructively address a difference of opinion with authors of recent papers. Readers are encouraged to point out potential flaws or discrepancies or to comment on exceptional studies published in the journal. Replication and refutation are cornerstones of scientific progress, and we welcome your comments.") My letter received a brusque rejection. Below I reproduce the letter for anyone interested in my reaction to the paper. (By the way, it's not just me. Other scientists whose work is targeted sometimes get the runaround on letters to the editor, too. For an amusing/astounding example, see "How to Publish a Scientific Comment in 1 2 3 Easy Steps" by Prof. Rick Trebino of Georgia Institute of Technology.[202])

Call me paranoid, but it seems to me that some top-notch journals are real anxious to be rid of the idea of irreducible complexity. Recall that last year *Genetics* published a paper purportedly refuting the difficulty of getting multiple required mutations by showing it's quick and easy in a computer—if one of the mutations is neutral (rather than harmful) and first spreads in the population. Not long before that, *PNAS* published a paper supposedly refuting irreducible complexity by postulating that the entire flagellum could evolve from a single remarkable prodigy-gene. Not long before that, *Science* published a paper allegedly refuting irreducible complexity by showing that if an investigator altered a couple of amino acid residues in a steroid hormone receptor, the receptor would bind steroids more weakly than the unmutated form. (That one also made the *New York Times!*) So, arguably picayune, question-begging, and just plain wrong results disputing IC find their way into front-line journals with surprising frequency. Meanwhile, in actual laboratory evolution experiments, genes are broken right and left as bacteria try to outgrow each other.

Well, at least it's nice to know that my work gives some authors a hook on which to hang results that otherwise would be publishable only in journals with impact factors of -3 or less. But if these are the best "refutations" that leading journals such as *PNAS* and *Science* can produce in more than a decade, then the concept of irreducible complexity is in very fine shape indeed.

To the Editor: Reducible Versus Irreducible Systems and Darwinian Versus Non-Darwinian Processes

THE RECENT paper by Clements et al.[203] illustrates the need for more care to avoid non sequiturs in evolutionary narratives. The authors intend to show that Darwinian processes can account for a reducibly complex molecular machine. Yet, even if successful, that would not show that such processes could account for irreducibly complex machines, which Clements et al. cite as the chief difficulty for Darwinism raised by intelligent design proponents like myself. Irreducibly complex molecular systems, such as the bacterial flagellum or intracellular transport system, plainly cannot sustain their primary function if a critical mechanical part is removed.[204] Like a mousetrap without a spring, they would be broken. Here the authors first postulate (they do not demonstrate) an amino acid transporter that fortuitously also transports proteins inefficiently.[205] They subsequently attempt to show how the efficiency might be improved. A scenario for increasing the efficiency of a pre-existing, reducible function, however, says little about developing a novel, irreducible function.

Even just as evidence for the applicability of Darwinian processes to reducibly complex molecular machines, the data are greatly over-interpreted. A Darwinian pathway is not merely one that proceeds by "numerous, successive, slight modifications"[206] but, crucially, one where mutations are random with respect to any goal, including the future development of the organism. If some mutations arise non-randomly, the process is simply not Darwinian. Yet the authors say nothing about random mutation. Their chief data are sequence similarities between bacterial and mitochondrial proteins. However, the presumably homologous proteins have different functions, and bind non-homologous proteins. What is the likelihood that, say, a Tim44-like precursor would forsake its complex of bacterial proteins to join a complex of other proteins? Is such an event reasonably likely or prohibitively improbable? Clements et al. do not provide even crude estimates, let alone rigorous calculations or experiments, and thus provide no support for a formally Darwinian process. Their only relevant data in this regard is their demonstration that a singly-mutated bacterial TimB can substitute for Tim14 in mitochondrial transport. While that is certainly an interesting result, rescuing a pre-existing, functioning system in the laboratory is not at all the same thing as building a novel system step-by-random-step in nature.

Biologists have long been wary of attempts to fill in our lack of knowledge of the history of life with imaginative reconstructions that go far beyond the evidence. As I have discussed,[207] extensive laboratory evolution studies over decades offer little support for the plausibility of such felicitous scenarios as Clements et al. propose. The authors may well be overlooking formidable difficulties that nature itself would encounter.

29. METHINKS NEW PNAS PAPER IS LIKE A WEASEL

"Methinks New *PNAS* Paper Is Like a Weasel," Evolution News
and Views, Discovery Institute, December 14, 2010.

A PAPER BY WILF AND EWENS RECENTLY PUBLISHED ONLINE IN THE *Proceedings of the National Academy of Sciences*, titled "There's Plenty of Time for Evolution,"[208] reads like a printed version of *Groundhog Day*, the classic movie where comedian Bill Murray keeps awakening to find it's the same day again. The paper's authors sniff at unnamed benighted folks who think there hasn't been enough time for (Darwinian) evolution to build the complexity we see in life. Not so, they protest. Why, all one has to do to see the light is to use the right mathematical model. In a model they describe, "After guessing each of the letters, we are told which (if any) of the guessed letters are correct, and then those letters are retained. The second round of guessing is applied only for the incorrect letters that remain after this first round, and so forth."

But this is no more than a mathematized version of Richard Dawkins's "Methinks it is like a weasel" analogy published in his 1986 book *The Blind Watchmaker*, where a string of letters is compared to that phrase in Dawkins's computer's memory, the letters that match are kept, and the ones that don't are randomly replaced until all letters match. But even Dawkins acknowledged in his book that the analogy "is misleading in important ways" because the results were judged by his computer "according to the criterion of resemblance to a distant ideal target... Life isn't like that." Well, little problems like "life isn't like that" apparently don't matter to some Darwinists who think every day is Groundhog Day.

30. IRREMEDIABLE COMPLEXITY

"Irremediable Complexity," Evolution News and
Views, Discovery Institute, August 22, 2011.

AN INTRIGUING "HYPOTHESIS" PAPER ENTITLED "HOW A NEUTRAL Evolutionary Ratchet Can Build Cellular Complexity,"[209] where the authors speculate about a possible solution to a possible problem, recently appeared in the journal *IUBMB Life*. It is an expanded version of a short essay called "Irremediable Complexity?"[210] published last year in *Science*. The authors of the manuscripts include the prominent evolutionary biologist W. Ford Doolittle.

The gist of the paper is this: The authors think that over evolutionary time, neutral processes would tend to "complexify" the cell. They call that theoretical process "constructive neutral evolution" (CNE). In an amusing analogy they liken cells in this respect to human institutions:

> Organisms, like human institutions, will become ever more "bureaucratic," in the sense of needlessly onerous and complex, if we see complexity as related to the number of necessarily interacting parts required to perform a function, as did Darwin. Once established, such complexity can be maintained by negative selection: the point of CNE is that complexity was not created by positive selection.[211]

In brief, the idea is that neutral interactions evolve serendipitously in the cell, spread in a population by drift, get folded into a system, and then can't be removed because their tentacles are too interconnected. It would be kind of like trying to circumvent the associate director of licensing delays in the Department of Motor Vehicles—can't be done.

The possible problem the authors are trying to address is that they think many systems in the cell are needlessly complex. For example, the spliceosome, which "splices" some RNAs (cuts a piece out of the middle of a longer RNA and stitches the remaining pieces together), is a huge conglomerate containing "five small RNAs (snRNAs) and >300 proteins, which must be assembled de novo and then disassembled at each of the many introns interrupting the typical nascent mRNA."[212] What's more, some RNAs don't need the spliceosome—they can splice themselves, without any assistance from proteins. So why use such an ungainly assemblage if a simpler system would do?

The authors think the evolution of such a complex is well beyond the powers of positive natural selection: "Even Darwin might be reluctant to advance a claim that eukaryotic spliceosomal introns remove themselves more efficiently or accurately from mRNAs than did their self-splicing group II antecedents, or that they achieved this by 'numerous, successive, slight modifications' each driven by selection to this end."[213]

Well, I can certainly agree with them about the unlikelihood of Darwinian processes putting together something as complex as the spliceosome. However, leaving aside the few RNAs involved in the spliceosome, I think their hypothesis of CNE as the cause for the interaction of hundreds of proteins—or even a handful—is quite implausible. (An essay skeptical of large claims for CNE, written from a Darwinian-selectionist viewpoint, has appeared recently[214] along with a response from the authors.[215])

The authors' rationale for how a protein drifts into becoming part of a larger complex is illustrated by Figure 1 of their recent paper (similar to the single figure in their *Science* essay). A hypothetical "Protein A" is imagined to be working just fine on its own, when hypothetical "Protein B" serendipitously mutates to bind to it. This interaction, postulate the authors, is neutral, neither helping nor harming the ability of Protein A to do its job. Over the generations Protein A eventually suffers a mutation which would have decreased or eliminated its activity. However, because of the fact that Protein B is bound to it, the mutation does not harm the activity of Protein A. This is still envisioned to be a neutral interaction by the authors, and organisms containing the Protein A-Protein B complex drift to fixation in the population. Then other mutations come along, co-adapting the structures of Protein A and Protein B to each other. At this point the AB complex is necessary for the activity of Protein A. Repeat this process several hundred more times with other proteins, and you've built up a protein aggregate with complexity of the order of the spliceosome.

Is this a reasonable hypothesis? I don't mean to be unkind, but the scenario is vague and undeveloped, and when examined critically it quickly loses plausibility. The first thing to note about the paper is that it contains absolutely no calculations to support the feasibility of the model. This is inexcusable. The mutation rates of various organisms—viral, prokaryotic, eukaryotic—are known to sufficient accuracy[216] that estimates of how frequently the envisioned mutations arrive could have been provided. The neutral theory of evolution is also well developed,[217] which would allow the authors to calculate how long it would take for the postulated neutral mutations to spread in a population. Yet no numbers—not even back-of-the-envelope calculations—are provided. Previous results by other researchers[218] have shown that the development of serendipitous specific binding sites between proteins would be expected to be quite rare, and to involve multiple mutations. Kimura[219] showed that fixation of a mutation by neutral drift would be expected to take a long, long time. Neither of these previous results bodes well for the authors' hypothesis.

The second thing to notice about the paper is that there is no experimental support for its hypothesis. The authors concede this where they comment, "Development of in vitro experimental systems with which to test CNE will be an important step forward in distinguishing complex biology that arose due to adaptation versus nonadaptive complexity, as part of a larger view to understand the interplay between neutral and adaptive evolution, such as the intriguing long-term evolution experiments of Lenski and coworkers."[220] Indeed, no such experimental evolutionary results have been reported to my knowledge, either by Lenski or by other workers (not in 2011 when the paper appeared, and not now nearly a decade later).

Besides the lack of support from calculations or experiments, the authors discuss no possible obstacles to the scheme. I certainly understand that workers want to accentuate the positive when putting a new model forward, but potential pitfalls should be pointed out so that other researchers have a clearer idea of the promise of the model before they invest time in researching it.

The first possible pitfall comes at the first step of the model, where a second protein is postulated to bind in a neutral fashion to a working protein. How likely is that step to be neutral? At the very least, we now have two proteins, A and B, that now have a large part of their surfaces obstructed that weren't before. Will this interfere with their activities? It seems there is a good chance. Second, simply by Le Chatelier's principle the binding of the two proteins must affect the free energies of their folded states. What's more, the flexibility of both proteins must be affected. Will these individual effects serendipitously cancel out so that the overall effect will be neutral? It seems like an awful lot to ask for without evidence.

In the next step of the model, Protein A is supposed to suffer a mutation that would have caused it to lose activity, but, luckily, when it is bound to Protein B it is stabilized enough so that activity is retained. What fraction of possible mutations to Protein A would fall in that range? It seems like a very specialized subfraction. Looking at the flip side, what fraction of mutations to Protein A and/or Protein B which otherwise would not have caused A to lose activity will now do so because of its binding to Protein B?

The last step of the model is the "co-adaptation" of the two proteins, where other, complementary mutations occur in both proteins. Yet this implies that the protein complex must suffer deleterious mutations at least every other step, provoking the "co-adaptive" mutation to fix in the population. Wouldn't these deleterious mutations be very unlikely to spread in the population?

Finally, multiply these problems all by a hundred to get a spliceosome. Or, rather, raise these problems to the hundredth power. But, then, why stop at a hundred? As the authors note approvingly: "Indeed, because CNE is a ratchet-like process that does not require positive selection, it will inevitably occur in self-replicating, error prone systems exhibiting sufficient diversity, unless some factor prevents it."[221]

Why shouldn't the process continue, folding in more and more proteins, until the cell congeals? I suppose the authors would reply, "Some factor prevents it." But might not that factor kick in at the first or second step? The authors give us no reason to think it wouldn't.

The CNE model, at least on the scale envisioned by the authors, faces other problems as well (for example, it would be a whole lot easier to develop binding sites for metal ions or metabolites that are present in the cell at much higher concentrations than most proteins), but I think this is enough to show it may not be as promising as the article would have one believe.

Besides the model itself, it is interesting to look at a professed motivation of the authors in proposing it. It may not have escaped your notice that "irremediable complexity" sort of sounds like "irreducible complexity." In fact, the authors put the model forward as their contribution to the good fight against "antievolutionists," whose "continued failure to consider CNE alternatives impoverishes evolutionary discourse and, by oversimplification, actually makes us more vulnerable to critiques by antievolutionists, who like to see such complexity as 'irreducible.'"[222]

So there you have it. The authors don't think Darwin can explain such complexity as is found in the proteasome, and they apparently rule out intelligent design. (By the way, when will these folks ever grasp the fact that intelligent design is not "antievolution"?) "Irremediable complexity" seems to be all that's left, no matter how unsupported and problematic it may be.

Although the authors seem not to notice, their entire model is built on a classic argument from ignorance, beginning with the definition of irremediable complexity: "'irremediable complexity': the seemingly gratuitous, indeed bewildering, complexity that typifies many cellular subsystems and molecular machines, particularly in eukaryotes."[223]

"Seemingly gratuitous." In other words, the authors don't know of a function for the complexity of some eukaryotic subsystems; therefore, they don't have functions. Well, the history of arguments asserting that something or other in biology is functionless is pretty grim. More, the history of assertions that even "simple" things (like, say, DNA, pre-1930) in the cell either don't have a function or are just supporting structures is abysmal. Overwhelmingly, progress in biology has consisted of finding new and ever-more-sophisticated properties of systems that had been thought simple. If apparently simple systems are much more complex than they initially seemed, I would bet heavily against the hypothesis that apparently complex systems are much simpler than they appear.

31. "Resurrected" Flagella Were Just Unplugged

"'Resurrected' Flagella Were Just Unplugged," Evolution
News and Views, Discovery Institute, March 3, 2015.

A RECENT PAPER IN *SCIENCE* CARRIES THE CLICKBAIT TITLE "EVOLU-tionary Resurrection of Flagellar Motility via Rewiring of the Nitrogen Regulation System."[224] I can't blame scientists in these days of comparatively low funding for trying to attract attention to their work. But the public will eventually grow jaded if misled readers keep finding jazzy labels stuck to picayune results.

A related cheerleading story in *The Scientist*[225] says that the workers didn't set out to investigate flagellar evolution. Rather, starting with the soil bacterium *Pseudomonas fluorescens*, which normally sports a fine, functioning flagellum, they intentionally deleted just the gene for the master switch protein (dubbed "FleQ") that controls flagellum development in order to see how the bug might deal with immobility while colonizing plants. (The genes for the dozens of other protein parts needed for the flagellum were all left intact.) They were surprised to see that the bacteria regained the ability to make flagella after being incubated on Petri dishes for a few days.

Their investigation of the once-again-mobile *P. fluorescens* showed a couple of genetic changes. Most frequently a point mutation (D228A) appeared in a protein called NtrB. That mutation had previously been shown[226] to cause the protein to stay on continuously (like breaking the controls of a chain saw so it can't be switched off). The ordinary role of NtrB is to chemically modify another protein, NtrC, under appropriate circumstance, which activates it. The mutant, constitutively active, unregulated NtrB kept NtrC continually in its active form, too.

Now, like FleQ (the master regulator switch for flagellar genes), active NtrC is itself a master switch that usually turns on genes involved in another pathway—nitrogen metabolism. It also turns out to be a homolog of FleQ. In other words, the two switch proteins were already structurally very similar. Apparently the unmutated NtrC already had some ability to cross-bind to the DNA control region that FleQ usually bound to, allowing the extra NtrC—produced when its regulator was broken—to flip the switch which turned on the pre-existing flagellum biosynthesis pathway.

That mutation turned on the flagellum pathway somewhat less effectively than occurs in the original, unmutated bacteria (ones that have intact FleQ). A subsequent mutation in the gene for the DNA-binding region of NtrC increased its ability to turn on the flagellum pathway to greater-than-normal. The authors suggest (but they didn't investigate) that the second mutation helped NtrC bind more tightly to the flagellum control region. Maybe so, but there's a fly in the ointment. A NtrC mutant that lost its entire DNA binding region also helped promote flagellum synthesis. Something funny is going on there.

Some other changes could take the place of the initial mutation in NtrB. Mutations which cause the loss of function of several other genes involved in nitrogen metabolism have the effect of keeping NtrB turned on all the time, too, which jacks up the concentration of NtrC and leads to the same result concerning flagella. Meanwhile, with the mutations in NtrB and NtrC, the poor bug lost the ability to control its genes for nitrogen metabolism. As a participant in the study said, "The bacteria that became much better at swimming were much worse at nitrogen regulation." But "sometimes the advantage can be so great that it's worth paying that cost because otherwise you die."

Here's an analogy for the work's relevance to the evolution of the flagellum. Suppose some guy told you that natural processes could make a functioning television. Intrigued, you say, great, show me. He takes a thousand working TV sets, unplugs them, and places the plugs next to electrical outlets. Eventually a mischievous racoon comes along, begins toying with one of the plugs and the nearby outlet and, in a stroke of dumb luck, plugs the television into the power outlet. See! the guy exclaims. What'd I tell you?! Natural process, functioning TV!

Hmm. Maybe *The Scientist* would describe such results as a "giant evolutionary leap" as they did with the flagellum paper, but I think most people would remain properly skeptical that natural processes had built a television, or even anything on the way to a television.

To recap, the first step in the path was a loss-of-function mutation, either directly to NtrB or to proteins that have the effect of keeping NtrB continually active. The second step was a mutation to NtrC, whose exact effect is unknown. These trigger the expression of a very complex, pre-existing pathway involving several dozen pre-existing proteins, leading to flagellum development. I myself would not characterize that as an "Evolutionary Resurrection." But I guess a paper titled "Crippled Bacteria Forced to Pay Heavy Evolutionary Cost Lest They Die" wouldn't wind up in *Science*.

32. LITTLE COMFORT FOR DARWINISTS

"From a 2011 Paper on Bacterial Flagella, Little Comfort for Darwinists,"
Evolution News and Views, Discovery Institute, November 21, 2014.

AN E-MAIL CORRESPONDENT POINTED OUT A 2011 PAPER TO ME, "Structural Diversity of Bacterial Flagellar Motors," published by the European Molecular Biology Organization.[227] The paper looks at bacterial flagella from a wide phylogenetic distribution using "electron cryotomography"—a pretty new technique that they say allows them to get greater detail more easily.

They write in the abstract that "while a conserved structural core was observed in all 11 bacteria imaged, surprisingly novel and divergent structures as well as different symmetries were observed surrounding the core." To my reading, though, the more striking fact is the first part of their sentence—that all flagella had the same necessary components in the same relative spatial relationships.

The second part of the sentence, that there are differences too, is interesting. But beyond showing somewhat altered shapes and unidentified regions of photographic density, the role of the novelties is unknown.

It seems to me the bigger story is the finding of a strongly conserved core occasionally tricked out with a few decorations. The authors mildly hint that there are renegades out there who would use the flagellum for nefarious purposes ("the occasional misrepresentation of the motor as an invariant, ideal machine"), which probably leads them to emphasize the novelty. But this study should offer little comfort to Darwinists.

33. Secret Obsessions

"New Paper on Flagellum Reveals Secret Obsessions," Evolution
News and Views, Discovery Institute, March 18, 2016.

SUPPOSE IN THE COURSE OF A PLEASANT CONVERSATION WITH A colleague you mentioned your vacation last year in Las Vegas. All of a sudden he starts ranting about Area 51—Vegas is only a few hours away, right? Did you see any lights in the sky? Any military vehicles heading north? You should stay at the Little A'Le'Inn motel like he has six times. You'll see some funny stuff there.

You'd probably back away slowly, smiling, wishing him a nice day.

That's the feeling I got after reading a couple of recent reports on science news sites. While describing an impressive piece of research on the bacterial flagellum and its variants, *New Scientist* could no longer contain itself: "Behold—the only known example of a biological wheel. Loved by creationists, who falsely think they are examples of 'intelligent design'... [T]he diversity of the motors and the fact that they have evolved many times in different bacterial lineages, scuppers the creationist view that the machinery is 'irreducibly complex.'"[228]

Wow!!!! Did you see those lights in the sky??!!!

You'd never guess from the breathless prose that the research paper (written by people whose work I commented on a few years ago) has nothing at all to do with irreducible complexity or intelligent design. Not only don't the phrases occur anywhere in the manuscript, the concepts don't show up either.

The dusty-sounding title, "Diverse High-Torque Bacterial Flagellar Motors Assemble Wider Stator Rings Using a Conserved Protein Scaffold,"[229] is apt for the work—an elegant, largely descriptive study of the structures of modern flagella from a few different kinds of bacteria that shows some are wider and more powerful, others narrower and less powerful. It reports no experiments that test whether random mutation and natural selection could explain even the variations of the molecular machines, let alone what they all have in common, which is considerable: "Despite differences in the organisms' swimming ability, the flagellar motor is composed of a conserved core of ~20 structural proteins. The mechanism of flagellar motility is conserved, with torque generated by rotor and stator components."

The paper does, however, show that if any one of dozens of proteins is knocked out, the ability of the cell to swim is lost, which of course is exactly what to expect from an irreducibly complex system.

Genetic Engineering & Biotechnology News (GEN) also lost it:

> The bacterial flagellum has been at the center of the thinly veiled creationism movement called intelligent design. Subscribers to this belief system have erroneously postulated that the flagellar motor system is "irreducibly complex" and could not have come about through Darwinian evolutionary mechanisms.... It is doubtful these findings will sway the opinion of its detractors, yet they do make it extremely more difficult for them to make their case.[230]

So, you see, those strange folks who think an extraordinarily sophisticated molecular machine points toward intelligent design will hold to their opinions no matter what. They can't be reasoned with, so you shouldn't even try.

Well, it's certainly clear that GEN and New Scientist have given up trying to reason on the subject.

One crazy person is a coincidence. Two are a trend. What's provoking some otherwise smart people into thinking this paper has anything at all to do with intelligent design, other than to reinforce its arguments? I think it's the same general factors that are responsible for much craziness in our world: fear and ignorance. The magazine staff works for and writes for people who fear intelligent design—either because they simply don't want it to be true or at least because they worry it will encourage the ignorant masses to question what they've been taught about life.

What's more, if you go by what they write, these folks are utterly clueless about what modern ID proponents actually argue. They seem to have gotten what opinions they have from perusing a New York Times story, or from glancing at press releases from scientific societies denouncing ID. No one gives any hint of having read a book by an ID proponent, or even of visiting a reliable ID website such as this one. Then they wonder why informed people don't think their arguments are persuasive.

Here are some elementary points they miss, put as simply as possible. Intelligent design is not about common ancestry. It doesn't matter to the ID argument whether life was "originally breathed into a few forms or into one," as Darwin wrote, or even into many. Nor does it matter whether the exact same molecular machine occurs in all organisms or if elegant variations on the theme are found in each separate family or genus or species.

It matters only whether the unabashedly purposeful structures can be seriously explained—in real scientific detail—as the result of unintelligent processes, as Darwinists have so far spectacularly failed to do, or whether intelligence was required to make them.

34. A REVIEW OF AVISE'S INSIDE THE HUMAN GENOME

"A Malodorous Argument for Darwinian Evolution," Evolution
News and Views, Discovery Institute, March 1, 2010.

UNIVERSITY OF CALIFORNIA EVOLUTIONARY BIOLOGIST JOHN Avise has penned a book, *Inside the Human Genome: A Case for Non-Intelligent Design,* and gotten it published by a top academic publishing house, Oxford University Press.[231] Avise, a member of the National Academy of Sciences, has for decades been a leading researcher in evolutionary and ecological genetics. He has written hundreds of research articles and over a dozen books. Clearly he has an impressive scientific mind.

Which makes it all the more astonishing that his new book shows all the intellectual savvy of a typical late-night college dorm room bull session. As his subtitle announces, Avise is anxious to show that, despite the claims of certain renegade biochemists, the molecular features of the human genome discovered by science in the recent past show no traces of intelligent design. They are chaotic, haphazard, a mess. Any designer with the smarts of at least, oh, say, John Avise, would have done a much better job.

Avise tries to steal three bases on a bunt. He claims that both Darwinian evolution and intelligent design can explain the functional parts of the genome, but only evolution can explain the dysfunctional parts (because a beneficent God would not have made those). So he points to what he deems to be poor design and—voilà!—that proves the most intricate, functional molecular machines arose by random mutation and natural selection. No actual separate demonstration of that is thought necessary.

In fact, Avise makes only the most cursory attempt to address the scientific argument for ID. His Chapter 5 is in large part devoted to answering (after a fashion) my *Darwin's Black Box.* Yet in the chapter Avise's only attempt to explain one of my book's examples of irreducible complexity is to cite Liu and Ochman's[232] dubious endeavor to tag all bacterial flagellar genes as descendants of one amazing prodigy gene. The rest of the chapter is pretty much hand-waving.

Avise's main theme is that genes can break, leading to genetic diseases. He has a figure outlining human chromosome 2 and the regions of this chromosome to which various diseases map, such as abetalipoproteinemia and Waardenburg syndrome. A nearby table lists genetic metabolic diseases compiled in *Mendelian Inheritance in Man*. His whole argument can pretty much be summed up in one brief quote: "Lesch-Nyhan syndrome hardly seems like the kind of outcome that would be countenanced by a loving all-powerful Diety [*sic*]."[233] In other words, the theological argument from evil—the same argument Darwin gave when he proclaimed that no beneficent God would allow wasp larvae to feed on the living bodies of caterpillars.

Well, it does not follow that, because the parts of my car can break, the car was not designed. Nor does it follow that the Ford motor company is evil. Of course Avise is making a brief against a "loving all-powerful" entity, which does not describe Ford. Yet, beginning with the Book of Job, throughout history philosophers and theologians have wrestled with the problem of evil. I'm no theologian, so I can't rigorously evaluate those arguments. But Avise is no theologian either and, despite writing an entire book that revolves around the problem of evil, he doesn't even attempt to engage those philosophical and theological arguments.

The bottom line is: if you're the kind of person who thinks that because some people smell bad due to a certain genetic disease (trimethylaminuria[234]) it follows that Darwinian processes made the eye, this is the book for you. Other folks will conclude that the academic standards of Oxford University Press have slipped a few notches.

PART THREE: DEBATING THE EDGE OF EVOLUTION

ALTHOUGH IT MIGHT HAVE SEEMED RELENTLESSLY DETAILED TO the uninitiated, *Darwin's Black Box* actually dealt with molecular machines and their "parts" in broad overview, much as a schematic drawing of a computer might represent its different parts as ovals or squares or triangles. But the parts of machines in our everyday world are themselves usually very carefully shaped and constructed. A simple triangle in a schematic drawing of a computer might itself represent an intricate piece of electronics. The gears of a watch are not just some bland "parts." Rather, the teeth of the gears have to be well-matched to meshing gears, have to be made of the right materials with the right hardness, be the right size to fit, and so on. The same goes for the parts of molecular machines. Their protein parts, too, have to mesh with other parts, be the right size, be the right strength, and more.

In thinking about the question of what unintelligent processes may or may not be able to account for in biology, it is necessary to go beyond a consideration of parts to a consideration of all necessary details. In terms of William Paley's famous example of a watch lying on a heath, we can be certain that the completed watch was designed. But is intelligence necessary to account even for such details as the teeth of the watch gears? In *Darwin's Black Box* and in my responses to critics of that book, I made the case that Darwin's mechanism of random mutation and natural selection surely can't produce irreducibly complex biological machines such as the bacterial flagellum. But can it even produce gear teeth? If it can do at least some things but not everything, can we discern an edge—a limit—to what we can reasonably expect unintelligent processes to accomplish in evolution? As it turns out, we can, and the evidence for where that edge rests is independent of the specific irreducible complexity arguments laid out in *Darwin's Black Box*. In other words, what follows stands on its own—a new argument in new territory. We're moving into what, until just a generation ago, was undiscovered country.

In seeking out an answer, the first thing is to acknowledge that it's a hard question to analyze from one's armchair. All manner of considerations and complications might enter into an answer. What's more, the answer may not be firm, but rather fuzzy. What's needed to answer it is less theorizing and more experimentation and observation. Enough with the bickering about how magnificent or hapless is Darwin's mechanism—let's instead *look and see* what it does do when we are watching. In other words, what do we observe actually happening in nature or in the lab when organisms encounter challenges in their environments?

Although that might be an obvious question for science to ask concerning evolution, for a long time it was extremely difficult to get an answer. That's because evolution occurs over many generations, and not to just a single organism but to a whole population. However, with advances in scientific techniques such as the ability to easily sequence DNA, and with long-term observation of microorganisms (which occur in astronomical numbers and reproduce quickly), a lot of pertinent data has been collected in the past twenty years.

In 2007 I wrote *The Edge of Evolution: The Search for the Limits of Darwinism*, which explored what was then available of this new data. There I reviewed data on resistance of the malaria microbe to chloroquine, the adaptation of the bacterium *E. coli* in lab experiments, the alteration of the human immunodeficiency virus (HIV) over time, and more. And on the basis of these observations and experiments with trillions of microbes over thousands of generations, I argued that unintelligent evolutionary processes were *very* limited indeed, much more limited than I had thought a decade earlier when I wrote *Darwin's Black Box*. When one looked at the actual molecular changes that were responsible for the helpful evolutionary changes, it became crystal clear that they were all very minor changes in pre-existing systems. Nothing remotely approaching the evolution of new molecular machinery, or even new parts, was seen. From this I argued that there is a strict limit to what Darwinian and other unintelligent evolutionary mechanisms can accomplish. Looked at from a different direction, there is good reason to conclude that purposeful intelligent design extends much more deeply into life than even I had thought.

The conclusion of *The Edge of Evolution* didn't go over with Darwinists any better than did the message of *Darwin's Black Box*. In fact, if anything their reaction was even more visceral. They were already braced from the decade of skirmishes over intelligent design, and immediately greeted the book with rhetorical guns blazing. This chapter includes my responses to the fire. Their original arguments can be easily discerned from my replies or, in most cases, can be tracked down on the internet with little trouble. As with the responses to *Darwin's Black Box*, I think the fragility of Darwin's theory is nowhere more easily seen than in the spectacle of highly intelligent partisans defending it so weakly.

35. EXCERPTS FROM THE EDGE OF EVOLUTION

These two excerpts give the gist of *The Edge of Evolution*. But if you haven't read the book yet, it's available at all major book stores and libraries.

EXCERPT FROM CHAPTER 1

COMMON DESCENT is what most people think of when they hear the word "evolution." It is the contention that different kinds of modern creatures can trace their lineage back to a common ancestor. For example, gerbils and giraffes—two mammals—are both thought to be the descendants of a single type of creature from the far past. And so are organisms from much more widely separated categories—buffalo and buzzards, pigs and petunias, yaks and yeast.

That's certainly startling, so it's understandable that some people find the idea of common descent so astonishing that they look no further. Yet in a very strong sense the explanation of common descent is also trivial. Common descent tries to account only for the *similarities* between creatures. It says merely that somehow shared features were there from the beginning—the ancestor had them. But all by itself, it doesn't try to explain how either the features or the ancestor got there in the first place, or why descendants differ. For example, rabbits and bears both have hair, so the idea of common descent says only that their ancestor had hair, too. Plants and animals both have complex cells with nuclei, so they must have inherited that feature from a common ancestor. But the questions of how or why are left hanging.

In contrast, Darwin's hypothesized mechanism of evolution—the compound concept of random mutation paired with natural selection—is decidedly more ambitious. The pairing of random mutation and natural selection tries to account for the *differences* between creatures. It tries to answer the pivotal question, what could cause such staggering transformations? How could one kind of ancestral animal develop over time into creatures as different as, say, bats and whales?

Let's tease apart that compound concept. First, consider natural selection. Like common descent, natural selection is an interesting but actually quite modest notion. By itself, the idea of natural selection says just that the more fit organisms of a species will produce more surviving offspring than the less fit. So, if the total numbers of a species stayed the same, over time the progeny of the more fit would replace the progeny of the less fit. It's hardly surprising that creatures that are somehow more fit—stronger, faster, hardier—would on average do better in nature than ones that were less fit—weaker, slower, more fragile.

By far the most critical aspect of Darwin's multifaceted theory is the role of random mutation. Almost all of what is novel and important in Darwinian thought is concentrated in this third concept. In Darwinian thinking, the only way a plant or animal becomes fitter than its relatives is by sustaining a serendipitous mutation. If the mutation makes the organism stronger, faster, or in some way hardier, then natural selection can take over from there, and help make sure its offspring grow numerous. Yet until the random mutation appears, natural selection can only twiddle its thumbs....

As we'll see throughout this book, genetic accidents can get life so far but no further. As earlier generations of scientists agreed, except at life's periphery the evidence for a pivotal role for Darwin's random mutations is terrible. For a bevy of reasons having little to do with science, this crucial aspect of Darwin's theory—the power of natural selection coupled to random mutation—has been grossly oversold to the modern public....

As a theory-of-everything, Darwinism is usually presented as a take-it-or-leave-it proposition. Either accept the whole theory, or decide that evolution is all hype and throw out the baby with the bath water. Both are mistakes. In dealing with an often-menacing nature we can't afford the luxury of elevating anybody's dogmas over data. The purpose of this book is to cut through the fog, to offer a sober appraisal of what Darwinian processes actually can and can't do, to find what I call *the edge of evolution*.

The Importance of the Pathway

In the real world, random mutation, natural selection, and common descent might all be completely true, and yet Darwinian processes still may not be an adequate explanation of life. In order to forge the many complex structures of life a Darwinian process would have to take numerous coherent steps, a series of beneficial mutations that successively build on each other, leading to the structure. In order to do so in the real world, rather than just in our imaginations, there must be a biological route to the structure that stands a reasonable chance of success in nature. In other words, variation, selection, and inheritance will only work if there is also a smooth evolutionary *pathway* leading from biological point A to biological point B....

In everyday life, the greater the distance between points A and B, and the more rugged the intervening landscape, the bleaker are the odds for success of a blindfolded walk, even—or perhaps especially—when following a simple-minded rule like "always climb higher; never back down." The same with evolution. In Darwin's day scientists were ignorant of many of the details of life, and so they could reasonably hope that evolutionary pathways would turn out to be short and smooth. But now we know better. The great progress of modern science has shown that life is enormously elegant and intricate, especially at its molecular foundation. That means that Darwinian pathways to many complex features of life are quite long and rugged. The problem for Darwin, then, as with a long, blindfolded stroll outdoors, is that in a rugged evolutionary landscape, random mutation and natural selection might just keep a species staggering down genetic dead-end alleys, getting stuck on the top of small anatomical hills, or wandering aimlessly over physiological plains, never even coming close to winning the biological pot of gold at a distant biological summit. If that is the case, then random mutation/natural selection would essentially be ineffective. In fact, the striving to climb any local evolutionary hill would actively prevent finding the peak of a distant biological mountain.

This point is crucial: if there is not a smooth, gradually rising, easily found evolutionary pathway leading to a biological system within a reasonable time, Darwinian processes won't work. In this book we'll examine just how demanding a requirement that is....

As a practical matter, how far apart do biological points A and B have to be, and how rugged the pathway between them, before random mutation and natural selection start to become ineffective? How can we tell when that point is reached? Where in biology is a reasonable place to draw the line marking the edge of evolution?

This book answers those questions.

EXCERPT FROM CHAPTER 7: THE TWO-BINDING-SITES RULE

Consider a hypothetical case where it would give an organism some advantage if a particular two of its proteins, which had been working separately, bound specifically to each other. Perhaps the two-protein complex would be able to perform some new task, or do an old task much better. The lesson from shape space is that, in order for the one to bind the other, we should expect to have to search through tens of millions of different mutant sequences before luckily happening upon one that would specifically stick with even modest strength, which would allow the two to spend even half of their time together. (This is likely the minimum necessary strength, enough to have a noticeable biological effect.) Since the mutation rate is so low—about one mutation at a particular site in a hundred million births—we would expect to have to slog through an enormous number of organisms before striking on that lucky one.

Let's make a rough calculation for the average number of organisms we would have to slog through to find a new protein-protein binding site. As I said, shape space tells us that about one in ten to a hundred million coherent protein binding sites must be sifted before finding one that binds specifically and firmly to a given target. The simplest way to alter a protein is by point mutation, where one amino acid is substituted for another at a position in a protein. There are twenty different kinds of amino acids found in proteins. That means that if just five or six positions changed to the right residues—the ones that would allow the two proteins to bind—then that would be an event of approximately the right frequency, since twenty multiplied by itself five or six times (20^5 or 20^6) is about three million or sixty million, respectively—relatively close to the ten to a hundred million different sites we need.

So one way to get a new binding site would be to change just five or six amino acids in a coherent patch in the right way. This very rough estimation fits nicely with studies that have been done on protein structure. Five or six amino acids may not sound like very much at first, since proteins are often made of hundreds of amino acids. But five or six amino acid substitutions means that reaching the goal requires *five or six coherent mutational steps*—just to get two proteins to bind to each other. As we saw in the last chapter, even *one* missing step makes the job much, much tougher for Darwin than when steps are continuous. If multiple steps are missing, the job becomes exponentially more difficult.

Let's consider one further wrinkle. Most amino acid changes in proteins diminish a protein's function. But about one-third of possible amino acid changes are like switching a 'k' for a 'c' in 'cat' or 'candy'; they can be accommodated without too much trouble. Such "neutral" changes can occur during evolution and spread around a population by chance. So let's suppose that of the five or six changes that have to happen to a protein to make a new binding site, a third of them are neutral. They could occur before the other key mutations, as a separate step, without harm. Although finding the right neutral changes would itself be an improbable step, we'll again err on the conservative side and discount the average number of neutral mutations from the average number of total necessary changes. That leaves three or four amino acid changes that might cause trouble if they occur singly. For the Darwinian step in question, they must occur together. Three or four simultaneous amino-acid mutations is like skipping two or three steps on an evolutionary staircase.

Although two or three missing steps doesn't sound like much, that's one or two more Darwinian jumps than were required to get chloroquine resistance in malaria. In Chapter 3 I dubbed that level a "CCC," a "chloroquine-complexity cluster" and showed that its odds were 1 in 10^{20} births. In other words (keeping in mind the roughness of the calculation): *Generating a single new cellular protein-protein binding site is of the same order of difficulty or worse than the development of chloroquine resistance in the malarial parasite.*

Now suppose that, in order to acquire some new, useful property, not just one but *two* new protein binding sites had to develop. A CCC requires, on average, 10^{20}, a hundred billion billion, organisms—more than the number of mammals that has ever existed on earth. So if other things were equal, then the likelihood of getting two new binding sites would be what we called in Chapter 3 a "double CCC"—the square of a CCC, or one in ten to the fortieth power. Since that's more cells than likely have ever existed on earth, such an event would not be expected to have happened by Darwinian processes in the history of the world. Admittedly, statistics are all about averages, so some freak event like this *might* happen—it's not ruled out by force of logic. But it is not bio-logically reasonable to expect it, or less likely events that occurred in the common descent of life on earth. In short, complexes of just three or more different proteins are beyond the edge of evolution. They are lost in shape space.

And the great majority of proteins in the cell work in complexes of *six* or more. Far beyond that edge.

36. RESPONSE TO DAWKINS IN THE NEW YORK TIMES

"Response to Richard Dawkins," Uncommon Descent (website), July 16, 2007.

HERE I RESPOND BRIEFLY TO RICHARD DAWKINS' REVIEW OF *THE Edge of Evolution* in the *New York Times*.[1] I must admit I was sur-prised that he agreed to do it. In the past Dawkins has said that on principle he would not interact with proponents of intelligent design, because that would give us publicity. I guess when the *New York Times* offers writing space, principles can be reconsidered.

Other Darwinist reviewers have blustered; Dawkins is the only one who has dripped venom. I will pass on replying to that. He makes just two substantive points in his review. The first is that the success of artifi-cial selection in things like dog breeding show the malleability of organ-isms, so why should Darwinian evolution be a problem? I already an-swered that point in my reply to Jerry Coyne. Briefly, it begs the question of what changes are occurring at the molecular level in those examples, whether simple ones or complex ones, and it begs the question of where the sophisticated molecular systems came from that we have learned control animal form and development. Dawkins seems quite reluctant to engage my argument at the molecular level; in his review he defers to other scientists for that. He himself gives the kind of argument that a nineteenth-century naturalist might give, before the elegance of the molecular foundation of life was discovered by modern biology.

Dawkins's second substantive point is that if I am right, then "Behe's calculations would at a stroke confound generations of mathematical geneticists, who have repeatedly shown that evolutionary rates are not limited by mutation. Single-handedly, Behe is taking on Ronald Fisher, Sewall Wright, J. B. S. Haldane, Theodosius Dobzhansky, Richard Lewontin, John Maynard Smith and hundreds of their talented co-workers and intellectual descendants."

It's a flattering thought, but incorrect. If I am right it would overturn virtually no theoretical work, simply because theoreticians have not dealt with the sorts of complex, functional systems I write about. For the most part, models have considered one or two simple mutations at a time, conceptually isolated from the real-life complexity of an actual cell or organism. That's necessary, because detailed models of complex systems would be intractable. Those (relatively) simple models can of course be very important and useful for things like predicting the spread of the sickle hemoglobin gene, or calculating from the number of neutral mutations the time since two species shared a common ancestor. But there is no theoretical evolutionary work on the production of molecular machinery.

(One of the luminaries Dawkins lists, John Maynard Smith, once briefly alluded to the kind of problem *The Edge of Evolution* deals with. In a letter to *Nature* in the early 1970s, Smith compared evolution of proteins to a word game where only one letter is allowed to be changed at a time, and misspellings are disallowed too. I cite Smith's paper in the book.)

At the end of his review Dawkins chides me for lack of peer-reviewed publications. Talk about the pot calling the kettle black. If Dawkins himself has many peer-reviewed research publications in the last few decades, he must be writing them under a pseudonym. Dawkins's hypocritical complaint makes a nice little example of Darwinian gate-keeping. The nebulous, wooly-minded scenarios Dawkins spins in his books, of the origins of bat echolocation, spider webs, and so on, have no real justification in peer-reviewed publications. Yet Dawkins is free to write

trade books without howls of protest from the scientific community because his stories fit the way many scientists want the world to be. But if
(ahem...) someone publishes a book critically analyzing the data from a
different perspective, the reaction is dramatically different.

37. RESPONSE TO SEAN CARROLL IN SCIENCE

"Response to Critics, Part 2: Sean Carroll," Uncommon
Descent (website), June 26, 2007.

ALMOST THE SAME DAY THAT *THE EDGE OF EVOLUTION* WAS OFFI-
cially released, *Science* published a long, lead review[2] by evolutionary developmental biologist Sean Carroll, whose own work I discuss
critically in Chapter 9. The review is three parts bluster to one part substance, which at least is more substance than Jerry Coyne's essay.

Here I'll ignore the bluster and deal with the substantive points.
Carroll first covers his rhetorical bases by warning readers that "Unfortunately, [Behe's] errors are of a technical nature and will be difficult for
lay readers, and even some scientists (those unfamiliar with molecular
biology and evolutionary genetics), to detect. Some people will be hood-
winked. My goal here is to point out the critical flaws in Behe's key arguments and to guide readers toward some references." So, you see, if Carroll's reasoning doesn't sound right, well, maybe that's because you, dear
reader, are too slow to understand him. If that's the case, you're supposed
to just take his word for it.

Unfortunately, his word is demonstrably questionable. He claims
that

> Behe's chief error is minimizing the power of natural selection to act
> cumulatively... Behe states correctly that in most species two adaptive
> mutations occurring instantaneously at two specific sites in one gene
> are very unlikely and that functional changes in proteins often involve
> two or more sites. But it is a non sequitur to leap to the conclusion, as

Behe does, that such multiple amino acid replacements therefore can't happen.

But I certainly do not say that multiple amino acid replacements "can't happen." A centerpiece of *The Edge of Evolution* is that it can and did happen. I stress in Chapter 3 that in the case of malarial resistance to chloroquine, multiple necessary mutations did happen in the membrane protein PfCRT. I also of course emphasize that it took a huge population size, one that would not be available to larger organisms. But Carroll seems uninterested in making distinctions.

Carroll cites several instances where multiple changes do accumulate gradually in proteins. (So do I. I discuss gradual evolution of antifreeze resistance, resistance to some insecticides by "tiny, incremental steps—amino acid by amino acid—leading from one biological level to another," hemoglobin C-Harlem, and other examples, in order to make the critically important distinction between beneficial intermediate mutations and detrimental intermediate ones.) But, as Carroll might say, it is a non sequitur to leap to the conclusion that all biological features therefore can gradually accumulate. Incredibly, he ignores the book's centerpiece example of chloroquine resistance, where beneficial changes do not accumulate gradually.

As a "second fatal blunder," he asserts I overlook proteins that bind to "short linear peptide motifs" of two or three amino acids. I'll get to that in a second. Notice, however, that here he is writing simply of a sub-class of protein binding sites, and never gets around to dealing with the question of how the majority of binding sites, those with interacting folded domains, developed. I assume that's because he has no answer.

Carroll lets his imagination run wild. He thinks it would be child's play for random processes to develop binding sites, at least for the sub-category of short peptide motif binding: "Very simple calculations indicate how easily such motifs evolve at random. If one assumes an average length of 400 amino acids for proteins and equal abundance of all amino acids, any given two-amino acid motif is likely to occur at random in every protein in a cell."

Wow, every protein in the cell will have a binding site! Methinks Carroll has just stumbled over an embarrassment of riches. If every protein (or even a large fraction of proteins) had such a binding site, then binding would essentially be non-specific. (It would be much like, say, the case of the digestive enzyme trypsin, which binds and cuts proteins wherever there is the amino acid lysine or arginine.) As I make clear in *The Edge of Evolution*, the problem the cell faces is not just to have protein binding sites (which could simply be large hydrophobic patches), but to bind specifically to the right partner.

In fact, if one takes the trouble to look up the references Carroll cites, one sees that a short amino acid motif is not enough for function in a cell. For example, Budovskaya et al.[3] show that the majority of proteins in the yeast *Saccharomyces cerevisiae* containing a motif recognized by a particular protein kinase were not phosphorylated by the enzyme. What does that mean? It just means that the simple motifs, while necessary for binding, are not sufficient. Other features of the proteins are necessary, too, features which Sean Carroll ignores.

In his enthusiasm Carroll seems not to have noticed that, as I discuss at great length in my book, no protein binding sites—neither short linear peptide motifs nor any other—developed in a hundred billion billion (10^{20}) malarial cells. Or in HIV. Or *E. coli*. Or in human defenses against malaria, save that of sickle hemoglobin. Like Coyne, Carroll simply overlooks observational evidence that goes against Darwinian views.

In fact, Carroll seems unable to separate Darwinian theory from data. He writes that "what [Behe] alleges to be beyond the limits of Darwinian evolution falls well within its demonstrated powers," and "Indeed, it has been demonstrated that new protein interactions and protein networks can evolve fairly rapidly and are thus well within the limits of evolution." Yet if one looks up the papers he cites, one finds no "demonstration" at all. Those papers show, respectively, that: A) different species have different protein binding sites (but, although the authors assume Darwinian processes, they demonstrate nothing about how the sites arose); or B) different species have different protein networks (but, again, the authors demonstrate nothing about how the networks arose). Like Jerry Coyne, Sean Carroll simply begs the question. Like Coyne, Carroll assumes whatever exists in biology arose by Darwinian processes. Apparently Darwinism has eroded Coyne's and Carroll's ability to separate data from theory.

In fact, the data I cite in *The Edge of Evolution* is a *real* demonstration. While we have studied them, in a truly astronomical number of chances, a variety of microbes developed precisely none of the sophisticated cellular mechanisms that Darwinist imaginations ascribe to random mutation and selection. That data demonstrates that random mutation doesn't explain the elegance of cellular systems.

38. BACK AND FORTH WITH SEAN CARROLL

"Back and Forth with Sean Carroll in *Science*," Uncommon Descent (website), October 17, 2007.

SCIENCE HAS PUBLISHED A LETTER[4] BY MYSELF RESPONDING TO SEAN Carroll's earlier review of *The Edge of Evolution*. In my letter I note, "In his unfavorable review of my book, *The Edge of Evolution*, Sean Carroll writes that 'Behe's chief error is minimizing the power of natural selection to act cumulatively,' and implies that I fail to discuss 'pyrimethamine resistance in malarial parasites...—a notable omission given Behe's extensive discussion of malarial drug resistance.'"

But, I demurred, I did write about pyrimethamine. Carroll admitted in *Science* right after my published letter that, well, yes, I did discuss pyrimethamine resistance, but his real concern was that I didn't give it the spin he wanted: "Behe did indeed discuss pyrimethamine resistance on pages 75 and 76 of his book. My criticism is that Behe omitted the clear evidence for the cumulative selection of multiple changes in the drug target protein in nature and that he invoked an altogether different and unsupported explanation in an attempt to bolster his main premise."[5]

Carroll's beef is that several papers he cites[6] have shown that, in the laboratory, in some respects intermediate mutations in the enzyme-target of pyrimethamine have better activity than the wild-type enzyme. But this data proves too much. If the mutations improve the enzyme in vitro, then that begs the question of why organisms with these mutations don't outcompete the wild type in nature, even in the absence of pyrimethamine.

One possibility, which plagues all in vitro work, is that perhaps the mutants have other, detrimental aspects, not measured in an assay, which makes the alteration a net burden in the wild. If that is the case, then the mutant enzyme might run rings around the wild-type enzyme when both are in a test tube in a lab, but could still be a bust in nature.

To see if a particular mutation in a particular enzyme helps an organism to survive in the wild, one has to show that it helps an organism to survive in the wild. None of the papers Carroll cites even tries to do that. On the other hand, the work I cite in my book looked at field studies of organisms in the wild. Workers wondered, "Because resistance to [pyrimethamine] can be conferred by a single point mutation, it was assumed that resistance could occur frequently. However, a recent population survey demonstrated a single origin of [resistant genes] in five countries: Thailand, Myanmar, PDR Lao, Cambodia, and Vietnam."[7]

It was hypothesized that multiple mutations in different genes might be required: "Because concurrent mutations in two different genes occur at reduced frequency, this would help explain the rarity with which resistance has evolved."[8]

(By the way, Hayton and Su also remark that, "Based on the mutant *pfcrt* haplotypes known so far, it is likely that simultaneous multipoint changes in *pfcrt* are necessary to confer [chloroquine resistance]."[9])

Carroll implies I'm somehow less than honest for passing on to readers the thinking of workers in the field in this area, while he passes off as near-conclusive ambiguous work done in vitro.

Toward the end of his lengthy letter Carroll remarks as an aside that "the complete disregard of a massive literature surrounding protein interactions" is "crucial to Behe's entirely unfounded conclusion." That's rich, considering that *Science* cut out the final paragraph of my letter responding to his review (*Science* allowed me about 200 words; they allowed Carroll about 500 words in response), a paragraph which read:

> Carroll writes, "It has been demonstrated that new protein interactions and protein networks can evolve fairly rapidly." If he is implying the changes occurred by Darwinian means, "demonstrated" is question-begging. The references he cites show only that differences exist in contemporary homologous protein sequences among various phyla, some functional, some not. How the functional differences arose—whether by random mutation and selection or not—is not addressed. On the other hand, studies I cite in my book show that, over thousands of generations, astronomical numbers of closely studied microorganisms failed to develop new protein interactions or networks.

If there is a "massive literature" on the evolution of protein-protein interactions which is pertinent to the questions I raise, Sean Carroll somehow failed to cite any of it in his review.

39. RESPONSE TO MILLER IN NATURE

"Response to Kenneth R. Miller," Uncommon Descent (website), July 11, 2007.

HERE I RESPOND TO THE UNFAVORABLE REVIEW OF *THE EDGE OF Evolution* by Kenneth R. Miller in *Nature*.[10] Like Sean Carroll, whose review in *Science* I discussed earlier, he employs much bluster. But Miller goes well beyond simple bluster. I overlooked Carroll's rhetoric and dealt only with his substantial arguments. This time I'll do things differently. Here I'll respond to Miller's substantive points. In the next installment we'll take a closer look at his style of argumentation.

After mentioning that de novo resistance to chloroquine is found roughly once in every 10^{20} malaria parasites, and quoting several sentences from *The Edge of Evolution* where I note, "On average, for humans to achieve a mutation like this by chance, we would need to wait a hundred million times ten million years," Miller writes:

Behe, incredibly, thinks he has determined the odds of a mutation "of the same complexity" occurring in the human line. He hasn't. What he has actually done is to determine the odds of these two exact mutations occurring simultaneously at precisely the same position in exactly the same gene in a single individual....

Behe obtains his probabilities by considering each mutation as an independent event, ruling out any role for cumulative selection, and requiring evolution to achieve an exact, predetermined result.

Miller makes the same mistake here that I addressed earlier when replying to Jerry Coyne's response. The number of one in 10^{20} is not a probability calculation. Rather, it is statistical data. It is perhaps not too surprising that both Miller and Coyne make that mistake, because in general Darwinists are not used to constraining their speculations with quantitative data. The fundamental message of *The Edge of Evolution*, however, is that such data are now available. Instead of imagining what the power of random mutation and selection might do, we can look at examples of what it has done. And when we do look at the best, clearest examples, the results are, to say the least, quite modest. Time and again we see that random mutations are incoherent and much more likely to degrade a genome than to add to it—and these are the positively-selected, "beneficial" random mutations.

Miller asserts that I have ruled out cumulative selection and required *Plasmodium falciparum* to achieve a predetermined result. I'm flattered that he thinks I have such powers. However, the malaria parasite does not take orders from me or anyone else. I had no ability to rule out or require anything. The parasite was free in the wild to come up with any solution that might help it, by any mutational pathway that was available. I simply reported the results of what the parasite achieved. In 10^{20} chances, it would be expected to have undergone huge numbers of all types of mutations—substitutions, deletions, insertions, gene duplications, and more. And in that astronomical number of opportunities, at best a handful of mutations were useful to it.

Miller makes two specific points. First, he comments, "Not only are each of these conditions unrealistic, but they do not apply even in the case of his chosen example. First, he overlooks the existence of chloroquine-resistant strains of malaria lacking one of the mutations he claims to be essential (at position 220). This matters, because it shows that there are several mutational routes to effective drug resistance." As I wrote in response to Coyne, however, my argument does not depend on any particular amino acid position being required, and in the paper Miller was referring to (Chen et al., *Antimicrobial Agents and Chemotherapy* 47 (2003): 3500–3505, apparently accidentally omitted in the *Nature* review, according to Coyne) other mutations are found in the malarial strain in which position 220 remained unchanged. Miller says this matters because there are several routes to drug resistance. It matters much less than he implies. Certainly, there may be several routes, maybe permutations of pathways, too. But whether or not there are several routes, the bottom line is that resistance arises only once for every 10^{20} parasites.

Miller continues: "Second, and more importantly, Behe waves away evidence suggesting that chloroquine resistance may be the result of sequential, not simultaneous, mutations (*Science* 298, 74–75; 2002), boosted by the so-called ARMD (accelerated resistance to multiple drugs) phenotype, which is itself drug induced."

If you read that paper, however, you find that it presents no evidence whatsoever for cumulative mutations; rather, it merely speculates about them. What's more, the paper makes no mention of the ARMD phenotype, and Miller says nothing about its relevance. Here Miller is simply throwing references and words around, but saying nothing meaningful.

40. Response to Miller, Continued

"Response to Kenneth R. Miller, Continued,"
Uncommon Descent (website), July 12, 2007.

In the first part of my response to Kenneth Miller's review, in which I addressed his substantive points, I ended by showing that a reference he cited did not contain the evidence he claimed it did. In this final part, I more closely examine Miller's tendentious style of argumentation.

Speaking of throwing around irrelevant references, Miller writes:

Telling his readers that the production of so much as a single new protein-to-protein binding site is "beyond the edge of evolution," [Behe] proclaims darwinian evolution to be a hopeless failure. Apparently he has not followed recent studies exploring the evolution of hormone-receptor complexes by sequential mutations (*Science* 312, 97–101; 2006), the 'evolvability' of new functions in existing proteins—studies on serum paraxonase (PON1) traced the evolution of several new catalytic functions (*Nature Genet.* 37, 73–76; 2005)—or the modular evolution of cellular signalling circuitry (*Annu. Rev. Biochem.* 75, 655–680; 2006).

Now, dear reader, when Miller writes of "protein-to-protein" binding sites in one sentence, wouldn't you expect the papers he cites in the next sentence would be about protein-to-protein binding sites? Well, although the casual reader wouldn't be able to tell, they aren't. *None* of the papers Miller cites involves protein-protein binding sites. The *Science* paper concerns protein-steroid-hormone binding; the *Nature Genetics* paper deals with the enzyme activity of single proteins; and the *Annual Reviews* paper discusses rearrangement of pre-existing protein binding domains. What's more, none of the papers deals with evolution in nature. They all concern laboratory studies where very intelligent investigators purposely rearrange, manipulate, and engineer isolated genes (not whole cells or organisms) to achieve their own goals. Although such studies can be very valuable, they tell us little about how a putatively blind, random evolutionary process might proceed in unaided nature.

Miller's snide comment, that apparently I haven't followed these developments, seems pretty silly, since it's so easy to find out that I followed them closely. You'd think he should have noticed that I cited the *Annual Reviews* article in *The Edge of Evolution* in Appendix D, which deals in detail with Wendell Lim's interesting work on domain swapping. You'd think he easily might have checked and seen that I was quoted in the

New York Times commenting on Joseph Thornton's *Science* paper when it first came out a year ago. You'd also think he'd then have to tell readers of the review why I thought the papers weren't pertinent. You'd be thinking wrong.

Much worse, Miller is as subtly misleading when writing about the substantive points of *The Edge of Evolution* as he is when making supercilious offhand comments. Miller writes, "Telling his readers that the production of so much as a single new protein-to-protein binding site is 'beyond the edge of evolution,' [Behe] proclaims darwinian evolution to be a hopeless failure." But the book says plainly that it is two, not one, binding sites that marks the edge of evolution. That was not an obscure point. Chapter 7 is entitled "The Two-Binding-Sites Rule"; Figure 7.4 has a line at two binding sites, with a big arrow pointing to it labeled "Tentative molecular edge of evolution." What's more, the book goes out of its way to say that Darwinism is certainly not a "hopeless failure," that there are important biological features it clearly can explain. That's why one chapter is called "What Darwinism Can Do."

Regrettably, that's Miller's own special style. He doesn't just sneer and thump his chest, as some other Darwinists do. He uses less savory tactics, too. His tactics include ignoring distinctions the author draws (cellular protein-protein binding sites vs. other kinds of binding sites), mischaracterizing an argument by skewing or exaggerating its claims ("so much as a single…"), and employing inflammatory, absolutist language ("[Behe] proclaims darwinian evolution to be a hopeless failure"). He turns the principle of charitable reading on its head. Instead of giving a text its best interpretation, he gives it the worst he can.

Call it the principle of malignant reading. He's been doing it for years with the arguments of *Darwin's Black Box*, and he continues it in this review. For example, despite being repeatedly told by me and others that by an "irreducibly complex" system I mean one in which removal of a part destroys the function of the system itself, Miller says, no, to him the phrase will mean that none of the remaining parts can be used for anything else—a straw man which can easily be knocked down. Unconscionably, he passes off his own tendentious view to the public as mine. People who look to Miller for a fair engagement of the arguments of intelligent design are very poorly served.

41. RESPONSE TO COYNE IN THE NEW REPUBLIC

"Response to Critics, Part 1: Jerry Coyne," Uncommon
Descent (website), June 24, 2007.

IN THE JUNE ISSUE OF THE *NEW REPUBLIC* JERRY COYNE WRITES AN exceptionally lengthy, 7500-word review of my book *The Edge of Evolution*.[11] Coyne is an eminent evolutionary biologist whose specialty is fruit fly genetics, which he employs as a tool to study speciation, his real interest. (He teamed up with University of Rochester biologist Allan Orr several years ago to write a book entitled *Speciation*.) Furthermore, he is a frequent contributor to the popular press, with articles and book reviews in the pages of the *Times Literary Supplement* and the *Guardian*. So not only does he have a sharp scientific mind, he can write clearly for a general audience, too. I knew of course that Coyne strongly dislikes intelligent design, but was hopeful as I first started his review that he would engage the book's arguments and offer thoughtful counterpoints, which could help sharpen my own thinking. (It's been my experience that the more thoughtfully Darwinism is engaged, and the more experimental evidence is brought to bear, the more its manifest problems come to light.)

Alas, it was not to be. The Coyne review is one very long mishmash of ad hominem, argument from authority, misunderstanding, and question begging. The ad hominem (questioning my motives, gratuitously citing folks who disagree with me without saying why that's pertinent to my argument, and so on) I will not reply to. The argument from authority is the most incomprehensible part of his essay. Alluding to my participation in the Dover, Pennsylvania court case of 2005, early in the review

Coyne writes, "More damaging than the scientific criticisms of Behe's work was the review that he got in 2005 from Judge John E. Jones III."

Wow, more damaging than scientific criticisms?! Leave aside the fact that the parts of the opinion Coyne finds so congenial (which are standard Darwinian criticisms of intelligent design) were actually written by the plaintiffs' lawyers and simply copied by the judge into his opinion. (Whenever the opinion discusses the testimony of any expert witness— for either side, whether scientists, philosophers, or theologians—the judge copied the lawyers' writing. Although such copying is apparently tolerated in legal circles, it leaves wide open the question of whether the judge even comprehended the abstruse academic issues discussed in his courtroom.) Frankly, it's astounding that a prominent academic evolutionary biologist like Coyne hides behind the judicial skirts of the former head of the Pennsylvania Liquor Control Board. If Coyne himself can't explain how Darwinism can cope with the challenges *The Edge of Evolution* cites, how could a non-scientist judge?

At some points in his review, it's hard to know whether Professor Coyne simply has a poor memory, or is so upset with the book that he gets confused. He writes, "For a start, let us be clear about what Behe now accepts about evolutionary theory. He has no problem with a 4.5-billion-year-old Earth, nor with evolutionary change over time... and that all species share common ancestors." "Now accepts"? I made that plain in *Darwin's Black Box* over ten years ago. Throughout the controversy of the past decade over ID, almost every time my work had been cited in a newspaper or journal, it has been noted that I think common ancestry is true. Yet apparently that comes as a surprise to Coyne.

His reasoning goes downhill from there. To hear him tell it, I "come clean" about an ancient earth and common descent—which I've always thought the evidence supported—because "there is simply too much evidence for any scientist to deny these facts without losing all credibility." But according to Coyne there is also simply too much evidence to deny that random mutation and selection can explain evolution. So how does he reconcile that, in his telling, I worry about my reputation in the one case but not the other? It's impossible to tell—I'm afraid his thinking is quite a muddle. Perhaps if he reflected a bit on why he's so upset, even though he acknowledges I agree with what he claims is the great bulk of evolutionary thinking, he would realize that the question of randomness versus design is actually the crucial point, both scientifically and otherwise. The rest are details.

Finally, in any future dictionary of logical fallacies, wherever there is an entry on the topic "begging the question," it's a safe bet there'll be a picture of Jerry Coyne next to it. He writes:

> Creationists equate the chance that evolution could produce a complex organism to the infinitesimal chance that a hurricane could sweep through a junkyard and randomly assemble the junk into a Boeing 747. But this analogy is specious. Evolution is manifestly not a chance process because of the order produced by natural selection—order that can, over vast periods of time, result in complex organisms looking as if they were designed to fit their environment. Humans, the product of non-random natural selection, are the biological equivalent of a 747, and in some ways they are even more complex.

So, you see, we know random mutation plus selection produces order in life because biology contains order! What's not to understand? There's plenty more of that kind of thinking, such as when he writes we know Darwinian processes can produce coherent results because the fossil record shows coherent changes!

The same question-begging is used to "answer" my argument on protein binding sites, but with a special twist. Writes Coyne: "In fact, interactions between proteins, like any complex interaction, were certainly built up step by mutational step.... This process could have begun with weak protein-protein associations that were beneficial to the organism. These were then strengthened gradually." So, reasons Coyne, we know protein binding sites developed gradually by random mutation because we know proteins have binding sites. So there!

The twist comes when Coyne claims, "Behe furnishes no proof, no convincing argument, that interactions cannot evolve gradually." So, apparently to Darwinists, contrary observational evidence doesn't count.

Or perhaps Coyne somehow overlooked Chapter 7, where I noted that in a hundred billion billion chances, no such interactions developed in malaria. Or in HIV. Or in ten trillion opportunities in *E. coli*. I guess he missed where I carefully reviewed the literature on new protein binding sites. Where I showed the disconnected nature of random mutation in Chapters 3 and 4. Well, I suppose if Coyne read *The Edge of Evolution* with his eyes firmly shut, then he could have missed those discussions.

42. BACK AND FORTH WITH JERRY COYNE

"Back and Forth with Jerry Coyne," parts 1–3,
Uncommon Descent (website), July 8–10, 2007.

UNIVERSITY OF CHICAGO EVOLUTIONARY BIOLOGIST JERRY COYNE has responded at TalkReason.org[12] to my reply to his review of *The Edge of Evolution* in *The New Republic*. Here I will respond back—not to everything he wrote (nor to other posts and replies on that website), but only to what I think are the more important points of his original response. Because it quickly gets awkward to include all of the context, I will just quote the portions of his response that I specifically address here. Readers who want to see the full back-and-forth should read his posted review and response.

PART I

Coyne: "It is clear from Behe's response on his Amazon blog to the negative reviews by Sean Carroll and myself of *The Edge of Evolution* that he really wants to score debating points, not to have a scientific discussion."

Me: I assure Professor Coyne that I want nothing more than a frank scientific discussion. From the tone of his original review in *The New Republic*, I felt the same way about him regarding debating points. The tone of his response is much more civil, which I appreciate.

Coyne: "Both Richard Dawkins (in his review of *The Edge of Evolution* in the *New York Times*) and myself have noted Behe's remarkable reluctance to submit his claims to peer-reviewed scientific journals. If Behe's theory is so world-shaking, and so indubitably correct, why doesn't he submit it to some scientific journals? (The reason is obvious, of course: his theory is flat wrong.)"

Me: Long ago I posted some of my correspondence with science journals on the Discovery Institute website. I urge readers to examine that, and decide if they agree with me. It is my conclusion, based on much experience, that broaching the topic of intelligent design in an evenhanded manner is intolerable to mainstream science journals. (On the other hand, philosophy of science journals are much more tolerant.)

As one science journal editor politely wrote to me, "As you no doubt know, our journal has supported and demonstrated a strong evolutionary position from the very beginning, and believes that evolutionary explanations of all structures and phenomena of life are possible and inevitable. Hence a position such as yours, which opposes this view on other than scientific grounds, cannot be appropriate for our pages."

In fact, if one is a known ID proponent as I am, even publishing simple, extensively qualified criticisms of aspects of Darwin's theory is extremely difficult, and a journal that does so gets pummeled by protest emails, as *Protein Science* did when it published a paper by David Snoke and myself.

Professor Coyne thinks he knows better about my not publishing in science journals. He writes, "The reason is obvious, of course: his theory is flat wrong." Well, of course I disagree. Here's a snippet from his review of *Darwin's Black Box* in *Nature* in 1996 that I think supports my view. There he wrote:

> There is no doubt that the pathways described by Behe are dauntingly complex, and their evolution will be hard to unravel. Unlike anatomical structures, the evolution of which can be traced with fossils, biochemical evolution must be reconstructed from highly evolved living organisms, and we may forever be unable to envisage the first proto-pathways. It is not valid, however, to assume that, because one man cannot imagine such pathways, they could not have existed.[13]

So even though "we may forever be unable to envisage" how unintelligent processes could produce some "dauntingly complex" system, Coyne is not willing to concede that maybe, just maybe, unintelligent processes did not produce them. Rather, ID proponents apparently are assigned the burden of proving that no one could even imagine a pathway. Good luck. What are the chances that a manuscript on intelligent design submitted to a science journal would be published if a fellow with his views, quite common in the science community, were a reviewer?

One journal editor wrote to me, "I am painfully aware of the close-mindedness of the scientific community to non-orthodoxy, and I think it is counterproductive." If the science community is close-minded to "ordinary" non-orthodoxy, it is implacably close-minded to the non-orthodoxy of intelligent design. As a practical matter, given the sociological realities of the relevant scientific community, the choice for an ID scientist such as myself is either to publish outside science journals or to not publish at all.

Coyne: "He questions whether Jones really understood intelligent design at all, or simply adopted the plaintiff's claims in the Dover case. In fact, it's palpably clear from Jones's written opinion that he saw right through Behe and his transparent creationism. And you can bet that if the verdict had gone in favor of Behe's side, he wouldn't be impugning Jones as 'the former head of the Pennsylvania Liquor Control Board.'"

Me: Professor Coyne should compare the written document signed by Judge Jones to the plaintiffs' lawyers "finding of fact" brief, given to him about a month before he issued his opinion. Here's a short excerpt. The lawyers' brief reads in part:

> The assertion that design of biological systems can be inferred from the "purposeful arrangement of parts" is based on an analogy to human design. According to Professor Behe, because we are able to recognize design of artifacts and objects, that same reasoning can be employed to determine biological design. Professor Behe testified that the strength of an analogy depends on the degree of similarity entailed in the two propositions. If this is the test, intelligent design completely fails.

Jones's opinion reads:

Indeed, the assertion that design of biological systems can be inferred
from the "purposeful arrangement of parts" is based upon an analogy
to human design. Because we are able to recognize design of artifacts
and objects, according to Professor Behe, that same reasoning can be
employed to determine biological design. Professor Behe testified that
the strength of the analogy depends upon the degree of similarity entailed in the two propositions; however, if this is the test, ID completely
fails.

As I said, whenever the opinion discusses expert testimony—on
either side, by scientists, philosophers, or theologians—Jones simply reproduced that text (sometimes very lightly copyedited) from the lawyers'
document. I myself could happily copy from, say, a scholarly book on
string theory, or Kant, or Aquinas, but my copying those words would
be no evidence that I understood them. Similarly, as I said, there is no
evidence Jones understood the academic issues discussed in his courtroom. Those who have hailed Jones as some sort of philosopher-king
have been badly misled.

Coyne has a fair point, that I probably wouldn't mention Jones's former political job at the helm of the Pennsylvania Liquor Control Board
if he had ruled differently. It's a frailty of human nature that one usually doesn't examine things too closely if they go your way. On the other
hand, Coyne and other Darwinists would almost certainly make the
same points I'm making now if some judge had ruled against them, and
simply copied defendants' lawyers' documents in his ruling. They would
certainly chide him for his lack of apparent understanding of their own
arguments.

PART 2

Coyne: "There is no evolutionary expectation that complex protein-protein interactions will evolve in a parasite adapting to a new drug."

Me: Darwinism purports to account for the complexity of cellular machinery which, along with much else, involves very many protein-protein
interactions. Yet if "there is no evolutionary expectation that complex

protein-protein interactions will evolve" in any particular circumstance, then for those skeptical of Darwinism, what independent reason is there to suppose the protein-protein interactions we do find in the cell evolved by random mutations? I can't think of any. So I and a lot of other people want to decide what Darwinian processes can do based on evidence, not supposition. And the evidence is decidedly against it.

It's myopic to view these results, as Coyne does, simply as "a parasite adapting to a new drug." Rather, they are data that bear directly on the question, "What can Darwinian processes do given an astronomical number of opportunities?" In the past, we did not have enough data to address that question. Now we do, and observational evidence indicates the answer is, "Not much at all." And as I show in the book, the results with malaria mirror results with *E. coli* and HIV, which are very different organisms in very different circumstances. In a truly enormous number of opportunities, nothing much of fundamental biochemical interest happened.

Coyne: "Behe's probability calculations, on which his entire argument rests, are flatly wrong because they assume that adaptation cannot occur one mutation at a time."

Me: Here is where Professor Coyne and other Darwinist reviewers really miss the boat and overlook the considerable power of the malaria results. The number I cite, one parasite in every 10^{20} for de novo chloroquine resistance, is not a probability calculation. Rather, it is a statistic, a result, a data point. (Furthermore, it is not my number, but that of the eminent malariologist Nicholas White.) I do not assume that "adaptation cannot occur one mutation at a time"; I assume nothing at all. I am simply looking at the results. The malaria parasite was free to do whatever it could in nature; to evolve resistance, or outcompete its fellow parasites, by whatever evolutionary pathway was available in the wild. Neither I nor anyone else were manipulating the results. What we see when we look at chloroquine-resistant malaria is pristine data—it is the best that random mutation plus selection was able to accomplish in the wild in 10^{20} tries.

Let me elaborate that last point. The fact that de novo chloroquine resistance is observed to be an event of frequency 1 in 10^{20} strongly implies that mutational events of greater frequency must be of little help, because events of greater frequency would have been expected to occur many times in the same time interval yet they are not observed to have been selected. For example, if a single point mutation such as K76T alone in PfCRT in the wild were sufficient to confer chloroquine resistance, then resistance would occur de novo in virtually every person treated with chloroquine, as it does in almost every person treated with atovaquone. In 10^{20} parasites that single mutation would have been expected to have occurred about 10^{10} times or more. What's more, every other possible single point mutation, at every position of the parasite's genome, would also be expected to have occurred roughly the same number of times. And enormous numbers of other types of mutations—deletions, insertions, gene duplications, and more—in every gene of the parasite, would also have occurred. The result: a very few mutations helped the parasite a bit; the overwhelming number of mutations did not help at all. Thus helpful mutations are exceedingly rare; perhaps in other cases they might be nonexistent.

Coyne: "The probability calculations are also wrong because Behe's argument is based on specifying a priori... the identical pair of mutations that occur in chloroquine-resistant malaria. He neglects the possibility (indeed, the certainty) that many other mutations that cause interactions between proteins and other molecules can also be adaptive."

Me: Coyne is wrong again, for the same reason. I did not specify "a priori" exactly which mutations had to occur to be adaptive. Was I somehow out in the wild in Africa and South America telling the parasite which mutations to try? The parasite was free in nature to do whatever it could. The results are not a priori; they are entirely a posteriori, observational data. Moreover, I did not "neglect the possibility" (let alone "the certainty") of anything. Nobody told the parasite to restrict mutations just to its *pfcrt* gene. If other mutations could have been adaptive, *Plasmodium falciparum* had 10^{20} chances in the wild to find them, to come up with whatever it could muster. In the malaria data, we simply observe the exceedingly modest results.

Incidentally, this bears on Coyne's comment on Miller's review that "one of the two mutations that Behe claims are 'required' for CQR is not actually required (Chen et al. 2003, reference accidentally omitted from Miller's piece)." If you read that paper you see that, yes, A220S is not found in some resistant strains, as it is in most. (By the way, I was always quite careful in my book to state that A220S had been found in most strains, because I was quite aware of the several exceptions.) However, one also reads that the strains missing A220S have several other, novel mutations, which may be playing a comparable role in them that the mutation at position 220 plays in most other strains. My argument does not depend on exactly which changes are needed in the protein. Rather, the important point is that multiple changes appear to be required for resistance in the wild.

And for the life of me, I don't see why that proposition—that two mutations might be needed for some adaptations, and that that would be a big evolutionary impediment—is being treated by Coyne and other Darwinists with such horror. It certainly has been discussed in the evolutionary literature in the past. In my book I quote Allan Orr remarking, "Given realistically low mutation rates, double mutants will be so rare that adaptation is essentially constrained to surveying—and substituting—one-mutational step neighbors. Thus if a double-mutant sequence is favorable but all single amino acid mutants are deleterious, adaptation will generally not proceed." All I have done is to point to an example of the situation he envisioned, to quantify it, and to argue that it's likely to be a fairly general phenomenon. Why the shock?

Part 3

Coyne: "The reviews by Ken Miller in *Nature* and Sean Carroll in *Science* cite several examples of the gradual origin of adaptations via the step-by-step accumulation of point mutations in proteins."

Me: Hardly. Read my responses to Sean Carroll's and Ken Miller's reviews. Carroll begs the question of what caused changes in proteins, and Miller points to work in which investigators manipulate proteins in the lab (not to work showing what happens in nature). Frankly, I'm quite encouraged by their citations. Leaving aside their blustering rhetoric, if that's the best that staunch, knowledgeable opponents of ID can come up with, even when writing in leading journals for a scientific audience, then I think the protein-protein binding site argument is on solid ground.

Coyne: "Finally, I note that Behe's 'response' completely ignores two devastating criticisms of his 'scientific' theory. First, as both Dawkins and I point out, if random mutations can't build complexity, how can they possibly have been so effective in artificial selection of plants and animals?"

Me: Because, of course, the genomes of many plants and animals already contain much developmental plasticity. Turning some existing genes or regulatory elements on or off, or tuning them up or down, or changing them slightly by simple, single mutations, can certainly affect the shapes and other properties of organisms somewhat. Artificial selection for such variants can easily explain dog breeds and such, as I noted in Chapter 9. But of course that begs the question of where the complex systems controlling the organisms' development came from.

Now, was their mention of artificial selection of plants and animals really a "devastating" criticism of my argument? Frankly, I'm a bit perplexed by this line of reply from Coyne and Dawkins. In *The Edge of Evolution*, and in *Darwin's Black Box* before it, I strongly emphasized that modern biology shows us that life is built upon intricate molecular systems, and that to understand the limits of random mutation and thus Darwin's theory, we have to concentrate our attention on the molecular level. I readily said that answers to questions about animal shape and other macrobiological properties would have to await elucidation of the molecular underpinnings of those properties. In *The Edge of Evolution* I argued that some biological levels (down to vertebrate class) could not be explained by Darwinism, but I based my argument entirely on advances in our knowledge of the complexity of the molecular developmental systems underpinning them. Even though I suspect Darwinism is also ineffective at lower biological levels, I stopped at the level of vertebrate class because, as I wrote, "at this point our reliable molecular data runs out, so a reasonably firm answer will have to await further research."

It does not engage my argument, then, for Professors Coyne and Dawkins simply to point out that varieties of dogs come in different shapes, sizes, and colors. To actually engage my argument, a reviewer has to argue from molecular properties.

Coyne: "Also, as I pointed out in my review, Behe asserts quite plainly in his book that the goal of the Designer was 'intelligent life.' I challenge him to provide a scientific rationale for this conclusion, which he failed to do in his response."

Me: I don't go into it in detail in the book (Chapter 10 was already too long), but in essence the scientific rationale is the arguments of Michael Denton in *Nature's Destiny* and Ward and Brownlee in *Rare Earth* (which I cite). Both of those books argue that the requirements in nature for intelligent life are much greater than for "simple" life, such as bacteria. If one takes the "purposeful arrangement of parts" as empirical (scientific) evidence of design, as I do, then one can recognize that the intricate arrangement of parts in nature needed for intelligent life points to that as a design goal.

43. GERT KORTHOF AND PSEUDOGENES

"Gert Korthof and Pseudogenes," Uncommon
Descent (website), October 15, 2007.

THE DUTCH BIOLOGIST GERT KORTHOF MAINTAINS A WEBSITE DEvoted to in-depth reviews of many books on evolution. Aside from often-insightful remarks, a delightful feature of his site is that he can write with great strength of feeling and yet not engage in insults or ad hominem remarks. He has posted an extensive review[14] of *The Edge of Evolution*.

He makes two main points. First, that while I profess to believe in both common descent and intelligent design, he sees an internal contradiction—there cannot be, he thinks, common descent if there is intelligent design, and vice versa. The second point is that he thinks I contradict what I wrote in *Darwin's Black Box* concerning the status of pseudogenes as evidence of common descent. I'll take these points in turn.

Korthof writes:

The "designed group" contains at least Kingdoms, Phyla, Classes (and maybe Orders, Families and Genera). Behe's "randomness group" contains at least species (and maybe genera, families and orders).

Explicit design appears to reach into biology to a certain level, to the level of the vertebrate class, but not necessarily further. (p. 220)

Apparently the vertebrate class is explicitly designed. That means that at a certain moment in the history of the earth the first vertebrate "was designed." The fundamental problem here is that the reason for invoking design is that natural processes are not sufficient to produce vertebrates. However, as soon as one single nonnatural event is invoked during the history of life, the genetic continuity of life is broken. Common Descent is based on the vertical (sometimes horizontal) transmission of genetic information. Without that genetic continuity, Common Descent breaks down.

Korthof is incorrect here. As I read him, Korthof is saying that the first moment that some vertebrate class appeared was the moment that it was designed. Thus there is a discontinuity that can't be classified as common descent. But as I tried to make clear, especially in the last chapter of my book, all design might have been built into the initial conditions of the universe and unfolded over time. I try to get this point across with the figure of the überphysicist who selects a certain one out of very many universes which will develop in just the way he wishes; he "activates it"; and from there on everything follows according to unbroken natural law. It may be that in that universe there are some apparently amazing coincidences and astronomically low-probability events [which were front-loaded into the initial set-up], but all events follow without further prodding after the initial activation. Thus, contra Gert Korthof, there can indeed be purposeful intelligent design and common descent.

Concerning the second point—the status of pseudogenes—Korthof writes:

> Here [in *Darwin's Black Box*] Behe argues against Ken Miller. Miller claimed that Intelligent Design cannot explain pseudogenes of hemoglobin in humans, because it would mean that "the designer made serious errors, wasting millions of bases of DNA on a blueprint full of junk. Evolution can explain them as nothing more than failed experiments." Indeed, it is true that in general one cannot conclude from structures with unknown function to no function at all, but the pseudogene is not an unknown structure, but a copy of a known functional gene with mutations which make it non-functional. So Behe's critique fails. Behe's second argument against pseudogenes as evidence for Common Descent is that "even if pseudogenes have no function, evolution has "explained" nothing about how pseudogenes arose" (DBB, 226) and his third is that "these chance events do not mean that the initial biochemical systems were not designed" (DBB, 228).
>
> My point is not to refute Behe's arguments, but simply point out the amazing and extraordinary fact that Behe in *Darwin's Black Box* dismissed exactly the same evidence that he now accepts without any explanation. In *The Edge* he simply states "a broken hemoglobin gene" and forgets that he stated 10 years ago that "this argument is unconvincing for three reasons." What was exactly wrong with his arguments in DBB?

This topic is actually pretty simple, but can get confusing if you take your eye off the ball. I take pseudogenes as good evidence of common descent, both in *Darwin's Black Box* and *The Edge*. However, I do not take pseudogenes as evidence that random mutation and natural selection could have produced the original functioning gene which eventually gave rise to a pseudogene. Kenneth Miller was trying to argue that pseudogenes are positive evidence against all intelligent design. In my mind that's like arguing that a broken-down car is evidence against intelligent design. Miller's argument was incorrect. But Korthof takes my rebuttal as arguing against common descent, which it wasn't meant to do. It was an argument that pseudogenes do not speak to the ability of Darwinian processes to make functioning genes.

I think a lot of folks get confused because they think that all events have to be assigned en masse to either the category of chance or to that of design. I disagree. We live in a universe containing both real chance and real design. Chance events do happen (and can be useful historical markers of common descent), but they don't explain the background elegance and functional complexity of nature. That required design.

44. SCIENCE, E. COLI, AND THE EDGE OF EVOLUTION

Excerpted and adapted from "Science, E. coli, and the Edge of Evolution," Discovery Institute, October 9, 2007, and "Science, E. coli, and the Edge of Evolution, Part 2," Discovery Institute, October 11, 2007.

AS I WROTE IN THE EDGE OF EVOLUTION, DARWINISM IS A MULTI-faceted theory, and to properly evaluate the theory one has to be very careful not to confuse its different aspects. Unfortunately, stories in the news and on the internet regularly confuse the facets of Darwinism, ignore distinctions made in The Edge of Evolution, or misstate the arguments of intelligent design. The disregard for critical distinctions blurs the issues badly. Here I briefly respond to two separate stories.

1) A few months ago an interesting paper in Science[15] by the group of Isabel Gordo demonstrated that beneficial mutations in E. coli were more frequent than had been thought. In fact, the authors remark that "we found a rate on the order of 10^{-5} per genome per generation, which is 1000 times as high as previous estimates, and a mean selective advantage of 1%." They show that the previous underestimates of the beneficial mutation rates were likely due to clonal interference—accumulation of beneficial mutations in large bacterial populations which then interfere

with each other to dominate the population, making beneficial mutations seem less frequent. Does this new result mean that Darwinian
evolution can construct molecular machinery much easier than thought?

No. While the result is interesting, readers of *The Edge of Evolution*
will not be very surprised by it. As I showed for mutations that help
in the human fight against malaria, many beneficial mutations actually
are the result of breaking or degrading a gene. Since there are so many
ways to break or degrade a gene, those sorts of beneficial mutations can
happen relatively quickly. For example, there are hundreds of different
mutations that degrade an enzyme abbreviated G6PD, which actually
confers some resistance to malaria. Those certainly are beneficial in the
circumstances. The big challenge for evolution, however, is not to degrade genes (Darwinian random mutations can do that very well!) but to
make the coherent, constructive changes needed to build new systems.
The bottom line is that the beneficial mutations reported in the new *Science* paper most likely are degradatory mutations, and so don't address
the challenges outlined in *The Edge of Evolution*.

2) On the Darwinian website The Panda's Thumb, a woman named
Abbie Smith questioned[16] whether results from HIV research actually
square with the claims I made that little fundamental change has occurred in the virus, even though it attains enormous populations sizes
and has a much increased mutation rate.

Although she calls herself a "pre-grad student," the tone of the post
is decidedly junior high, the tone of someone who is trying hard to compete with all the other Mean Girls on that unpleasant website. I'll pass
over all that and try to stick to the substance.

Her post mainly concerns a small protein coded for by HIV-1 called
Vpu. She first points out that the amino acid identity between the homologous chimp SIV protein with HIV Vpu is 39%, much less than
that of other homologous viral proteins, and she seems to regard that
fact by itself as remarkable. Yet the alpha and beta chains of human hemoglobin are only about 44% identical, and have virtually superimpos-
able structures and very similar functions. The fact that the chimp and

human versions of VPU have 39% identity indicates they are structurally virtually identical. That doesn't seem like a fundamental change to me.

She goes on to write that Vpu acts to degrade CD4 molecules by binding to them and recruiting the pathway that degrades CD4. Unfortunately, she seems not to have read the beginning of Chapter 8 of *The Edge of Evolution* ("Objections to the Edge"), where I make some careful distinctions:

> This chapter makes some important distinctions and addresses potential objections. It considers counter-arguments to my attempt to define the edge of evolution—not philosophical ones, about the "other side" of that boundary, but technical and logical ones about the line itself....
>
> Another, more important point to note is that I'm considering just cellular proteins binding to other cellular proteins, not to foreign proteins. *Foreign proteins injected into a cell by an invading virus or bacterium make up a different category.* The foreign proteins of pathogens almost always are intended to cripple a cell in any way possible. Since there are so many more ways to break a machine than to improve it, this is the kind of task at which Darwinism excels. Like throwing a wad of chewing gum into a finely tuned machine, it's relatively easy to clog a system—much easier than making the system in the first place. Destructive protein-protein binding is much easier to achieve by chance. [my emphasis]

So the example she chose is from exactly the category that I excluded in the above paragraph. My exclusion isn't arbitrary. As I wrote, there are many more ways to cripple a machine than to build one, so destructive Darwinian processes can appear to accomplish more. Yet *The Edge of Evolution* is concerned with how molecular machinery is constructed, not destroyed. One can't ignore such critical distinctions and make progress. But, in my experience, many Darwinists overlook important differences.

She goes on to list several other properties of Vpu, but, while interesting, none at all are what one should call "fundamental" changes. For example, she notes, the HIV Vpu has several sites that are negatively charged by virtue of being phosphorylated. She continues: "Yet some

SIVcpz Vpus have only one [phosphorylation] site, and instead utilize a simple string of negatively charged amino acids in place of the second site. Different ways of performing similar tricks with totally different amino acids. I think that's biochemically significant as well."

Well, I disagree. I don't think that's biochemically fundamental at all. In each case one has a blob of negative charge. Since the mutation rate of HIV is so extremely high, kinase sites are likely replaced every day in some virus with multiple glutamates or aspartates and vice versa. She also points out that some virus Vpu subtypes have altered the location of modification sites or acquired signals to localize protein in particular subcellular compartments. But again, because of the virus's extremely high mutation rate, such sites would be expected to come and go frequently. As I emphasized in *The Edge of Evolution*, HIV's enormous numbers and very high mutation rate cause immense variation. The question, however, is to what extent the immense variation has produced novel virus systems or machinery? And, as I indicated, the answer is very little. Butler et al.[17] remark under the subheading "Biological Consequences of HIV Diversity": "With such breadth of genetic diversity among HIVs, one might expect significant biological differences between the clades. Although interesting variations can be seen, much of the data concerning biological implications of HIV diversity is contradictory."

Plenty of differences do exist, and some are "interesting," but not all that great.

Darwinists overlook the considerable power of the example of the relatively minor changes in HIV. There have been a truly astronomical number of copies produced in just the past fifty years or so. And because of its much increased mutation rate, it has undergone in the past half-century as many of some kinds of mutations as all the cells have undergone in the history of the world. If Darwinism had the power that its boosters claim, we should expect to see truly fundamental changes. Yet despite the enormous number of opportunities, only minor changes have appeared. That is very strong evidence of the strict limits on what Darwinian processes can accomplish.

45. RESPONSE TO IAN MUSGRAVE ON HIV

In *The Edge of Evolution* I discussed the very modest changes that have occurred in the human immunodeficiency virus HIV over the past decades—even though such viruses occur in huge numbers and mutate much faster than plant or animal DNA. This brought some complaints from a few HIV researchers that the virus can too evolve! I responded in the essay above and the one below. More recent work on the Ebola virus strongly supports the contention that, while these viruses are of course very dangerous, there is a very definite, very near edge to their evolution.

"Response to Ian Musgrave's 'Open Letter to Dr. Michael Behe,'" parts 1–5, Uncommon Descent (website), November 12–16, 2007.

IN THIS POST I REPLY TO PROFESSOR IAN MUSGRAVE'S "OPEN LETTER to Dr. Michael Behe" on the Panda's Thumb blog.[18] Musgrave wrote: Dear Dr. Behe,

I have recently read your response to Abbie Smith's article on the HIV-1 protein VPU. Ms Smith showed how Vpu's recently evolved viroporin activity directly contradicts your statement that HIV has evolved no new biding [sic—*binding*] sites since it entered humans (*The Edge of Evolution*, page 143 and figure 7.4, page 144). I was greatly disappointed in your response. I must admit to having a special involvement in this case. Firstly, I drew the illustrations for Ms Smith's article, and its follow up. But secondly, as a member of my professional association's education committee, I am directly concerned with the support and nurturing of the new generations of enquiring minds, those that we will pass the torch of enquiry on to when we retire. It is in this regard that your response [sic—*is*] very disturbing. It is almost the exact opposite of what a concerned scientist and science communicator should have done.

It was bad enough that you chose to ignore her for over two months and then did not do her the courtesy of replying on her blog. It was bad enough that you chose to start by belittling her and playing the "I'm a Professor and she is a mere student" card (conveniently ignoring the fact that she actually works on HIV). This is particularly egregious in science, where we pay attention to the evidence and logic of an argument, rather than the letters after an author's name. Doubly so if we wish to guide young scientists into a demanding profession.

Dear Dr. Musgrave,

I find your letter disingenuous. The tone of Abbie Smith's post was insulting, jut-jawed, and puerile:

+ "I'm ERV. This is my dog, Arnold Schwarzenegger. And this is my friend, Vpu. I presume you and Vpu haven't met, as you recently repeated in an interview with *World* magazine the same sentiment you gurgled ad nauseam in '*Edge of Evolution*.'"

+ "Ah, Michael Behe, you might try to talk your way around Vpu now… Sorry, you'll find no escape with that limp-wristed, ad hoc parry."

+ "Ah, Michael Behe, you might try to talk your way around Vpu NOW by saying, "Vpu might be *new* new in HIV-1, but it's not *NEW* *new* new.""

+ "This is just one of a billion plus examples of lazy Creationists taking advantage of the ignorance of their followers."

As far as I'm concerned, if a complete stranger sends me a message with a sneering tone like that, she can go soak her head. I had no intention of replying to Smith's post at all; I did so only after I received requests from other folks who wanted me to reply.

At no point in my reply was I "belittling her and playing the 'I'm a Professor and she is a mere student card'" as you allege. The only reference to Smith's tone I made was the following sentence: "Although she calls herself a 'pre-grad student' the tone of the post is decidedly junior high school, of someone who is trying hard to compete with all the other Mean Girls on that unpleasant website." I think a re-reading of her post shows that my evaluation was quite judicious. After that passage I addressed only the science, not her sneering tone.

Frankly, Professor Musgrave, I find your concern "as a member of my professional association's education committee" for "the support and nurturing of the new generations of enquiring minds" to be unconvincing. One of the very basic prerequisites for education is to be able to engage in civil discourse, especially with people whose views are different from your own. It is clear to me that Smith has not yet mastered that skill. To the extent that you consider yourself one of her professional mentors, you have failed in your responsibilities.

PART 2

Musgrave: "But by far the worst, you ignored her core argument. That in the space of a decade HIV-1 Vpu developed a series of binding sites that made it a viroporin, a multisubunit structure with a function previously absent from HIV-1.

It is not clear to me why you call that Smith's "core argument." In her post, her writing meanders quite a bit; it's hard for me to glean what she thinks is most important. After sneering a bit at me, Smith began her post by asserting that Vpu is a "new" gene (even though it is found in SIVcpz across several primate families, as her own citations show).[19] She then spent several paragraphs making the point that Vpus from chimps and humans share only 37% sequence identity. That was followed by the declaration, "Turns out a LOT of evolution has been going on in HIV-1 since it was transferred to humans 50–60 years ago." So, rather naturally I think, I concluded that she thought the simple fact of 37% sequence identity was compelling evidence for the power of Darwinian evolution. In reply, I pointed out that proteins with similar structures and functions have such sequence identity (trypsin and chymotrypsin are ~40% identical), so mere degree of sequence identity means little.

After that point she went on to speak of the function of Vpu to assist in degrading CD4 (again, not the "core argument" you assert). She claimed that this involved the evolution of two protein-protein binding sites. In my reply to her, I pointed out that in my book I placed viral protein-cellular proteins binding in a different category (more about that later), so I didn't reply further to that. However, I now think that reply wasn't the best one. Rather, I should have simply pointed out that Smith's references themselves show that "Vpu-mediated CD4 down-regulation and degradation is conserved among highly divergent SIVcpz strains."[20] Now, why are we even talking about the evolution of a process which has been conserved from chimps and other primates to humans? Where was it shown that the Vpu-CD4 degradation process evolved two protein binding sites? We have primate viral ancestors that have that function and human viruses that have that function. What was the point here again?

She then has two paragraphs on what you call the "core" argument. I'll get back to those, because right after those paragraphs she reverts to discussing CD4 degradation. She writes: "HIV-1 Vpu requires two casein kinase II sites.... Yet some SIVcpz Vpus have only one CKII site, and instead utilize a simple string of negatively charged amino acids in place of the second site. Different ways of performing similar tricks with totally different amino acids. I think that's biochemically significant as well."

I disagree with her assessment; I think this is a trivial biochemical change given HIV's mutation rate. Casein kinase II sites have the consensus sequence (S/T-X-X-D/E).[21] (Notice one of the consensus amino acids is already acidic.) Now, since the mutation rate of HIV is about 10^{-4}, and since there are 10^9–10^{10} viruses in an infected individual, that means any particular double mutation would happen in every individual with AIDS every day. Thus in every hundred persons with HIV, every day there would occur at least one virus with that consensus sequence mutated to, say D-D-X-E. (That would allow for three nucleotide changes, since two are needed to go from S/T to D/E.) If all that was necessary at the site was a blob of negative charge to replace the phosphate negative charge, you'd get that every day in a group of a hundred infected individuals. Smith thinks "that's biochemically significant." I think she hasn't done her math. I view that as sequence drift—the feature stays the same (negatively charged residues), but the sequence drifts within limits.

The same goes for the handful of other changes that impress her: "For instance, Subtype C Vpu's are characteristically longer than the others, have key phosphorylation sites shifted, have an extra CKII site, and its tertiary structure is totally different (Subtype B Vpu's have an M_r of 43,000 in an SDS-PAGE gel, while Subtype C is 34,000)."

Since casein kinase II sites consist of just two specific amino acid residues (separated by any two residues), phosphorylation sites can be shifted, created, or destroyed daily as easily as the calculations from the last paragraph show. By themselves they are no more significant than the simple drift in the sequence of proteins. The fact that the Vpu-fusion proteins molecular weight shifts from 43,000 to 34,000 in SDS-PAGE is difficult to interpret. It may or may not mean anything significant about the protein's native structure.

PART 3

IN MY reply to Smith I quoted from a review[22] which asked the question why, with so much genetic variation, do we just see "interesting varia-tions" in biological properties. Smith, replying to me on her blog in high dudgeon, quotes the next paragraph of the review which details some of those interesting variations:

> The long terminal repeat region (LTR) of the HIV genome regulates transcription and viral replication, acting as a promoter responsive to the viral Tat protein. Although all subtypes share the same LTR func-tion, they differ with respect to LTR sequence structure, basal activity and response to cytokines and transcription factors [95]. The majority of HIV-1 group M subtypes contain two nuclear factor binding sites (NF-kB). A minority of subtype C contain an extra NF-kB that may promote replication in the presence of TNF-and chronic immune ac-tivation [135].

Smith remarks, "You know, that kinda looks like my essay, except looking at LTRs instead of Vpu." She means that she views these as sig-nificant as the changes in Vpu she pointed out in her original essay. I agree with her, but think that biochemically none of them are all that significant. (One possible point of miscommunication is that I'm concentrating on the biochemical machinery of the virus, while Smith is more concerned with epidemiological factors. As she writes: "I'm trying to figure out the physiological and epidemiological impact of the changes. I'm hoping that I can figure out a genetic determinate of transmission, narrowing down the epitopes a potential vaccine would need to contain. Everybody is trying to figure this kind of thing out now in the HIV research world." That focus is certainly critically important for public health, but it's not the same question as how do complex biochemical systems arise.)

One must remember that because of HIV's mutation rate of 10^{-4}, and because of its large population size (10^9–10^{10} per individual), multiple mutations in a single virus will occur much more frequently than for cells. For example, in general, all possible double point mutations will occur in a single virus in a single individual every day. All possible triple point mutations will occur in a population of a hundred individuals every month or so. All possible quadruple point mutations will occur in a population of a million individuals every month. All this means that multiple amino acid mutations or nucleotide changes in signaling sequences are easily achieved in HIV. Because of their much lower mutation rate, however, cells cannot match those numbers.

In the paragraph she quotes above from the review, since the transcription factor binding sites are rather short, since HIV has such a large mutation rate, and since duplicating a binding site is certainly not an unusual feat, all of these are what I would call "genome drift"—minor variations on a theme, not something to crow about after 10^{20} replications and a greatly enhanced mutation rate. In my view Smith and other Darwinists are much too easily impressed. As the review mentions after enumerating these changes: "Although there are clear differences in LTR sequences and basal replication capacity among subtypes, the influence of these molecular level changes on specific subtype epidemics and the global spread of the virus remain uncertain."[23]

In other words, even though it may affect epidemiological properties unpredictably, the frantic variation of HIV seems to occur within a very restricted envelope of biochemical possibilities.

In her original post Smith writes further, "It turns out that one of the biochemical differences is that Subtype B Vps have a Golgi retention signal in the second alpha-helix of the cytoplasmic domain."

Ian Musgrave mentions this too: "HIV-1 Vpu has a new binding site, YRKL in the cytoplasmic alpha helical section, not present in SIVcpz Vpu, which efficiently targets Vpu to the Golgi complex, making the degradation process more efficient. Can you please explain why the appearance of a new targeting motif in HIV-1 Vpu is NOT an example of a new binding site."

But, although not identical, SIV variants have similar sequences at the same location:[24] YKRN; WKRN; WRQI; etc. Given the mutation rate of HIV of 10^{-4}, and the fact that an infected individual carries $10^8 - 10^9$ viral copies, the conversion of an SIV variant into the Golgi retention signal YRKL could take place with relative ease: perhaps one such de novo mutant per year in a population of ten thousand organisms. If it gave a selective advantage, there would be no problem for this arising by Darwinian means.

At this point I should perhaps remind Dr. Musgrave that the title of my book is *The Edge of Evolution*. In it I explain that Darwinian processes can do some things, but not others, and I try to find a rough dividing line. As I emphasize, that means one has to make distinctions between categories. A virus like HIV, with its small genome size and much greater mutation rate, has to be considered differently from cells with their larger genome sizes and lower mutation rates. As a rule of thumb, HIV can acquire two specific point mutations as easily as a cell can acquire one. And with its great population size, it would be child's play for HIV to alter many signaling sequences. To answer Dr. Musgrave's question, I wouldn't list this as a new binding site, not because it doesn't bind a cellular protein, but because, as I explicitly state in the book, I place viral protein-cellular protein interactions in a separate category. My book concerns cellular protein-protein binding sites (or new viral-viral sites).

PART 4

AND NOW let's talk about Dr. Musgrave's "core argument," that subsequent to the virus leaping to humans from chimps Vpu developed the ability to act as a viroporin, allowing the leakage of cations which helps release the virus from the cell membrane. Yes, I'm perfectly willing to concede that this does appear to be the development of a new viral protein-viral protein binding site, one which I overlooked when writing about HIV. So the square point in Figure 7.4 representing HIV should be placed on the Y axis at a value of one, instead of zero, and Table 7.1 should list one protein-binding site developed by HIV instead of zero.

One should, however, also make some distinctions with this example. First, although there apparently are five or so copies of Vpu in the viroporin complex, that does not mean that five binding sites developed. Only one new binding site need develop for one area of a protein which binds to a different area of the same protein, to form a homogeneous complex with, say, C5 symmetry. That is all that is required for a circularly symmetric structure to form. Second, the viroporin is not some new molecular machine. There is no evidence that it exerts its effect in, say, an ATP- or energy-dependent manner. Rather, similar to other viroporins, the protein simply forms a passive leaky pore or weak channel.[25] This situation is probably best viewed as a foreign protein degrading the integrity of a membrane, rather than performing some positive function.

And third, I explicitly pointed out in Chapter 8 of *The Edge of Evolution* that HIV had undergone enough mutating in past decades to form all possible viral-viral binding sites, but commented that apparently none of them had been helpful. This I discussed as the "principle of restricted choice":

> A third reason for doubt is the overlooked problem of restricted choice. That is, not only do new protein interactions have to develop, there has to be some protein available that would actually do some good. Malaria makes about 5,300 kinds of proteins. Of those only a very few

help in its fight against antibiotics, and just two are effective against chloroquine. If those two proteins weren't available or weren't helpful, then, much to the joy of humanity, the malarial parasite might have no effective evolutionary response to chloroquine. Similarly, in its frantic mutating, HIV has almost certainly altered its proteins at one point or another in the past few decades enough to cover all of shape space. *So new surfaces on HIV proteins would have been made that could bind to any other viral protein in every orientation.* Yet of all the many molecules its mutated proteins must have bound, none seem to have helped it; no new protein-protein interactions have been reported. Apparently the choice of proteins to bind is restricted only to unhelpful ones.[26] [my emphasis]

So Dr. Musgrave's "core argument" turns out to be a decidedly double-edged sword. Yes, I agree that one overlooked protein-protein interaction developed, leading to a leaky cell membrane. However, in the past fifty years many, many more potential viral protein-viral protein interactions must have also developed but not been selected because they did the virus little good. That, dear readers, is "restricted choice," a very large contributor to the edge of evolution.

PART 5

FINALLY, DR. Musgrave objects to my placing viral protein-cellular protein interactions in a separate category from cellular protein-cellular protein interactions. As noted in my response essay immediately above, in Chapter 8 of *The Edge of Evolution* I had written:

Another, more important point to note is that I'm considering just cellular proteins binding to other cellular proteins, not to foreign proteins. Foreign proteins injected into a cell by an invading virus or bacterium make up a different category. The foreign proteins of pathogens almost always are intended to cripple a cell in any way possible. Since there are so many more ways to break a machine than to improve it, this is the kind of task at which Darwinism excels. Like throwing a wad of chewing gum into a finely tuned machine, it's relatively easy to clog a system—much easier than making the system in the first place. Destructive protein-protein binding is much easier to achieve by chance.[27]

Musgrave protests, "This is simply not true, either generally or in the particular case of Vpu. Importantly, your statement shows that you do not understand what Vpu does. Vpu down regulates the surface protein CD4 (the viroporin activity is something separate related to viral release). It does not 'gum-up' CD4, it specifically binds to it...."

Dr. Musgrave misunderstands my point. I did not say that Vpu acted as a nonspecific wad of chewing gum. Rather, my point is a general one, that, initially, when during the course of evolution a viral or bacterial protein is injected for the first time into a cell, it is encountering a new environment, one it hasn't adapted to before, and which hasn't adapted to it. In that case there are many "targets of opportunity" for the foreign protein. If by serendipity it sticks to some cellular protein to a certain degree, and that association interferes to a degree with the function of the cellular protein, then that will likely benefit the virus, and the association can be strengthened by mutations to the viral protein over time. After a while the association can become very specific, as Vpu's has.

On the other hand, consider cellular proteins. Cellular proteins must continually exist in a confined space, dense with many other cellular proteins, and so they are normally selected to not bind to most other cellular proteins. In other words, for eons the surfaces of cellular proteins have been honed so as to not interact with almost any other protein in a very concentrated cellular milieu. (Work has shown protein surfaces also adapt to particular sub-cellular compartments.[28]) As the textbook *Biochemistry* by Voet and Voet puts it, "Proteins, because of their multiple polar and nonpolar groups, stick to almost anything; anything, that is, but other proteins. This is because the forces of evolution have arranged the surface groups of proteins so as to prevent their association under physiological conditions. If this were not the case, their resulting nonspecific aggregation would render proteins functionally useless."[29]

How does this affect the situation when, in some future environment, it might be beneficial for two cellular proteins that had worked separately, to associate, to develop a new protein-protein binding site? From considerations of shape space,[30] it seems in general it would be very difficult. Work on antibodies and other proteins shows that a protein which is "tolerant" of a second protein (that is, does not bind it) will be tolerant of mutants of that protein, until a relatively large number of amino acids have changed.[31] It seems likely that this will be the case for cellular proteins too.[32] If so, that means that two proteins which had previously tolerated each other over eons in a cell would have to undergo multiple mutations to acquire the ability to specifically bind to each other.

That's why I put cellular protein-protein interactions in a different category from foreign protein-cellular protein interactions.

46. Waiting Longer for Two Mutations

"Waiting Longer for Two Mutations," Discovery Institute, March 9, 2009.

AN INTERESTING PAPER APPEARED RECENTLY IN AN ISSUE OF THE journal *Genetics*, "Waiting for Two Mutations: With Applications to Regulatory Sequence Evolution and the Limits of Darwinian Evolution."[33] As the title implies, it concerns the time one would have to wait for Darwinian processes to produce some helpful biological feature (here, regulatory sequences in DNA) if two mutations are required instead of just one. It is a theoretical paper, which uses models, math, and computer simulations to reach conclusions, rather than empirical data from field or lab experiments, as *The Edge of Evolution* does. The authors, Rick Durrett and Deena Schmidt, declare in the abstract of their manuscript that they aim "to expose flaws in some of Michael Behe's arguments concerning mathematical limits to Darwinian evolution." Unsurprisingly (bless their hearts), they pretty much do the exact opposite. Since the journal *Genetics* publishes letters to the editors (most journals don't), I sent a reply to the journal, which they published.[34]

In their paper (as I write in my reply):

They develop a population genetics model to estimate the waiting time for the occurrence of two mutations, one of which is premised to damage an existing transcription-factor-binding site, and the other of which creates a second, new binding site within the nearby region from a sequence that is already a near match with a binding site sequence (for example, 9 of 10 nucleotides already match).

The most novel point of their model is that, under some conditions, the number of organisms needed to get two mutations is proportional not to the inverse of the square of the point mutation rate (as it would be if both mutations had to appear simultaneously in the same organism), but to the inverse of the point mutation rate times the square root of the point mutation rate (because the first mutation would spread in the population before the second appeared, increasing the odds of getting a double mutation). To see what that means, consider that the point mutation rate is roughly one in a hundred million (1 in 10^8). So if two specific mutations had to occur at once, that would be an event of likelihood about 1 in 10^{16}. On the other hand, under some conditions they modeled, the likelihood would be about 1 in 10^{12}, ten thousand times more likely than the first situation. Durrett and Schmidt compare the number they got in their model to my literature citation that the probability of the development of chloroquine resistance in the malarial parasite is an event of order 1 in 10^{20}, and they remark that it "is 5 million times larger than the calculation we have just given." The implied conclusion is that I have greatly overstated the difficulty of getting two necessary mutations. Below I show that they are incorrect.

Serious Problems

Interesting as it is, there are some pretty serious problems in the way they applied their model to my arguments, some of which they owned up to in their reply, and some of which they didn't. When the problems are fixed, however, the resulting number is remarkably close to the empirical value of 1 in 10^{20}.[35] I will go through the difficulties in turn.

The first problem was a simple oversight. They were modeling the mutation of a ten-nucleotide-long binding site for a regulatory protein in

DNA, so they used a value for the mutation rate that was ten times larger than the point mutation rate. However, in the chloroquine-resistance protein discussed in *The Edge of Evolution*, since particular amino acids have to be changed, the correct rate to use is the point mutation rate. That leads to an underestimate of a factor of about thirty in applying their model to the protein. As they wrote in their reply, "Behe is right on this point." I appreciate their agreement here.

The second problem has to do with their choice of model. In their original paper they actually developed models for two situations—for when the first mutation is neutral, and for when it is deleterious. When they applied it to the chloroquine-resistance protein, they unfortunately decided to use the neutral model. However, it is very likely that the first protein mutation is deleterious. As I wrote discussing a hypothetical case in Chapter 6 of *The Edge of Evolution*:

> Suppose, however, that the first mutation wasn't a net plus; it was harmful. Only when both mutations occurred together was it beneficial. Then on average a person born with the mutation would leave fewer offspring than otherwise. The mutation would not increase in the population, and evolution would have to skip a step for it to take hold, because nature would need both necessary mutations at once.... The Darwinian magic works well only when intermediate steps are each better ("more fit") than preceding steps, so that the mutant gene increases in number in the population as natural selection favors the offspring of people who have it. Yet its usefulness quickly declines when intermediate steps are worse than earlier steps, and is pretty much worthless if several required intervening steps aren't improvements.

If the first mutation is indeed deleterious, then Durrett and Schmidt applied the wrong model to the chloroquine-resistance protein. In fact, if the parasite with the first mutation is only 10% as fit as the unmutated parasite, then the population-spreading effect they calculate for neutral mutations is pretty much eliminated, as their own model for deleterious mutations shows. What do the authors say in their response about this possibility? "We leave it to biologists to debate whether the first PfCRT mutation is that strongly deleterious." In other words, they don't know;

it is outside their interest as mathematicians. (Again, I appreciate their candor in saying so.) Assuming that the first mutation is seriously deleterious, then their calculation is off by a factor of 10^4. In conjunction with the first mistake of thirty-fold, their calculation so far is off by five-and-a-half orders of magnitude.

MAKING A STRING OF ONES

THE THIRD problem also concerns the biology of the system. I'm at a bit of a loss here, because the problem is not hard to see, and yet in their reply they stoutly deny the mistake. In fact, they confidently assert it is I who am mistaken. I had written in my letter that "their model is incomplete on its own terms because it does not take into account the probability of one of the nine matching nucleotides in the region that is envisioned to become the new transcription-factor-binding site mutating to an incorrect nucleotide before the tenth mismatched codon mutates to the correct one."

They retort, "This conclusion is simply wrong since it assumes that there is only one individual in the population with the first mutation." That's incorrect. Let me explain the problem in more detail.

Consider a string of ten digits, either 0 or 1. We start with a string that has nine 1s, and just one 0. We want to convert the single 0 to a 1 without switching any of the 1s to a 0. Suppose that the switch rate for each digit is one per hundred copies of the string. That is, we copy the string repeatedly, and, if we focus on a particular digit, about every hundredth copy or so that digit has changed. Okay, now cover all of the numbers of the string except the 0, and let a random, automated procedure copy the string, with a digit-mutation rate of one in a hundred. After, say, seventy-nine copies, we see that the visible 0 has just changed to a 1. Now we uncover the rest of the digits. What is the likelihood that one of them has changed in the meantime? Since all the digits have the same mutation rate, then there is a nine in ten chance that one of the other digits has already changed from a 1 to a 0, and our mutated string still does not match the target of all 1s. In fact, only about one time out of ten will we uncover the string and find that no other digits have changed except the visible digit. Thus the effective mutation rate for transforming the string with nine matches out of ten to a string with ten matches out of ten will be only one tenth of the basic digit-mutation rate. If the string is a hundred long, the effective mutation rate will be one-hundredth the basic rate, and so on. (This is very similar to the problem of mutating a duplicate gene to a new selectable function before it suffers a degradative mutation, which has been investigated by Lynch and coworkers.[36])

So, despite their self-assured tone, in fact on this point Durrett and Schmidt are "simply wrong." And, as I write in my letter, since the gene for the chloroquine resistance protein has on the order of a thousand nucleotides, rather than just the ten of Durrett and Schmidt's postulated regulatory sequence, the effective rate for the second mutation is several orders of magnitude less than they thought. Thus with the, say, two orders of magnitude mistake here, the factor-of-thirty error for the initial mutation rate, and the four orders of magnitude for mistakenly using a neutral model instead of a deleterious model, Durrett and Schmidt's calculation is a cumulative seven-and-a-half orders of magnitude off. Since they had pointed out that their calculation was about five million-fold (about six-and-a-half orders of magnitude) lower than the empirical result I cited, when their errors are corrected the calculation agrees pretty well with the empirical data.

AN IRRELEVANT EXAMPLE

Now I'd like to turn to a couple of other points in Durrett and Schmidt's reply which aren't mistakes with their model, but which do reflect conceptual errors. As I quote above, they state in their reply, "This conclusion is simply wrong since it assumes that there is only one individual in the population with the first mutation." I have shown above that, despite their assertion, my conclusion is right. But where do they get the idea that "it assumes that there is only one individual in the population with the first mutation"? I wrote no such thing in my letter about "one individual." Furthermore, I "assumed" nothing. I merely cited empirical results from the literature. The figure of 1 in 10^{20} is a citation from the literature on chloroquine resistance of malaria. Unlike their model, it is not a calculation on my part.

Right after this, in their reply Durrett and Schmidt say that the "mistake" I made is a common one, and they go on to illustrate "my" mistake with an example about a lottery winner. Yet their own example shows they are seriously confused about what is going on. They write:

> When Evelyn Adams won the New Jersey lottery on October 23, 1985, and again on February 13, 1986, newspapers quoted odds of 17.1 trillion to 1. That assumes that the winning person and the two lottery dates are specified in advance, but at any point in time there is a population of individuals who have won the lottery and have a chance to win again, and there are many possible pairs of dates on which this event can happen.... The probability that it happens in one lottery 1 year is ~1 in 200.

No kidding. If one has millions of players, and any of the millions could win twice on any two dates, then the odds are certainly much better that somebody will win on some two dates than that Evelyn Adams win on October 23, 1985 and February 13, 1986. But that has absolutely nothing to do with the question of changing a correct nucleotide to an incorrect one before changing an incorrect one to a correct one, which is the context in which this odd digression appears. What's more, it is not the type of situation that Durrett and Schmidt themselves modeled. They asked the question, given a particular ten-base-pair regulatory sequence, and a particular sequence that is matched in nine of ten sites to the regulatory sequence, how long will it take to mutate the particular regulatory sequence, destroying it, and then mutate the particular near-match sequence to a perfect-match sequence? What's even more, it is not the situation that pertains in chloroquine resistance in malaria. There several particular amino acid residues in a particular protein (PfCRT) have to mutate to yield effective resistance. It seems to me that the lottery example must be a favorite of Durrett and Schmidt's, and that they were determined to use it whether it fit the situation or not.

MULTIPLYING RESOURCES

THE FINAL conceptual error that Durrett and Schmidt commit is the gratuitous multiplication of probabilistic resources. In their original paper they calculated that the appearance of a particular double mutation in humans would have an expected time of appearance of 216 million years, if one were considering a one-kilobase region of the genome. Since the evolution of humans from other primates took much less time than that, Durrett and Schmidt observed that if the DNA "neighborhood" were a thousand times larger, then lots of correct regulatory sites would already be expected to be there. But, then, exactly what is the model? And if the relevant neighborhood is much larger, why did they model a smaller neighborhood? Is there some biological fact they neglected to cite that justified the thousand-fold expansion of what constitutes a "neighborhood," or were they just trying to squeeze their results post hoc into what a priori was thought to be a reasonable time frame?

When I pointed this out in my letter, Durrett and Schmidt did not address the problem. Rather, they upped the stakes. They write in their reply that "there are at least 20,000 genes in the human genome and for each gene tens if not hundreds of pairs of mutations that can occur in each one." The implication is that there are very, very many ways to get two mutations. Well, if that were indeed the case, why did they model a situation where two particular mutations—not just any two—were needed? Why didn't they model the situation where any two mutations in any of 20,000 genes would suffice? In fact, since that would give a very much shorter time span, why did the journal *Genetics* and the reviewers of the paper let them get away with such a miscalculation?

The answer of course is that in almost any particular situation, almost all possible double mutations (and single mutations and triple mutations and so on) will be useless. Consider the chloroquine-resistance mutation in malaria. There are about 10^6 possible single amino acid mutations in malarial parasite proteins, and 10^{12} possible double amino acid mutations (where the changes could be in any two proteins). Yet only a handful are known to be useful to the parasite in fending off the antibiotic, and only one is very effective—the multiple changes in PfCRT. It would be silly to think that just any two mutations would help. The vast majority are completely ineffective. Nonetheless, it is a common conceptual mistake to naively multiply postulated "helpful mutations" when the numbers initially show too few.

A Very Important Point

Here's a final important point. *Genetics* is an excellent journal; its editors and reviewers are top-notch; and Durrett and Schmidt themselves are fine researchers. Yet, as I show above, when simple mistakes in the application of their model to malaria are corrected, it agrees closely with empirical results reported from the field that I cited. This is very strong support that the central contention of *The Edge of Evolution* is correct: that it is an extremely difficult evolutionary task for multiple required mutations to occur through Darwinian means, especially if one of the mutations is deleterious. And, as I argue in the book, reasonable application of this point to the protein machinery of the cell makes it very unlikely that life developed through a Darwinian mechanism.[37]

47. The Old Enigma

"The Old Enigma," parts 1–3, Uncommon Descent
(website), March 31–April 2, 2009.

WHEN *THE EDGE OF EVOLUTION* WAS FIRST PUBLISHED, SOME Darwinist reviewers sneered that the problem it focused on—the need for multiple mutations to form some protein features (such as binding sites), where intermediate mutations were deleterious—was a chimera. There were no such things, they essentially said. University of Wisconsin geneticist Sean Carroll, reviewing the book for *Science*, stressed examples where intermediate mutations were beneficial. (I never said there weren't such cases, and discussed several in the book.) In the same vein, University of Chicago evolutionary biologist Jerry Coyne assured readers of the *New Republic*, "In fact, interactions between proteins, like any complex interaction, were certainly built up step by mutational step... This process could have begun with weak protein-protein associations that were beneficial to the organism. These were then strengthened gradually..." The take-home message of the reviews for the public and for scientists in other fields was the same: Nothing to see here, folks. Move along. No problem here.

Contrast those assurances with a recent paper that addresses "the old enigma of the evolution of complex features in proteins that require two or more mutations." Those words (reminiscent of the title of my 2004 paper in *Protein Science* with David Snoke, "Simulating Evolution by Gene Duplication of Protein Features that Require Multiple Amino Acid Residues," which is cited by the recent paper) were written by the prominent bioinformatician (and no friend of ID) Eugene Koonin in his review of the Whitehead et al. paper, "The Look-ahead Effect of Phenotypic Mutations."[38] (Reviews are published along with papers on the *Biology Direct* web site.)

Old enigma? Old enigma? Who knew that evolving just a couple of interactive amino acid residues was a long-standing mystery? Someone should tell Carroll and Coyne....

I will discuss the specifics of the paper below. But let me first drive home this point. The development of protein features, such as protein-protein binding sites, that require the participation of multiple amino acid residues is a profound, fundamental problem that has stumped the evolutionary biology community until the present day (and continues to do so, as I explain below). It is a fundamental problem because all proteins exert their effects by physically binding to something else, such as a small metabolite or DNA or other protein, and require multiple residues to do so. The problem is especially acute for protein-protein interactions, since most proteins in the cell are now known to act as teams of a half-dozen or more, rather than individually. Yet if one can't explain how specific protein-protein interactions developed, then it is delusional to claim that we can explain how anything that depends on them developed, such as the molecular machinery of the cell. It's like saying, "We understand perfectly well how a car could evolve; we just don't know how the pieces could get fit together." If such a basic requirement for putting together complex systems is not understood, nothing is understood. Keep this in mind the next time you hear a blithe Darwinian tale about the undirected evolution of the cilium or bacterial flagellum.

PART 2

KOONIN IS clearly very impressed with the new paper, which he calls "brilliant" and "a genuinely important work that introduces a new and potentially major mechanism of evolution." His enthusiasm is a good indication that the problem is a major one, and that no other papers exist which deal effectively with it.

So what is the paper (a theoretical, mathematical-modeling study) about? When a mutationless gene is transcribed and translated into a protein, errors can creep in. It turns out that these error rates are much higher than for copying DNA. Using published mutation rates, Whitehead et al. estimate that one in ten standard-sized proteins will contain an error; that is, they will contain an amino acid that is not coded for by the gene. The authors call these "phenotypic mutations." Inherited changes that occur in the DNA are called "genotypic mutations."

Now, the idea is this. Suppose an organism needs two mutations to acquire some new feature, such as a disulfide bond. Further suppose that a single organism in a population that initially has neither of the mutations acquires just one of the necessary mutations in its DNA. Because of phenotypic mutations, this single organism will also contain some copies of the protein that have the second mutation. If the selective benefit of these phenotypic mutations is proportional to their concentration, as the authors suppose, then that organism may have an advantage over other organisms with no mutations. In a sense, the authors say, evolution can look a step ahead, so the authors dub this the "look-ahead effect." The reviewer Eugene Koonin agrees that the paper "in a sense, overturns the old adage of evolution having no foresight. It seems like, even if non-specifically and unwittingly, some foresight might be involved."

As the authors and one of the other referees note, this is pretty reminiscent of something called the "Baldwin effect," which was first proposed in the nineteenth century. The authors contend that there are subtle differences between the Baldwin effect and the look-ahead effect. Yet, regardless of who deserves priority for the idea, I don't think the look-ahead effect contributes much at all to solving the problem of multiple mutations. In my own opinion, the idea of the paper is certainly clever, but Koonin vastly overestimates its importance. It offers virtually no help in solving the "old enigma," as I explain below.

First, the effect is quite minor at best. Since, based on transcriptional and translational mutation rates, the fraction of proteins with the correct phenotypic mutation is expected to be about one-hundredth of one percent (10^{-4}) of the total number of protein copies, the presumed selective effect will be only 10^{-4} times the selective effect of the double genotypic mutant. So if the double genotypic mutant had a selective advantage of 0.1 (a pretty substantial value), the phenotypic look-ahead mutant would have an advantage of just 10^{-5}. If the double genotypic mutant has less of an advantage, the look-ahead has proportionately less. Because of this, the effect would be helpful only for large population sizes: too small of a population and there is no effect, because the mutation is effectively neutral. One can construct situations in which the selective advantage of a particular double genotypic mutant would be enormous (for example, if it conferred antibiotic resistance) so the look-ahead effect would be greater, but positing the general occurrence of such situations in nature amounts to special pleading.

It's also important to realize that the authors of the paper purposely did not consider mitigating factors in their analysis. As they wrote, "The goal of our analysis was to demonstrate that the look-ahead effect is theoretically possible, and as such, we intentionally excluded confounding factors for the sake of clarity." Other possible important effects that weren't considered in the model include the influence of the first genotypic mutation on the stability of the spectrum of proteins with pheno-typic mutations, effects of the mutations on translation rates, and so on. It is certainly understandable to simplify a model as much as possible for an initial investigation. However, any confounding effects will only diminish the strength of an already-weak influence.

PART 3

SECOND, THE authors assume that, in the absence of phenotypic muta-tions, the first genotypic mutation would be strictly neutral. That is, the selection coefficient for the first mutation is very, very close to zero. It turns out that this is a critical feature. If the first mutation were slightly positive itself (without considering look-ahead) then it could be selected on its own, and the look-ahead effect makes little difference. On the oth-er hand, if the first mutation is slightly negative (including look-ahead), then it will not be positively selected and, again, the effect makes es-sentially no difference. It is only in a very restricted range of selection coefficients that any significant influence will be seen.

A related point is the question: Except for purposes of illustration, why should the look-ahead effect be conceptually separated from every-thing else that goes into the selection coefficient? Clearly any mutation can have many effects, from stabilizing (or destabilizing) the structure of a protein to increasing (or decreasing) its interaction with other pro-teins, to favorably (or unfavorably) affecting the energy budget of a cell, and so on. All of the effects can influence whether the mutation is fa-vorable overall or not, so why separate out look-ahead? If, considering all influences, a particular mutation is favorable because offspring with the mutation survive with higher probability, then that is represented by a positive selection coefficient; if unfavorable, a negative coefficient. It is dubious to subdivide survival due to a particular mutation into tiny parts.

Third, the look-ahead effect is manifestly a double-edged sword. Consider the sequence of the protein one mutation before it reached what we previously called the "unmutated state"—that is, the sequence of the protein that was fixed in the population right before it reached the sequence that was two mutations from the highly favorable form. We can call it "sequence minus one." Now suppose a mutation appears in the DNA of one cell that would take us to the starting sequence (call it "sequence zero") if it spread and became fixed in the population. The next mutation (call it "sequence plus one") can appear in this individual cell as a phenotypic, look-ahead mutation. The final mutation ("sequence plus two"), which has the highly selectable feature, does not appear even as a phenotypic mutation in this cell. But now suppose that sequence plus one was not strictly neutral without look-ahead, but somewhat deleterious (as most protein mutations are). Then, because of the look-ahead effect, sequence zero will be selected against, and the probability that the population ever develops sequence zero will be much lower.

The take-home point is that, although looking ahead might help the final step a bit if the penultimate mutation is otherwise strictly neutral, the look-ahead effect will actively inhibit the development of a multimutation feature if one of the steps in a mutational pathway is somewhat deleterious. And the more deleterious it is, the more effectively the path is blocked. In a rugged adaptive landscape, the look-ahead effect is as likely to hurt as to help. In other words, it is a net of zero. So Darwinism remains great at "seeing" the immediately next step, but it has no reliable power to see beyond.

Finally and most importantly, recall the central message of *The Edge of Evolution*: To have a good idea of what Darwinian evolution can do, we no longer need to rely solely on speculative models, which may overlook or misjudge aspects of biology that nature would encounter. We already have good data in hand. We already have results that should constrain models. Over many thousands of generations, astronomical numbers of malarial cells seem not to have been able to take advantage of the look-ahead effect or anything else to build new, coherent molecular machinery. All that's been seen in that system in response to antibiotics are a few point mutations. In tens of thousands of generations, with a cumulative population size in the trillions, no coherent new systems have been seen in the fascinating work of Richard Lenski on the laboratory evolution of *E. coli*. Instead, even beneficial mutations have turned out to be degradative ones, where previously functioning genes are deleted or made less effective. And that's the same result as has been seen in the human genome in response to selective pressure due to malaria—a number of degraded genes or regulatory elements, and no new machinery.

Theoretical models must be constrained by data. If models don't reproduce what we do know happens in adaptive molecular evolution, then they are wholly unreliable in telling us anything about what we don't know. Unless a model can also reproduce empirical results such as those cited just above, it should be regarded as fanciful.

48. To Traverse a Maze, It Helps to Have a Mind

"To Traverse a Maze, It Helps to Have a Mind," Evolution News and Views, Discovery Institute, November 7, 2012.

A PAPER APPEARED RECENTLY IN *SCIENCE* THAT REMINDED ME OF one of my favorite toys when I was a kid—a rolling-ball maze. Over the years I had a few different varieties, including two- and three-dimensional ones. The basic gist is that a person has to twist and turn the toy to roll a ball through a plastic, transparent maze from the entrance to the exit. Of course there are a lot of dead ends and blind alleys, so it's pretty tricky, at least at first. Once you learn the path, it becomes trivial.

Näsvall et al. do the same with a protein, manipulating experimental conditions to roll it around dead ends and have it come out in the place they want.[39] Although the printed paper itself and an accompanying commentary by Elizabeth Pennisi[40] paint the results as an advance in understanding evolution, that's so only if evolution has eyes and a mind like a kid solving a maze. The investigators' exceptionally intelligent manipulations are relegated to the online supplemental materials.

Reading a brief part of the supplemental Materials and Methods section, entitled "Selection for bifunctional HisA mutants," is enough to see the absurdity of taking the results as a model for undirected Darwinian evolution. The authors write that they plated various independent cultures of a bacterial strain missing a protein needed to make the essential amino acid tryptophan on minimal-media+histidine plates "to select any mutant that could produce tryptophan without a TrpF enzyme. We included histidine in the medium because previous studies have demonstrated that mutations that confer TrpF activity to Thermotoga maritima HisA results in loss of HisA activity." They continue:

> The expression of the his operon (including hisA) is regulated by an attenuation mechanism that regulates the amount of read-through of a transcriptional terminator before the first structural gene of the operon according to the availability of charged histidinyl-tRNA. As this results in very low expression of hisA in medium containing histidine, we included a mutation which removes the transcriptional terminator, thereby leading to derepressed transcription of the his operon even in the presence of histidine.

In other words:

+ They deleted an enzyme that previous work showed could likely be replaced by another existing enzyme.
+ They added the necessary nutrient histidine because previous work showed that mutations conferring an ability to make tryptophan destroyed the ability to make histidine.
+ The added histidine would have shut off production of the protein, so they removed the genetic control element to keep it in production.
+ Later, once they found mutations to produce tryptophan, they removed histidine from the medium to encourage the production of mutations restoring histidine synthesis.

Roll the ball to the left to avoid one obstacle, roll it backward to avoid another, turn the maze over to drop the ball into the next corridor.... Needless to say, this ain't how unaided nature works—unless nature is guiding events toward a goal.

49. Misusing Protistan Examples

Nothing shows the threadbare nature of Darwin's theory more than when official science organizations try to defend it.

"Misusing Protistan Examples to Propagate Myths About Intelligent Design," Evolution News and Science Today, Discovery Institute, February 15, 2010.

The *Journal of Eukaryotic Microbiology* recently published several papers from a workshop sponsored by the International Society of Protistologists titled "Horizontal Gene Transfer and Phylogenetic Evolution Debunk Intelligent Design." So here we have a respected scientific society, presumably planning a workshop months in advance, and finally laying out their considered case for why intelligent design fails. As you might imagine, I was most anxious to read about it. So, what was the content of the papers? Rather than scholarly treatments, the manuscripts read like press releases from the National Center for (Darwinian) Science Education. So the introductory essay by Avelina Espinosa[41] tells us that ID has "creationist beginnings," claims that I say "evolution" is "impossible," and places in my mouth the phrase "design creationism." (I have never uttered that phrase except to disparage it.) There was about as much scholarship in the papers as you'd get from a typical politician.

The first of the full articles concerned itself mostly with common descent, which I have always said I think is correct, and which in any case is not an issue of intelligent design. Another article, however, briefly dealt with my case from *The Edge of Evolution*, that some adaptations are likely to require multiple mutations, and thus be very improbable.[42] In "Using Protistan Examples to Dispel the Myths of Intelligent Design,"[43] University of Georgia professor Mark Farmer and Wadsworth Center researcher Dr. Andrea Habura start off sloppily: "According to Behe (2007), the odds that mutations required to impart chloroquine resistance in *Plasmodium* could arise naturally are so impossibly long that they lie beyond what he considers 'The Edge of Evolution.'" But the book clearly states that chloroquine resistance in *Plasmodium* did arise naturally, by Darwinian processes. I go on to argue it took very many malarial parasites to chance upon resistance, and that pointed to a limit for Darwinian evolution for more complex mutations, or for populations with smaller numbers than Plasmodium, but I clearly said the opposite of what Farmer and Habura ascribe to me. I said that chloroquine resistance arose naturally. That doesn't give a reader confidence that the authors concern themselves much with the details of an argument.

Farmer and Habura think I am wrong that multiple mutations were necessary in the protein PfCRT to confer chloroquine resistance on malaria. They think only one will suffice. What's more, they claim there are experiments to show that. They cite two papers. But neither paper even tries to test whether a single mutation in PfCRT confers chloroquine resistance. Lakshmanan et al.[44] show that if they remove one particular mutation (K76T) from a mutant protein that carried a half-dozen or so other mutations (compared to the wild-type protein), the protein no longer confers chloroquine resistance. That experiment shows the K76T is necessary; however, it does not show it is sufficient by itself, as Farmer and Habura thought. The same goes for the second paper. In their methods section Jiang et al.[45] write that "Parasite 106/1K76 [a chloroquine-sensitive strain that does not have the K76T mutation] has six mutations found typically in Southeast Asian CQR parasite... except a key mutation at PfCRT 76 position.") Thus both these papers show that K76T is necessary, but neither shows it to be sufficient. To do so one would have to test the K76T in the wild-type, unmutated background.

To recap, several years after *The Edge of Evolution* was published a scientific society held a workshop to demonstrate the book's errors. Yet they couldn't even get the book's argument straight, and the experimental work they cited against my argument is not even pertinent to it. Apparently the design argument drives some scientists so much to distraction that they lose their normally robust powers of reasoning.

50. EVOLUTION AND THE EBOLA VIRUS

"Evolution and the Ebola Virus: Pacing a Small Cage," Evolution
News and Science Today, Discovery Institute, October 24, 2014.

I RECEIVED A REQUEST TO COMMENT ON THE EVOLUTION OF THE EB-
ola virus. The best, most recent work, which sequenced and compared
ninety-nine entire virus genomes, was published in *Science* last month:
"Genomic Surveillance Elucidates Ebola Virus Origin and Transmission during the 2014 Outbreak."[46] It came at a very high cost—five co-
authors actually died of Ebola in the course of the work.

The paper is mostly concerned with tracing the timing and spread
of the virus. The authors show that the virus is acquiring mutations very
rapidly (several hundred compared to a virus from an earlier outbreak in
1976), as is to be expected for RNA viruses, which have a much higher
mutation rate than DNA viruses. None of the mutations was evaluated
for functionality/selective value (which would be quite a job to do), but
all seem to be point mutations—no gene duplications or anything fancy.
In fact, the larger virus family of which Ebola is a part, the filoviruses,
seems to be pretty much unchanged in its basic genomic structure and
organization over the past ten thousand years.

Ebola's high rate of mutation is similar to what John Sanford has
demonstrated for the H1N1 virus that caused the influenza pandemic
after World War I. He makes a compelling case that the accumulating
mutations there were degradatory, could not be eliminated easily by selection, and eventually caused the virus's extinction in 2009.

Even small mutations in Ebola could have important epidemiological and medical effects if, for example, they affect the virus's ability to
bind to various cell receptor proteins. Yet, despite furious mutation,
there's no evidence that much of anything is changing at the basic gene
or biochemical level for the Ebola virus in particular or the filoviruses in
general. It seems to be another instance of what one group evocatively
described as RNA viruses "pacing a small cage"—dangerous, but confined within definite limits.

Part Four: Richard Lenski on the Edge

In the late 1980s Richard Lenski, a professor of biology at Michigan State University, began the longest-running laboratory evolution experiment to date. In a small flask he grew *E. coli* bacteria for twenty-four hours. During that time the bugs underwent six to seven generations, doubling in population each generation, finally producing hundreds of millions of cells. The next day he took one percent of the culture, added it to fresh culture broth, and let it grow for another day—another six to seven generations, more hundred millions. Lenski has repeated that ritual for over twenty-five years, and the *E. coli* have evolved for over 50,000 generations under his watchful gaze. That's 50,000 generations of random mutation of bacterial DNA, fierce competition among trillions of cells to leave the most descendants, and pitiless natural selection of the most fit. What do the results show us about Darwinism and other undirected evolutionary mechanisms?

51. Excerpt from The Edge of Evolution

I first reported on the ongoing state of Lenski's work in *The Edge of Evolution*.

Michael J. Behe, *The Edge of Evolution: The Search for the Limits of Darwinism* (New York: Free Press, 2007), excerpt from Chapter 7.

The studies of malaria and HIV provide by far the best direct evidence of what evolution can do. The reason is simple: numbers. The greater the number of organisms, the greater the chance that a lucky mutation will come along, to be grabbed by natural selection. But other results with other organisms can help us find the edge of evolution, especially laboratory results where evolutionary changes can be followed closely. The largest, most ambitious, controlled laboratory evolutionary study was begun more than a decade ago in the laboratory of Professor Richard Lenski at Michigan State University. Lenski wanted to follow evolution in real time. He started a project to watch the unfolding of cultures of the common gut bacterium *Escherichia coli*. *E. coli* is a favorite laboratory organism that has been studied by many scientists for more than a century. The bug is easy to grow and has a very short generation span of as little as twenty minutes under favorable conditions. Like those of *P. falciparum*, *H. sapiens*, and HIV, the entire genome of *E. coli* has been sequenced.

Unlike malaria and HIV, which both have to fend for themselves in the wild, and fight tooth and claw with the human immune system, the *E. coli* in Lenski's lab were coddled. They had a stable environment, daily food, and no predators. But doesn't evolution need a change in the environment to spur it on? Shouldn't we expect little evolution of *E. coli* in the lab, where its environment is tightly controlled? No and no. One of the most important factors in an organism's environment is the presence of other organisms. Even in a controlled lab culture where bacteria are warm and well-fed, the bug that reproduces fastest or outcompetes others will dominate the population. Like gravity, Darwinian evolution never stops.

But what does it yield? In the early 1990s Lenski and coworkers began to grow *E. coli* in flasks; the flasks reached their capacity of bacteria after about six or seven doublings. Every day he transferred a portion of the bugs to a fresh flask. There have now been over thirty thousand generations of *E. coli* born and died in Lenski's lab, roughly the equivalent of a million years in the history of humans. In each flask the bacteria would grow to a population size of about five hundred million. Over the whole course of the experiment, perhaps ten trillion, 10^{13}, *E. coli* have been produced. Although ten trillion sounds like a lot (it's probably more than the number of primates on the line from chimp to human), it's virtually nothing compared to the number of malaria cells that have infested the earth. In the past fifty years there have been about a billion times as many of those as E. *coli* in the Michigan lab, which makes the study less valuable than our data on malaria.

Nonetheless, the E. *coli* work has pointed in the same general direction. The lab bacteria performed much like the wild pathogens: a host of incoherent changes have slightly altered pre-existing systems. Nothing fundamentally new has been produced. No new protein-protein interactions, no new molecular machines. Like thalassemia in humans, some large evolutionary advantages have been conferred by breaking things. Several populations of bacteria lost their ability to repair DNA. One of the most beneficial mutations, seen repeatedly in separate cultures, was the bacterium's loss of the ability to make a sugar called ribose, which is a component of RNA. Another was a change in a regulatory gene called *spoT*, which affected en masse how fifty-nine other genes work, either increasing or decreasing their activity. One likely explanation for the net good effect of this very blunt mutation is that it turned off the energetically costly genes that make the bacterial flagellum, saving the cell some energy. Breaking some genes and turning others off, however, won't make much of anything. After a while, beneficial changes from the experiment petered out. The fact that malaria, with a billion-fold more chances, gave a very similar pattern to the more modest studies on E. *coli*, strongly suggests that that's all Darwinism can do.

52. New Work by Richard Lenski

"New Work by Richard Lenski," Evolution News and
Views, Discovery Institute, October 21, 2009.

A NEW PAPER[1] FROM RICHARD LENSKI'S GROUP HAS APPEARED IN *Nature* and has garnered a fair amount of press attention. The new paper continues the grand experiment that Lenski has been publishing about lo these many years—allowing a culture of the bacterium E. *coli* to continuously grow and evolve under his close observation. The only really new thing reported is a technical improvement—these days one can have the entire genome of E. coli "re-sequenced" (that is, determine the sequence of the entire DNA of the particular E. coli you're working with) for an affordable cost. (There are companies which will do it for a fee.) So Lenski and collaborators had the whole genomes—each and every nucleotide—sequenced of the E. coli that they have been growing for the past twenty years. Since they froze away portions of their bacterial culture at different times along the way, they now have the exact sequences of the evolving culture at many time points, from inception to 2,000 generations to 10,000 to 40,000. Thus, they can know exactly which mutations appeared when—an almost-complete paper trail. Very, very cool!

From that information they identify a few dozen mutations which they say are likely beneficial ones. That is almost certainly true, but what they don't emphasize is that many of the beneficial mutations are degradative—that is, they eliminate a gene or its protein's function. About half of the mutations they initially identified in previous work, but some they report here for the first time. They don't discuss what the new ones do (they may not yet know), but odds are high that most of them also are degradative, causing proteins either to stop working or to work less well. In any event, there is no indication that any of these are on their way to building some complex new system.

Interestingly, in this paper they report that the E. coli strain became a "mutator." That means it lost at least some of its ability to repair its DNA, so mutations are accumulating now at a rate about seventy times faster than normal. Lenski had reported years earlier that a number of other lines of the evolving population (they started with twelve separate cultures) had become mutators, too.[2] So it seems that loss of ability to repair DNA is a common occurrence under these conditions.

Lenski is a very good self-promoter (no criticism intended; that's a good thing—scientists have to interest other people in their work), and he always accentuates the positive. So if a gene is blasted to bits by a mutation, he talks cheerfully about how it is a beneficial change that helps the bacterium grow faster. One has to dig hard into the data to see that the bacterium is losing genetic information. In press coverage for this paper, he avows that a "new dynamic relationship was established" in the bacterium's evolution, and one has to read the details of the paper to find out that this is due to a degradative mutation that compromises its normal ability to repair its DNA.

Despite his understandable desire to spin the results his way, Lenski's decades-long work lines up wonderfully with what an ID person would expect—in a huge number of tries, one sees minor changes, mostly degradative, and no new complex systems. So much for the power of random mutation and natural selection. For his work in this area we should be very grateful. It gives us solid results to point to, rather than having to debate speculative scenarios.

53. "EVOLVABILITY" AND TORTUOUS DARWINIAN PATHWAYS

"Richard Lenski, 'Evolvability,' and Tortuous Darwinian Pathways,"
Evolution News and Views, Discovery Institute, April 18, 2011.

SEVERAL PAPERS ON THE TOPIC OF "EVOLVABILITY" HAVE BEEN PUBlished relatively recently by the laboratory of Richard Lenski.[3] Most readers will quickly recognize Lenski as the Michigan State microbiologist who has been growing cultures of E. coli for over twenty years in order to see how they would evolve, patiently transferring a portion of each culture to new media every day, until the aggregate experiment has now passed 50,000 generations. I'm a huge fan of Lenski et al.'s work because, rather than telling just-so stories, they have been doing the hard laboratory work that shows us what Darwinian evolution can and likely cannot do.

The term "evolvability" has been used widely and rather loosely in the literature for the past few decades. It usually means something like the following: a species possesses some biological feature which lends itself to evolving more easily than other species that don't possess the feature, so that the lucky species will tend to adapt and survive better than its rivals over time. The kind of feature that is most often invoked in this context is "modularity." That word itself is often used in a vague manner. As I wrote in *The Edge of Evolution*, "Roughly, a module is a more-or-less self-contained biological feature that can be plugged into a variety of contexts without losing its distinctive properties. A biological module can range from something very small (such as a fragment of a protein), to an entire protein chain (such as one of the subunits of hemoglobin), to a set of genes (such as *Hox* genes), to a cell, to an organ (such as the eyes or limbs of *Drosophila*)."[4]

Well, Lenski and coworkers don't use "evolvability" in that sense. They use the term in a much broader sense: "Evolutionary potential, or evolvability, can be operationally defined as the expected degree to which a lineage beginning from a particular genotype will increase in fitness after evolving for a certain time in a particular environment."[5] To put it another way, in their usage "evolvability" means how much an organism will increase in fitness over a defined time starting from genotype A versus starting from genotype B, no matter whether genotypes A and B have any particular identifiable feature such as modularity or not.

Lenski's group published a very interesting paper last year showing that the more defective a starting mutant was in a particular gene (*rpoB*, which encodes a subunit of RNA polymerase), the more "evolvable" it was.[6] That is, more-crippled cells could gain more in fitness than less-crippled cells. But none of the evolved crippled cells gained enough fitness to match the uncrippled parent strain. Thus it seemed that more-crippled cells could gain more fitness simply because they started from further back than less-crippled ones. Compensatory mutations would pop up somewhere in the genome until the evolving cell was near to its progenitor's starting point. This matches the results of some viral evolution studies where some defective viruses could accumulate compensatory mutations until they were similar in fitness to the starting strain, whether they began with one-tenth or one-ten-billionth of the original fitness.[7]

In a paper published a few weeks ago the Michigan State group took a somewhat different experimental tack.[8] They isolated a number of cells from relatively early in their long-term evolution experiment. (Every 500th generation during the 50,000-generation experiment Lenski's group would freeze away the portion of the culture which was left over after they used a part of it to seed a flask to continue the growth. Thus they have a very complete evolutionary record of the whole lineage, and can go back and conduct experiments on any part of it whenever they wish. Neat!) They saw that different mutations had cropped up in different early cells. Interestingly, the mutations which gave the greatest advantage early on had become extinct after another 1,000 generations. So Lenski's group decided to investigate why the early very-beneficial mutations were nonetheless not as "evolvable" (because they were eventually outcompeted by other lineages) as cells with early less-beneficial mutations.

The workers examined the system thoroughly, performing many careful experiments and controls. (I encourage everyone to read the whole paper.) The bottom line is that they found that changing one particular amino acid residue in one particular protein (called a "topoisomerase," which helps control the "twistiness" of DNA in the cell), instead of a different amino acid residue in the same protein, interfered with the ability of a subsequent mutation in a gene (called *spoT*) for a second protein to help the bacterium increase in fitness. In other words, getting the "wrong" mutation in topoisomerase—even though that mutation by itself did help the bacterium—prevented a mutation in *spoT* from helping. Getting the "right" mutation in topoisomerase allowed a mutation in *spoT* to substantially increase the fitness of the bacterium.

The authors briefly discuss the results (the paper was published in *Science*, which doesn't allow much room for discussion) in terms of "evolvability," understood in their own sense.[9] They point out that the strain with the right topoisomerase mutation was more "evolvable" than the one with the wrong topoisomerase mutation, because it outcompeted the other strain. That is plainly correct, but does not say anything about "evolvability" in the more common and potentially much more important sense of an organism possessing modular features that help it evolve new systems. "Evolvability" in the more common sense has not been tested experimentally in a Lenski-like fashion.

In my own view, the most interesting aspect of the recent Lenski paper is its highlighting of the pitfalls that Darwinian evolution must dance around, even as it is making an organism somewhat more fit.[10] If the "wrong" advantageous mutation in topoisomerase had become fixed in the population (by perhaps being slightly more advantageous or more common), then the "better" selective pathway would have been shut off completely. And since this phenomenon occurred in the first instance where anyone had looked for it, it is likely to be commonplace. That should not be surprising to anyone who thinks about the topic dispassionately. As the authors note, "Similar cases are expected in any population of asexual organisms that evolve on a rugged fitness landscape with substantial epistasis, as long as the population is large enough that multiple beneficial mutations accumulate in contending lineages before any one mutation can sweep to fixation." If the population is not large enough, or other factors interfere, then the population will be stuck on a small peak of the rugged landscape.

This fits well with recent work by Lenski's and others' laboratories, showing that most beneficial mutations actually break or degrade genes, and also with work by Thornton's group showing that random mutation and natural selection likely could not transform a steroid hormone receptor back into its homologous ancestor, even though both have very similar structures and functions, because the tortuous evolutionary pathway would be nearly impossible to traverse.[11] The more that is learned about Darwin's mechanism at the molecular level, the more ineffectual it is seen to be.

54. MORE FROM LENSKI'S LAB, STILL SPINNING FURIOUSLY

"More from Lenski's Lab, Still Spinning Furiously," Evolution
News and Views, Discovery Institute, January 30, 2012.

RECENTLY A NEW PAPER BY RICHARD LENSKI AND COLLEAGUES[12] appeared in *Science* with, as usual, commentary in the *New York Times*. Lenski's lab must own a red phone with a direct line to the Gray Lady.

The gist of the paper is that a certain bacteriophage (a virus that infects bacteria) called "lambda" gained the ability to bind a different protein on the surface of its host, the bacterium *E. coli*, than the protein it usually binds. The virus has to bind to the cell's surface as a prelude to invading it. The protein it normally binds is called LamB. Lenski's lab, however, used a bacterial strain that had turned off the production of LamB in 99% of *E. coli* cells but, crucially, 1% of cells still produced the protein.

Thus the virus could still invade some cells, reproduce, and not go extinct. Under these conditions the viral binding protein (called "J") underwent several mutations, apparently to better bind LamB in the fewer cells that produced it. Then, surprisingly, after the viral gene gained a fourth mutation, the viral J protein acquired the ability to bind a different protein on *E. coli*, called OmpF. Now the virus could use OmpF as a platform for invading the cell. Since all cells made OmpF, the virus was no longer restricted to invading just the 1% of cells that made LamB, and it prospered. The workers repeated the experiment multiple times, and frequently got the same results.

As always, the work of the Lenski lab is solid and interesting, but is spun to make it appear to support Darwinian evolution more than it does. As the authors acknowledge, this is certainly not the first time a lab has evolved a virus to grow on a different strain of host. In a review of mine[13] published in the *Quarterly Review of Biology*, there is a section entitled "Evolution Experiments with Viruses: Adapting to a New Host" discussing just that topic.

In general, viruses have been shown to be able to adapt to bind to related host cells that have similar surface features. In almost all cases the virus uses the same binding protein, and the same (mutated) binding site to attach to the new host cell. This also seems to be the case with Lenski's new work. As stated above, the first several mutations apparently strengthen the ability of the J protein to bind to the original site, LamB, while the fourth mutation allows it to bind to OmpF.

As Lenski and his co-authors state, however, the mutated viral J protein can still bind to the original protein, LamB, which strongly suggests the same binding site (that is, the same location on the J protein) is being used. It turns out that both LamB and OmpF have similar three-dimensional structures, so that strengthening the binding to one fortuitously led to binding to the other.

In my review[14] I discussed why this should be considered a "modification of function" event rather than a gain-of-function one. The bottom line is that the results are interesting and well done, but not particularly novel, nor particularly significant.

To me, the much more significant results of the new paper, although briefly mentioned, were not stressed as they deserved to be. The virus was not the only microbe evolving in the lab. The E. coli also underwent several mutations. Unlike for lambda, these were not modification-of-function mutations—they were complete loss-of-function mutations.

The mechanism the bacterium used to turn off LamB in 99% of cells to resist initial lambda infection was to mutate and destroy its own gene locus called malT, which is normally useful to the cell. After acquiring the fourth mutation the virus could potentially invade and kill all cells. However, E. coli itself then mutated to prevent this, too. It mutated by destroying some genes involved in importing the sugar mannose into the bacterium. It turns out that this "mannose permease" is used by the virus to enter the interior of the cell. In its absence, infection cannot proceed.

So at the end of the day there was left the mutated bacteriophage lambda, still incompetent to invade most *E. coli* cells, plus mutated *E. coli*, now with broken genes which remove its ability to metabolize maltose and mannose. It seems Darwinian evolution took a small step sideways and two big steps backwards.

55. MULTIPLE MUTATIONS
NEEDED FOR E. COLI

"Multiple Mutations Needed for *E. Coli*," Uncommon
Descent (website), June 9, 2008.

An INTERESTING PAPER HAS JUST APPEARED IN THE *PROCEEDINGS of the National Academy of Sciences*, "Historical Contingency and the Evolution of a Key Innovation in an Experimental Population of *Escherichia coli*."[15] It is the "inaugural article" of Richard Lenski, who was recently elected to the National Academy. Lenski, of course, is well known for conducting the longest, most detailed lab evolution experiment in history, growing the bacterium *E. coli* continuously for about twenty years in his Michigan State lab. For the fast-growing bug, that's over 40,000 generations!

I discuss Lenski's fascinating work in Chapter 7 of *The Edge of Evolution*, pointing out that all of the beneficial mutations identified from the studies so far seem to have been degradative ones, where functioning genes are knocked out or rendered less active. So random mutation much more easily breaks genes than builds them, even when it helps an organism to survive. That's a very important point. A process which breaks genes so easily is not one that is going to build up complex coherent molecular systems of many proteins, which fill the cell.

In his new paper Lenski reports that, after 30,000 generations, one of his lines of cells has developed the ability to utilize citrate as a food source in the presence of oxygen. (*E. coli* in the wild can't do that.) Now, wild *E. coli* already have a number of enzymes that normally use citrate and can digest it (it's not some exotic chemical the bacterium has never seen before). However, the wild bacterium lacks an enzyme called a "citrate permease" which can transport citrate from outside the cell through the cell's membrane into its interior. So all the bacterium needed to do to use citrate was to find a way to get it into the cell. The rest of the machinery for its metabolism was already there. As Lenski put it, "The only known barrier to aerobic growth on citrate is its inability to transport citrate under oxic conditions."[16]

Other workers (cited by Lenski) in the past several decades have also identified mutant *E. coli* that could use citrate as a food source. In one instance the mutation wasn't tracked down.[17] In another instance a protein coded by a gene called *citT*, which normally transports citrate in the absence of oxygen, was overexpressed.[18] The overexpressed protein allowed *E. coli* to grow on citrate in the presence of oxygen. It seems likely that Lenski's mutant will turn out to be either this gene or another of the bacterium's citrate-using genes, tweaked a bit to allow it to transport citrate in the presence of oxygen. (At the time I originally wrote this essay, he hadn't yet tracked down the mutation, but see my next essay below.)

The major point Lenski emphasizes in the paper is the historical contingency of the new ability. It took trillions of cells and 30,000 generations to develop it, and only one of a dozen lines of cells did so. What's more, Lenski carefully went back to cells from the same line he had frozen away after evolving for fewer generations and showed that, for the most part, only cells that had evolved at least 20,000 generations could give rise to the citrate-using mutation. From this he deduced that a previous, lucky mutation had arisen in the one line, a mutation which was needed before a second mutation could give rise to the new ability. The other lines of cells hadn't acquired the first, necessary, lucky, "potentiating"[19] mutation, so they couldn't go on to develop the second mutation that allows citrate use. Lenski argues that this supports the view of the late Stephen Jay Gould that evolution is quirky and full of contingency. Chance mutations can push the path of evolution one way or another, and if the "tape of life" on earth were rewound, it's very likely evolution would take a completely different path than it has.

I think the results fit a lot more easily into the viewpoint of *The Edge of Evolution*. One of the major points of the book was that if only one mutation is needed to confer some ability, then Darwinian evolution has little problem finding it. But if more than one is needed, the probability of getting all the right ones grows exponentially worse. "If two mutations have to occur before there is a net beneficial effect—if an intermediate state is harmful, or less fit than the starting state—then there is already a big evolutionary problem."[20] And what if more than two are needed? The task quickly gets out of reach of random mutation.

To get a feel for the clumsy ineffectiveness of random mutation and selection, consider that the workers in Lenski's lab had routinely been growing *E. coli* all these years in a soup that contained a small amount of the sugar glucose (which they digest easily), plus about ten times as much citrate. Like so many cellular versions of Tantalus, for tens of thousands of generations trillions of cells were bathed in a solution with an abundance of food—citrate—that was just beyond their reach, outside the cell. Instead of using the unreachable food, however, the cells were condemned to starve after metabolizing the tiny bit of glucose in the medium—until an improbable series of mutations apparently occurred. As Lenski and coworkers observe: "Such a low rate suggests that the final mutation to Cit^+ is not a point mutation but instead involves some rarer class of mutation or perhaps multiple mutations. The possibility of multiple mutations is especially relevant, given our evidence that the emergence of Cit^+ colonies on MC plates involved events both during the growth of cultures before plating and during prolonged incubation on the plates."[21]

In *The Edge of Evolution* I had argued that the extreme rarity of the development of chloroquine resistance in malaria was likely the result of the need for several mutations to occur before the trait appeared. Even though the evolutionary literature contains discussions of multiple mutations,[22] Darwinian reviewers drew back in horror, acted as if I had blasphemed, and argued desperately that a series of single beneficial mutations certainly could do the trick. Now here we have Richard Lenski affirming that the evolution of some pretty simple cellular features likely requires multiple mutations.

If the development of many of the features of the cell required multiple mutations during the course of evolution, then the cell is beyond Darwinian explanation. I show in *The Edge of Evolution* that it is very reasonable to conclude they did.

56. LENSKI, CITRATE, AND BIOLOGOS

"Rose-Colored Glasses: Lenski, Citrate, and BioLogos," Evolution
News and Views, Discovery Institute, November 13, 2012.

READERS OF MY POSTS KNOW THAT I'M A BIG FAN OF PROFESSOR Richard Lenski, a microbiologist at Michigan State University and member of the National Academy of Sciences. For the past few decades he has been conducting the largest laboratory evolution experiment ever attempted. Growing *E. coli* in flasks continuously, he has been following evolutionary changes in the bacterium for over 50,000 generations (which is equivalent to roughly a million years for large animals). Although Lenski is decidedly not an intelligent design proponent, his work enables us to see what evolution actually does when it has the resources of a large number of organisms over a substantial number of generations. Rather than speculate, Lenski and his coworkers have observed the workings of mutation and selection. For this, we ID proponents should be very grateful.

In a manuscript published a few years ago in the *Quarterly Review of Biology*,[23] I discussed laboratory evolution results from the past four decades up to that point, including Lenski's. His laboratory had shown clearly that random mutation and selection improved the bacterium with time, as measured by the number of progeny it could produce in a given time. He demonstrated without doubt that beneficial mutations exist and can spread quickly in a population of organisms. However, once Lenski's lab eventually identified the mutations at the DNA level (a difficult task), many of the beneficial mutations turned out to be, surprisingly, degradative ones. In other words, breaking or deleting some pre-existing genes or genetic regulatory elements so that they no longer worked actually helped the organism under the conditions in which it was grown. Other beneficial mutations altered pre-existing genes or regulatory elements somewhat.

What conspicuously was not seen in his work were beneficial mutations that resulted from building what I dubbed new Functional Coded elemenTs, or "FCTs." Roughly, an FCT is a sequence of DNA that affects the production or processing of a gene or gene product (see my review for a more rigorous definition). In short, improvements had been made by breaking existing genes, or fiddling with them in minor ways, but not by making new genes or regulatory elements. From this information I formulated "The First Rule of Adaptive Evolution": *Break or blunt any functional coded element whose loss would yield a net fitness gain.* To say the least, the First Rule is not what you would expect from a process, such as Darwinian evolution, which is touted as being able to build amazingly sophisticated molecular machinery.

Before my review was published, the Lenski lab observed a mutant strain in their experiments that could metabolize citrate in the presence of oxygen, which unmutated *E. coli* cannot do.[24] (Importantly, however, the bacterium can metabolize citrate in the absence of oxygen.) This allowed the mutated bacterium to outcompete its relatives, because the growth medium contained a lot of citrate, as well as oxygen. It was an intriguing result, and was touted as a major innovation, but at the time Lenski's lab was unable to track down at the DNA level the exact mutations that caused the change.

Now they have. In a recent publication in *Nature*[25] they report the multiple mutations that confer and increase the ability to transport citrate in an atmosphere containing oxygen. They divide the mutations conceptually into three categories: 1) potentiation; 2) actualization; and 3) refinement. "Actualization" is the name they give to the mutation that first confers a weak ability to transport citrate into the laboratory *E. coli*. (It turns out that the bacterium is lacking only a protein to transport citrate into the cell in the presence of oxygen; all other enzymes needed to further metabolize citrate are already present.) The gene for the citrate transporter, *citT*, that works in the absence of oxygen is directly upstream from the genes for two other proteins that have promoters that are active in the presence of oxygen. A duplication of a segment of this region serendipitously placed the *citT* gene next to one of these promoters, so the *citT* gene could then be expressed in the presence of oxygen. Gene duplication is a type of mutation that is known to be fairly common, so this result, although requiring a great deal of careful research to pin down, is unsurprising.

Over time the mutant got better at utilizing citrate, which the authors called "refinement." Further work showed this was due to multiple duplications of the mutant gene region, up to 3–9 copies. Again, gene duplication is a fairly common process, so again it is unsurprising. In another experiment Lenski and coworkers showed that increasing the concentration of the citrate transporter gene was sufficient in itself to account for the greater ability of E. coli to grow on citrate. No other mutations were needed.

A more mysterious part of the whole process is what the group called "potentiation." It turns out that the original E. coli they began with decades ago could not benefit from the gene duplication that brought together a *citT* gene with an oxygen-tolerant promoter. Before it could benefit, a preliminary mutation had to occur in the bacterium somewhere other than the region containing the citrate-metabolism genes. Exactly what that mutation was, Lenski and coworkers were not able to determine. However, they examined the bacterium for mutations that may contribute to potentiation, and speculated that "a mutation in *arcB*, which encodes a histidine kinase, is noteworthy because disabling that gene upregulates the tricarboxylic acid cycle." (They tried, but were unable to test this hypothesis.) In other words, the "potentiation" may involve degradation of an unrelated gene.

Lenski's lab did an immense amount of careful work and deserves much praise. Yet the entirely separate, $64,000 question is: What do the results show about the power of the Darwinian mechanism? The answer is, they do not show it to be capable of anything more than what was already known. For example, in my review of lab evolution experiments I discussed the work of Zinser et al.[26] where a sequence rearrangement brought a promoter close to a gene that had lacked one. I also discussed experiments such as Licis and van Duin[27] where multiple sequential mutations increased the ability of an FCT. Despite Lenski's visually startling result—where a usually clear flask became cloudy with the overgrowth of bacteria on citrate—at the molecular level nothing novel occurred.

Another person who follows Lenski's results closely is Dennis Venema, chair of the Biology Department at Trinity Western University and contributor to the BioLogos website. Founded by Francis Collins, BioLogos defends the compatibility of Darwinian science and Christian theology. I agree that the Darwinian mechanism (rightly understood) is theoretically compatible with Christian theology. However, I also think Darwinism is grossly inadequate on scientific grounds. A number of BioLogos writers think it is adequate, and attempt to defend it against skeptics of Darwinism, most especially against proponents of intelligent design such as myself.

In several posts at BioLogos,[28] Professor Venema compared the results of Lenski's current citrate work to arguments I had made in my *QRB* review and in *The Edge of Evolution*. Whereas I had argued that there was a limit to the number of unselected (either detrimental or neutral) mutations we could reasonably expect an undirected, Darwinian process to have at its disposal in building a complex system, Venema thought Lenski's recent work showed that limit to be exceeded. Furthermore, while I had pointed out that none of the mutations seen in Lenski's work up to the date of the review was a gain-of-FCT, Venema wrote that the newly published citrate mutations constituted such a feature.

I disagree on both counts. The gene duplication which brought an oxygen-tolerant promoter near to the *citT* gene did not make any new functional element. Rather, it simply duplicated existing features. The two FCTs comprising the oxygen-tolerant citrate transporter locus—the promoter and the gene—were functional before the duplication and functional after. I had written in my review that one type of mutation that could be categorized as a gain-of-FCT was gene duplication with subsequent sequence modification, to allow the gene to specialize in some task. Venema thinks the mutation observed by Lenski is such an event. He has overlooked the fact that there was no subsequent sequence modification; a segment of DNA simply tandemly duplicated, bringing together two pre-existing FCTs. (It is true that the sequence of the protein coded by the duplicated gene includes a fragment from one of the nearby genes, but there is no evidence nor reason to think that the fused fragment is necessary for the activity of the protein.) In my review I classify that as a modification-of-function event. An example of a true gain-of-FCT by duplication cited in my review was the work of Olsthoorn and van Duin[29] where a 14-nucleotide duplication led to the formation of new functional coded elements (it did not simply repeat pre-existing elements), so it is not just a modification-of-function mutation. The citrate mutation did nothing like that.

Venema counts the number of mutations needed to get a fully functioning citrate-importing function in Lenski's work, and arrives at roughly a half-dozen. Unfortunately, several of those are tandem duplications of the weak $citT$ transporter gene, which clearly are selectable, beneficial mutations. In arriving at the limits to Darwinism, I emphasized that that mechanism would certainly work if gradually increasing, serial, beneficial mutations could do the job. Thus such mutations do not count in estimating the limit. Only required deleterious and neutral mutations count against the limit to Darwinian evolution. Venema argues that perhaps all of complex functional biology could be reached by gradual, beneficial mutations. Well, bless his optimistic heart, but the data give us no reason to think that, because gradually increasing one protein's total cellular activity by sequential gene duplication is successively beneficial, all routes to complex systems involving multiple distinct elements will be. Quite the opposite, as I have often argued.

Professor Venema also counts several "potentiating" mutations as contributing to the system. But whatever those mutations are, they are not part of the citrate metabolic system itself. Rather, they are at most part of the genetic background. If Lenski and coworkers' speculations are correct,[30] at least one of the potentiating mutations degrades an unrelated gene, and thus itself counts as a loss-of-FCT mutation. When counting the mutations contributing to the edge of evolution for building a feature, only the ones directly involved in the feature are counted, not ones that indirectly contribute to a receptive genetic background (which are legion). Thus, unlike Venema, I count perhaps three to four mutations—the original duplication placing the oxygen-tolerant promoter near the $citT$ gene, plus several rounds of duplication of that region. All of the mutations are modification-of-function ones in the classification system I described. I should add that there is no reason to think that Darwinian processes cannot produce gain-of-FCT mutations, and I reviewed several such events. But they are greatly outnumbered by loss-of-FCT and modification-of-function beneficial mutations.

In my own view, in retrospect, the most surprising aspect of the oxygen-tolerant $citT$ mutation was that it proved so difficult to achieve. If, before Lenski's work was done, someone had sketched for me a cartoon of the original duplication that produced the metabolic change, I would have assumed that would be sufficient—that a single step could achieve it. The fact that it was considerably more difficult than that goes to show that even skeptics like myself overestimate the power of the Darwinian mechanism.

57. Twenty-Five Years and Counting

"Lenski's Long-Term Evolution Experiment: Twenty-five Years and Counting,"
Evolution News and Views, Discovery Institute, November 21, 2013.

THE CURRENT ISSUE OF *SCIENCE* CARRIES A FOUR-PAGE PANEGYric[31] highlighting the career of Richard Lenski on the occasion of the twenty-fifth anniversary of the beginning of his long-term evolution experiment. In 1988 Lenski started what then seemed a slightly wacky project—to let cultures of the bacterium *Eschericia coli* grow continuously under his watchful gaze in his lab at Michigan State University. Every day he or one of a parade of grad students and postdocs would transfer a small portion of the culture into fresh media in a new test tube, allowing the bacteria to grow six to seven generations per day. Twenty-five years later the culture—a cumulative total of trillions of cells—has been going for an astounding 58,000 generations and counting. As the article points out, that's equivalent to a million years in the lineage of a large animal such as humans. Combined with an ability to track down the exact identities of bacterial mutations at the DNA level, that makes Lenski's project the best, most detailed source of information on evolutionary processes available anywhere, dwarfing rival lab projects and swamping field studies. That's an achievement well worth celebrating.

Still, the important question to ask is, what exactly has this venerable project shown us about evolution? The study has addressed some narrow points of peculiar interest to evolutionary population geneticists, but for proponents of intelligent design the bottom line is that the great majority of even beneficial mutations have turned out to be due to the breaking, degrading, or minor tweaking of pre-existing genes or regulatory regions.[32] There have been no mutations or series of mutations identified that appear to be on their way to constructing elegant new molecular machinery of the kind that fills every cell. For example, the genes making the bacterial flagellum are consistently turned off by a beneficial mutation (since flagella aren't needed in stirred flasks, and apparently it saves cells energy used in constructing flagella). The suite of genes used to make the sugar ribose is the uniform target of a destructive mutation, which somehow helps the bacterium grow more quickly in the laboratory. Degrading a host of other genes leads to beneficial effects, too.

The *Science* story references a new paper from Lenski's lab[33] showing that the bacterial strain continues to improve its growth rate. The chief talking point of the paper is that the rate of improvement follows a curve that will not max out—improvements would continue indefinitely, although at an ever-slowing rate. The natures of the newer beneficial mutations, however, are not reported—whether they, too, are degradative changes, or minor, sideways changes, or truly constructive changes. (I know which way I'll bet....)

In one supplementary figure the authors show that the increasing growth rate is built on some previously known, beneficial-yet-degradative mutations. Earlier this year Lenski's lab[34] identified a mutation that built on a previous mutation, too, which may prefigure what kind of changes the unidentified mutations in the current paper will turn out to be. Over the course of the project several of the dozen separate strains developed what is called a "mutator" phenotype. In English, that means that the cell's ability to faithfully copy its DNA is degraded, and its mutation rate has increased some 150-fold. As Lenski's work showed, that's due to a mutation (dubbed mutT) that degrades an enzyme that rids the cell of damaged guanine nucleotides, preventing their misincorporation into DNA. Loss of function of a second enzyme (mutY), which removes mispaired bases from DNA, also increases the mutation rate when it occurs by itself. However, when the two mutations, mutT and mutY, occur together, the mutation rate decreases by half of what it is in the presence of mutT alone—that is, it is seventy-five-fold greater than the unmutated case.

Lenski is an optimistic man, and always accentuates the positive. In the paper on mutT and mutY, the stress is on how the bacterium has improved with the second mutation. Heavily unemphasized is the ominous fact that one loss-of-function mutation is "improved" by another loss-of-function mutation—by degrading a second gene. Anyone who is interested in long-term evolution should see this as a baleful portent for any theory of evolution that relies exclusively on blind, undirected processes.

58. Citrate Hype—Now Deflated

"Richard Lenski and Citrate Hype—Now Deflated," Evolution
News and Views, Discovery Institute, November 17, 2016.

DISHONESTY COMES IN DEGREES, FROM THE WHITE LIE TOLD TO
spare another's feelings to criminal fraud for one's own financial
gain. Somewhere in the middle lies hype in science. Certainly a bit of
innocent, accentuate-the-positive spinning of research results can help
a scientist catch people's attention. Unfortunately, that can escalate into
hucksterism that seriously exaggerates the importance of the work.

Most scientists aren't even tempted to try it, because most areas of
research aren't sexy enough to pull it off. It is a problem, however, for
those who work on topics that catch the news media's attention: cures for
cancer; cloning; grand theories of the universe; and, of course, evolution.

Which brings us to Michigan State's Richard Lenski. As longtime
readers of *Evolution News and Views* well know, to study evolution, for
more than twenty-five years Lenski's lab has continuously grown a dozen
lines of the bacterium *E. coli* in small culture flasks, letting them replicate for six or seven generations per day and then transferring a portion
to fresh flasks for another round of growth. The carefully monitored
cells have now gone through more than 60,000 generations, which is
equivalent to over a million years for a large animal such as humans. (It's
dubbed the Long-Term Evolution Experiment—LTEE.) As I've written
before, the work itself is terrific. However, the implications of the work
are often blown seriously out of proportion by a cheerleading science
news media eager for stories to trumpet.

In 2008 Lenski's group reported that after more than fifteen years
and 30,000 generations of growth, one of the *E. coli* cell lines suddenly
developed the ability to consume citrate, which for technical reasons had
been present in the liquid culture medium. Later work by the Michigan
State team showed the ability was due to the duplication and rearrangement of a gene for a protein that normally imports citrate into the cell,

but only when no oxygen is present. The mutation allowed the protein to work when oxygen was present, as it was throughout the LTEE.

It was an interesting, if modest, result—a gene had been turned on under conditions where it was normally turned off. But the authors argued it might be pretty important. In their paper they wrote that the mutant's ability could be the result of "historical contingency"—that is, a rare, serendipitous event that might alter the course of evolution. They also remarked that, since an inability to use citrate in the presence of oxygen had been a characteristic used to help define *E. coli* as a species, perhaps the mutation marked the beginning of the evolution of a brand new species.

One scientist who thought the results were seriously overblown was Scott Minnich, professor of microbiology at the University of Idaho and—full disclosure—a colleague of mine as a Fellow at Discovery Institute's Center for Science and Culture. Minnich knew that decades ago the microbiologist Barry Hall had isolated an *E. coli* mutant that could also use citrate after only a few weeks of growth (also cited in Lenski's paper), and that other studies had shown mutants could be isolated rapidly if they were selected directly—that is, if they were grown where the only available food source was the selecting substrate such as citrate, rather than a mixture of the selecting substrate plus glucose, as in Lenski's experiment.

So Minnich's lab re-did the work under conditions he thought would be more effective.[35] The bottom line is that they were able to repeatedly isolate the same mutants Lenski's lab did as easily as falling off a log—within weeks, not decades. In an accompanying commentary highlighting the Idaho group's paper in the *Journal of Bacteriology*, the prominent UC Davis microbiologist John Roth and his colleague Sophie Maisnier-Patin agreed that Lenski's "idea of 'historical contingency' may require reinterpretation."[36]

Richard Lenski was not pleased. Although in a response on his blog[37] he acknowledged up front that the Idaho group's science was "fine and interesting," he insisted that yes, the mutation was indeed historical-

ly contingent. Roth and Maisner-Patin's comments to the contrary supposedly represented "a false dichotomy." After all, historical contingency
just "means that history matters," and whether Lenski's cells developed
the mutation clearly depended on how they had been treated in his lab.
Ipso facto, it was contingent.

But of course it's vacuous to say simply that "history matters." Any
near-certain outcome can be prevented if necessary conditions for it to
occur aren't present. A ball will always roll down a hill—unless someone puts a barrier in front of it. The fact that the Minnich lab easily
and repeatedly obtained the same results with multiple bacterial strains
and growth conditions shows the results are not some special example of
historical contingency, if that phrase has any nontrivial meaning at all.
Rather, it's a humdrum, repeatable result.

Lenski also tried to split hairs over the question of speciation. He
faulted Minnich for writing skeptically of Lenski's citrate mutation,
"This was interpreted as a speciation event." Lenski countered that in
their initial paper his group had only been wondering out loud if the
mutant would "eventually become" a distinct species. It's a *process*, not an
event, you see. But Minnich's group had cited two publications in their
paper that backed up their take on things. The first was a review paper where Lenski himself described the experiment and then remarked
coyly, "That sounds a lot like the origin of species to me. What do you
think?"[38] (Wink, wink, nudge, nudge.) The second was Elizabeth Pennisi's puff piece[39] in *Science* on the LTEE in 2013 where she wrote (presumably after consulting with Lenski) "because one of *E. coli's* defining
characteristics is the inability to use citrate for energy in the presence of
oxygen, the citrate-consuming bacteria could be seen as a new species." If
Lenski plays fast and loose with the public's perceptions of his work, he
shouldn't complain when he's called on it.

In a disgraceful move, Lenski impugned Scott Minnich's character.
Since he's a "fellow of the Discovery Institute" sympathetic with intelligent design, the skeptical discussion in Minnich's paper (which underwent thorough peer review by an excellent journal that chose to high-

light it with commentary from eminent scientists) "suggests an ulterior nonscientific motive." (Apparently Lenski himself can speculate about all sorts of grand possibilities, ulterior-motive free.) You see, the Idaho scientists had the temerity to write, "A more accurate, albeit controversial, interpretation of the LTEE is that E. coli's capacity to evolve is more limited than currently assumed."

Well, perhaps someone personally involved in the work might see unending possibilities. But what should an objective observer call a situation where the exact same mutations occur time and time again?—Limitless? Where a problem has no other solution except the one found?—Flexible? Where deletion of either of the genes (citT or dctA) involved in the mutation prevents citrate utilization, as Minnich's group showed?—Resourceful? Where none of the other thousands of genes in the cell can substitute?—Inventive? Where even the easily obtainable mutation has apparently been of little use in nature?—Earth-shaking?

With regard to citrate evolution, the Minnich lab's results have revealed E. coli to be a one-trick pony. And, as I've written previously, in other respects Lenski's own work has shown that E. coli evolves in his lab overwhelmingly by loss-of-function and decrease-of-function mutations that are damaging.

If that isn't "more limited than currently assumed," it's close enough. The take-home lesson is that, although the unvarnished work itself is great, the hype surrounding the LTEE has seriously misled the public and the scientific community. It's far past time that a pin was stuck in its balloon.

59. Citrate Death Spiral

"Citrate Death Spiral," Evolution News and Science Today, Discovery Institute, June 17, 2020.

Michigan State University biologist Richard Lenski and collaborators have just published a terrific new paper in the journal eLife.[40] Anyone who wants to see a crystal-clear example of the inherent, unavoidable, fatal difficulties that the Darwinian mechanism itself poses for unguided evolution should read it closely.

The paper concerns the further evolution of a widely discussed mutant strain of the bacterium *E. coli* discovered during the course of Lenski's Long-Term Evolution Experiment (LTEE). The LTEE is his more-than-three-decades-long project in which *E. coli* was allowed to grow continuously in laboratory flasks simply to observe how it would evolve.[41] As I've written before, almost all of the beneficial mutations that were discovered to have spread through the populations of bacteria in the LTEE were ones that either blunted pre-existing genes (decreasing their previous biochemical activity) or outright broke them.[42]

There seemed, however, to be one interesting exception.[43] One morning after more than 30,000 generations of bacterial growth, one flask of *E. coli* (out of twelve separate flasks that Lenski maintained for comparison and replication's sake) seemed cloudier than the other eleven flasks. That indicated substantially more bacteria than usual had grown in the nutrient broth. After much hard laboratory work, Lenski's group showed that a region of the prodigious bacterium's DNA that was close to a gene coding for a citrate transporter (that is, a protein whose job is to bring external, dissolved citrate into the cell; citrate is a common chemical that cells metabolize) had duplicated.[44] The duplication mutation placed the control region of a different gene next to that of the citrate transporter.

Here's why that helped. The citrate-transporter gene's natural regulator causes the gene to be turned off whenever oxygen is around, as it was under the normal laboratory growth conditions at MSU. The second regulator, however, allows the gene it controls to be turned on in the presence of oxygen. The mutation that placed a copy of the regulator of the second gene next to the citrate gene then allowed the citrate gene to be turned on in the presence of oxygen, too. Since for technical purposes there was a lot of dissolved citrate in the nutrient broth, the mutant *E. coli* could import and metabolize ("eat") the citrate, which was unavailable to nonmutants. With all that extra food, the mutant grew like crazy, quickly surpassing nonmutants.

The novel result was widely reported, and the conjecture was floated that perhaps the mutant was on its way to forming a new species.[45] As I wrote in *Darwin Devolves*, however, other, much more ominous, genetic results should have tempered any optimism. For example, the citrate mutant had accumulated many of the same beneficial-but-degradative mutations that had previously spread through the population—the new mutation did not, could not, repair the previously damaged genes. And later work showed that several more broken genes had been selected in the mutant, apparently to help it metabolize citrate more efficiently.[46]

The new paper now reports on 2,500 generations of further evolution of the citrate mutant, in nutrient media that contains either citrate alone or citrate plus glucose (as for earlier generations). As always with the Lenski lab, the research is well and thoroughly done. But the resulting *E. coli* is one sick puppy. Inside the paper they report, "The spectrum of mutations identified in evolved clones was dominated by structural variation, including insertions, deletions, and mobile element transpositions." All of those are exceedingly likely to break or degrade genes. Dozens more genes were lost. The citrate mutant tossed genetic information with mindless abandon for short-term advantage.

In a particularly telling result, the authors "serendipitously discovered evidence of substantial cell death in cultures of a Cit^+ clone sampled from... the LTEE at 50,000 generations." In other words, those initial random "beneficial" citrate mutations that had been seized on by natural selection tens of thousands of generations earlier had led to a death spiral. The death rate of the ancestor of the LTEE was ~10%; after 33,000 generations it was ~30%; after 50,000, ~40%. For the newer set of experiments, the death rate varied for different strains of cells in different media, but exceeded 50% for some cell lines in a citrate-only environment. Indeed, the authors identified a number of mutations—again, almost certainly degradative ones—in genes for fatty acid metabolism that, they write with admirable detachment, "suggest adaptation to scavenging on dead and dying cells."

The degraded *E. coli* was eating its dead.

LESSONS TO DRAW

LET ME emphasize: the *only* result from the decades-long, 50,000-plus generation *E. coli* evolution experiment that even *seemed* at first blush like it had a bit of potential to yield a novel pathway in the bacterium has resulted instead in spectacular *devolution*. As Lenski and co-authors wrote in *Science* in 2019 in their dismissive review of *Darwin Devolves* (which focused strongly on the clear degradation occurring over the course of the LTEE):

> There are indeed many examples of loss-of-function mutations that are advantageous, but Behe is selective in his examples. He dedicates the better part of chapter 7 to discussing a 65,000-generation *Escherichia coli* experiment, emphasizing the many mutations that arose that degraded function—an expected mode of adaptation to a simple laboratory environment, by the way—while dismissing improved functions and deriding one new one as a "sideshow." (Full disclosure: The findings in question were published by coauthor Richard Lenski.) [47]

The "one new" function was the citrate mutation. I had called it a "side show" in *Darwin Devolves* precisely because the *E. coli* of the LTEE were accumulating degradative mutations much faster than any mutations that might with charity be called constructive:

> Interesting as it is, the ambiguous citrate mutation that started the hoopla is a side show. The overwhelmingly important and almost completely unnoticed lesson is that genes are being degraded left and right, both when they directly benefit the bacteria and when they do so indirectly in support of another mutation. The occasional, particularly noticeable modification-of-function or gain-of-FCT mutation can't turn back the tide of damaging and loss-of-FCT ones. [48]

The more molecular evolutionary work that rolls in, the more the above conclusion becomes a mere truism.

In their new paper the Lenski group rightfully points out the great strength of its experimental evolutionary system: "It is rarely feasible to examine evolution in action as organisms invade, colonize, and adapt to a new niche in nature, especially with independently evolving replicates and control populations. In this study, we investigated how *E. coli* variants with the new ability to grow aerobically on citrate adapted to a novel, citrate-only resource environment in the laboratory."[49]

Exactly. That means the LTEE gives us our clearest insight into the general effects of Darwin's mechanism, which continues to operate no matter what other processes may also be occurring (I'm looking at you, Extended Evolutionary Synthesis), no matter whether an organism is in a lab or in the wild, no matter which kind of organism—whether microbe, plant, or animal—is subject to its tender care. So, thanks to the Lenski group, we know that devolution is relentless—it never rests. In good times and bad, if a change in a species could help it adapt more closely to its environment, degradative mutations will arrive most quickly by far to offer their assistance. And, of course, under selective pressure a species has no choice but to accept helpful-but-degradative ones, even if that eventually leads to the species languishing. Thanks in very large part to the fine work done over decades at Michigan State we can now be certain that, like the citrate-eating *E. coli*, as an explanation for the great features of life Darwin's theory itself is in a death spiral.

PART FIVE: JOSEPH THORNTON MAKES EVEN EASY PROBLEMS HARD FOR DARWIN

R ICHARD LENSKI HAS BEEN PROBING THE LIMITS TO DARWINIAN evolution in the best way—by simply growing bacteria and watching what happens. Yet there are other valuable approaches, too. In my view the most rigorous and cleverest ones have been directed by Joseph Thornton, now a professor at the University of Chicago. Thornton has studied the abilities of similar modern proteins from different branches of life and compared them to their presumed ancient common ancestor protein—which he constructed (he calls it "resurrected") in his own lab after comparing the modern ones. He got surprising results.

60. ON IRREDUCIBLE COMPLEXITY

I'm afraid the following essay has an exasperated tone. At the time Thornton's newly published work was getting the full finally-someone-put-those-stupid-IDers-in-their-place treatment, with a special commentary in *Science* about how the amazing results "solidly refute all parts of the intelligent design arguments" and a write-up in the *New York Times*. A *Times* reporter even called me to gloat. To me the results seemed as though someone who showed a splinter could be removed from the wooden base of a mousetrap and the mousetrap still work was claiming the problem of irreducible complexity had been solved. In other words, the puffing up of a trivial result.

"On Molecular Exploitation and the Theory of Irreducible Complexity," Discovery Institute, January 1, 2006.

THE STUDY BY BRIDGHAM ET AL.[1] PUBLISHED IN THE APRIL 7 ISSUE of *Science* is the lamest attempt yet—and perhaps the lamest attempt that's even possible—to deflect the problem that irreducible complexity poses for Darwinism.

The bottom line of the study is this: the authors started with a protein which already had the ability to strongly interact with three kinds of steroid hormones (aldosterone, cortisol, and "DOC" [11-deoxycorticosterone]). After introducing several simple mutations, the protein interacted much more weakly with all of those steroids. In other words, a pre-existing ability was decreased.

That's it! The fact that this extremely modest and substantially irrelevant study is ballyhooed with press releases, a commentary in *Science* by Christoph Adami,[2] and forthcoming stories in the mainstream media, demonstrates the great anxiety some folks feel about intelligent design.

In the study the authors wished to see if two related modern proteins called the glucocorticoid (GR) receptor and mineralocorticoid receptor (MR) could be derived from a common ancestral protein. Using clever analysis, the authors made a protein that they thought represented the ancestral protein. That protein binds several structurally similar hormones, as does modern MR. They then introduced two amino acid changes into the protein which are found in modern GR. The two changes caused the ancestral protein to bind the different kinds of hormones anywhere from ten- to a thousandfold more weakly. That protein bound aldosterone about threefold more weakly than cortisol. The authors note that modern GR (in tetrapods) also binds aldosterone more weakly than cortisol. So perhaps, the thinking goes, an ancestral gene that could bind both hormones duplicated in the past, one copy accumulated those two mutations to become the modern GR, and the other copy became modern MR.

Here are a number of comments in response:

1) This continues the venerable Darwinian tradition of making grandiose claims based on piddling results. There is nothing in the paper

that an ID proponent would think was beyond random mutation and natural selection. In other words, it is a straw man.

2) The authors (including Christoph Adami in his commentary) are conveniently defining "irreducible complexity" way, way down. I certainly would not classify their system as anywhere near IC. The IC systems I discussed in *Darwin's Black Box* contain multiple, active protein factors. Their "system," on the other hand, consists of just a single protein and its ligand. Although in nature the receptor and ligand are part of a larger system that does have a biological function, the piece of that larger system they pick out does not do anything by itself. In other words, the isolated components they work on are not irreducibly complex.

3) In the experiment just two amino acid residues were changed! No new components were added, no old components were taken away.

4) Nothing new was produced in the experiment; rather, the pre-existing ability of the protein to bind several molecules was simply weakened. The workers begin their experiments with a protein that can strongly bind several structurally very similar steroids, and they end with a protein that at best binds some of the steroids tenfold more weakly. (Figure 4C.)

5) Such results are not different from the development of antibiotic resistance, where single amino acid changes can cause the binding of a toxin to a particular protein to decrease (for example, warfarin resistance in rats, and resistance to various AIDS drugs). Intelligent design proponents happily agree that such tiny changes can be accomplished by random mutation and natural selection.

6) In the "least promising" intermediate (designated L111Q) the protein has essentially lost its ability to bind any steroid. In the "most promising" intermediate protein (the one that has just the S106P alteration) the protein has lost about 99% of its ability to bind DOC and cortisol, and lost about 99.9% of its ability to bind aldosterone. (Figure 4C.)

7) Although the authors imply (and Adami claims directly) that the mutated protein is specific for cortisol, in fact it also binds aldosterone with about half of the affinity. (Compare the red and green curves in the

lower right-hand graph of Figure 4C.) What's more, there actually is a much larger difference (about thirtyfold) in binding affinity for aldosterone and cortisol with the initial, ancestral protein than for the final, mutated protein (about twofold). So the protein's ability to discriminate between the two ligands has decreased by tenfold.

8) One would think that the hundredfold decrease in the ability to bind a steroid would at least initially be a very detrimental change that would be weeded out by natural selection. The authors do not test for that; they simply assume it wouldn't be a problem, or that the problem could somehow be easily overcome. Nor do they test their speculation that DOC could somehow act as an intermediate ligand. In other words, in typical Darwinian fashion the authors pass over with their imaginations what in reality would very likely be serious biological difficulties.

9) The fact that such very modest results are ballyhooed owes more, I strongly suspect, to the antipathy that many scientists feel toward ID than to the intrinsic value of the experiment itself.

10) In conclusion, the results (and even the imagined-but-problematic scenario) are well within what an ID proponent already would think Darwinian processes could do, so they won't affect our evaluation of the science. But it's nice to know that *Science* magazine is thinking about us!

61. Nature Publishes Paper on the Edge of Evolution

"*Nature* Publishes Paper on the Edge of Evolution," Evolution News and Views, Discovery Institute, September 30, 2009.

NATURE HAS PUBLISHED AN INTERESTING PAPER RECENTLY WHICH places severe limits on Darwinian evolution. This is the first of three essays discussing it.

The manuscript, from the laboratory of Joseph Thornton at the University of Oregon, is titled "An Epistatic Ratchet Constrains the Direction of Glucocorticoid Receptor Evolution."[3] The work is interpreted by its authors within a standard Darwinian framework. Nonetheless, like the important work over the years of Michigan State's Richard Lenski on laboratory evolution of E. *coli*, which has shown trillions of bacteria evolving under selection for tens of thousands of generations yielding just broken genes and minor changes, the new work demonstrates the looming brick wall which confronts unguided evolution in at least one system. And it points strongly to the conclusion that such walls are common throughout all of biology.

In the paper Bridgham et al. continue their earlier work on steroid hormone receptor evolution. Previously they had constructed in the laboratory a protein which they inferred to be the ancestral sequence of two modern hormone receptors abbreviated GR and MR.[4] They then showed that if they changed two amino acid residues in the inferred ancestral receptor protein into ones which occur in GR, they could change its binding specificity somewhat in the direction of modern GR's specificity. (All the work was done on molecules in the laboratory. No measurements were made of the selective value of the changes in real organisms in nature. Thus, any relevance to actual biology is speculative.) They surmised that a gene duplication plus sequence diversification could have given rise to MR and GR. As I wrote in a comment at the time, that was interesting work, and the conclusion was reasonable, but the result was exceedingly modest and well within the boundaries that an intelligent design proponent like myself would ascribe to Darwinian processes. After all, the starting point was a protein which binds several steroid hormones, and the ending point was a slightly different protein that binds the same steroid hormones with slightly different strengths. How hard could that be?

Well, it turns out that Darwinian evolution can have a lot of trouble accomplishing even that simple task, or at least its opposite. In the new paper the authors try the reverse experiment. They begin with the more modern hormone receptor (which is more restrictive in the steroids it binds) and ask whether a Darwinian process could get the ancestral activity back (which is more permissive). Their answer is no, it couldn't. They show that a handful of amino acid residues in the more recent re-

ceptor would first have to be changed before it could act as the ancestral form is supposed to have done, and that is very unlikely to occur.
In other words, the new starting point is also a protein which binds a
steroid hormone, and the new desired ending point is also a slightly different protein that binds steroid hormones. How hard could that be?
But it turns out that Darwinian processes can't reach it, because several
amino acids would have to be altered before the target activity kicked in.

A number of points can be drawn from this fine work:

+ The central point of *The Edge of Evolution*[5] was that if several
  amino acids of a protein must be changed before a certain
  selective effect is available, then that is effectively beyond the
  reach of Darwinian processes. Bridgham et al.[6] confirm that
  conclusion. (As an aside, it would make a good project for a
  sociologist of science to ask why the same conclusion is met with
  howls of protest when presented by a Darwinian skeptic such as
  myself, but garners praise when presented by someone else.)

+ There is no reason to think the protein studied by Bridgham
  et al.[7] is unusual in its difficulty of developing a binding site for
  even a relatively closely related substance. In fact, in the absence
  of strong opposing data, that should be the default, reasonable
  assumption.

+ That same reasonable assumption counts strongly against any
  two unrelated proteins easily developing a binding site for each
  other.

+ That reasonable assumption therefore negates all woolly
  Darwinian evolutionary scenarios where critical protein binding
  sites are assumed without justification to pop up when needed
  (such as, say, in the building of multiprotein structures like the
  cilium or flagellum).

+ Thus, the work strongly supports *The Edge of Evolution*'s
  conclusion that Darwinian processes are highly unlikely to have
  built the complex molecular machinery of the cell.

62. NATURE PAPER REACHES "EDGE OF EVOLUTION"

"*Nature* Paper Reaches 'Edge of Evolution' and Finds Darwinian Processes Lacking," Evolution News and Views, Discovery Institute, October 6, 2009.

NATURE HAS RECENTLY PUBLISHED AN INTERESTING PAPER WHICH places severe limits on Darwinian evolution. The manuscript, from the laboratory of Joseph Thornton at the University of Oregon, is titled, "An Epistatic Ratchet Constrains the Direction of Glucocorticoid Receptor Evolution."[8] The work is interpreted by its authors within a standard Darwinian framework, but the results line up very well with arguments I made in *The Edge of Evolution*. This is the second of three essays discussing it.

Using clever synthetic and analytical techniques, Bridgham et al. show that the more recent hormone receptor protein that they synthesized, a GR-like protein, can't easily revert to the ancestral structure and activity of an MR-like protein because its structure has been adjusted by selection to its present evolutionary task, and multiple amino acid changes would be needed to switch it back. That is a very general, extremely important point that deserves much more emphasis. In all cases—not just this one—natural selection is expected to hone a protein to suit its current activity, not to suit some future, alternate function. And that is a very strong reason why we should not expect a protein performing one function in a cell to easily be able to evolve another, different function by Darwinian means. In fact, the great work of Bridgham et al. shows that it may not be do-able for Darwinian processes even to produce a protein performing a function very similar to that of a homologous protein.

Before reading their paper, even I would have happily conceded for the sake of argument that random mutation plus selection could convert an MR-like protein to a GR-like protein and back again, as many times as necessary. Now, thanks to the work of Bridgham et al., even such apparently minor switches in structure and function are shown to be quite problematic. It seems Darwinian processes can't manage to do even as much as I had thought.

(As an aside into the circus world of popular-level debates on Darwinism, the work of Bridgham et al. nicely shows the fallacy of the anti-ID retort that, say, a mousetrap is not irreducibly complex because, if the catch and holding bar are removed, parts of it can still be used as a tie clip. The same principle that holds for cellular machinery would hold for all machinery. Something that was shaped by selection to work as a tie clip would not look like an ancestor of a mousetrap, or easily be converted to one by random changes plus selection. If you look at images of tie clips on the internet, none of them resemble mousetraps—except for those purposely designed by folks who were arguing against irreducible complexity.)

Another point worth driving home in this post concerns the frequently encountered argument that, well, just because one kind of protein can't develop a useful binding site or selectable property easily doesn't mean that some other kind of cellular protein can't. (In keeping with their Darwinian framework, Bridgham et al. seem to allude to this.) After all, there are thousands to tens of thousands of kinds of proteins in a typical cell. If one of them is ruled out, the reasoning goes, many more possibilities remain.

This argument, however, is specious. For any given evolutionary task, the number of proteins in the cell that are candidates for helpful mutations is almost always very limited. For example, as I discussed in *The Edge of Evolution*, out of thousands of malaria proteins, mutations in only a handful are helpful to the parasite in its fight against chloroquine, and only one is really effective—the mutations in the PfCRT protein. Ditto for the human proteins that can mutate to help resist malaria—there's just a handful. In the case of the hormone receptors discussed by Bridgham et al.,[9] one can note that, out of ten thousand vertebrate proteins, the one that gave rise to a new steroid hormone receptor was an already-existing steroid hormone receptor. This should be quite surprising to folks who believe the many-proteins argument, because the steroid receptor was outnumbered 10,000 to 1 by other protein genes, yet it won the race to duplicate and form a new functional receptor. If all things were equal, we should be very surprised by that. But of course, not all things are equal. The reason the receptor duplicated to give rise to a closely related receptor is that no other protein in the cell is likely to be able to do so in a reasonable amount of evolutionary time.

The bottom line is that, for a given evolutionary task, at best only a handful of proteins will likely be helpful to evolve, and at worst none may help. To calculate the probability of, say, a helpful protein-protein interaction developing in response to any particular selective pressure, it's mistaken to gratuitously multiply odds by the total number of proteins in a cell. Combined with the point made by Bridgham et al.[10] that even tiny structural/functional changes may not be achievable by random mutation/selection, these considerations pretty much squelch the likelihood of Darwinian processes doing much of significance during evolution.

63. DOLLO'S LAW AND THE SYMMETRY OF TIME

"Dollo's Law, the Symmetry of Time, and the Edge of Evolution,"
Evolution News and Views, Discovery Institute, October 12, 2009.

NATURE HAS RECENTLY PUBLISHED AN INTERESTING PAPER WHICH places severe limits on Darwinian evolution. The manuscript, from the laboratory of Joseph Thornton at the University of Oregon, is titled "An Epistatic Ratchet Constrains the Direction of Glucocorticoid Receptor Evolution."[11] The work is interpreted by its authors within a standard Darwinian framework, but the results line up very well with arguments I made in *The Edge of Evolution*. This is the last of three essays discussing it.

Bridgham et al. are interested in the reversibility of evolution, and discuss their results in terms of something called "Dollo's law." Louis Dollo, an early twentieth-century paleobiologist, was interested in discerning phylogenies. He maintained that one could always distinguish ancestral forms from descendant forms. Stephen Jay Gould[12] commented that Dollo's "law" was not an empirical observation, but rather a postulate which he felt was necessary to properly construct phylogenies. Over the years the meaning of "Dollo's law" transmogrified. In modern usage, the phrase has come to mean that complex traits, once lost, do not re-evolve in the same lineage. For example, whales do not re-evolve gills, even though they are aquatic and ultimately—via a series of land-living intermediates according to current phylogenetic theory—descended from fish, because gills are a lost, complex trait in that lineage.

Dollo's law is taken with a grain of salt by many biologists, and apparent exceptions to the law have been noted (cited in Bridgham et al.). Nonetheless, although Dollo's law isn't very reliable at the organismal level, maybe it can do better at the molecular level. Bridgham et al. wanted to test that idea:

> Evolutionary reversibility represents a strong test of the importance of contingency and determinism in evolution. If selection is limited in its ability to drive the reacquisition of ancestral forms, then the future outcomes available to evolution at any point in time must depend strongly on the present state and, in turn, on the past. Ready reversibility, in contrast, would indicate that natural selection can produce the same optimal form in any given environment, irrespective of history. The evolutionary reversibility of a protein can be evaluated at three levels: molecular sequence, protein function, and the structural/mechanistic underpinnings for that function. The latter is most relevant to understanding the roles of contingency and determinism in evolution.... True reversal, involving restoration of the ancestral phenotype by the ancestral structure-function relations, would indicate that the forms of functional proteins can evolve deterministically, irrespective of contingent historical events.[13]

After experimentally supporting the claim that a GR-like protein would be very unlikely to revert to an MR-like ancestral form by Darwinian means, they concluded: "We predict that future investigations, like ours, will support a molecular version of Dollo's law: as evolution proceeds, shifts in protein structure-function relations become increasingly difficult to reverse whenever those shifts have complex architectures, such as requiring conformational changes or epistatically interacting substitutions."

I think the experimental work of Bridgham et al. is great, and I think their interpretive reasoning is fine as far as it goes. But it is severely pinched by their Darwinian framework; their results point to much more.

Just as for some general laws of physics, there is nothing inherently time-asymmetric about generic random mutation and selection. So there should be nothing particularly special about evolving back in history versus forward. The only thing that would be "special" about going back is that you can (potentially) know which way you had come, so you can see if the steps can be retraced, as Bridgham et al. did. However, the huge roadblock that the authors discovered for one homologous protein converting to another by Darwinian processes did not have to be in the past—the roadblock could as easily have been in the future. If the GR-like protein had come first in history, then no MR-like protein would likely have arisen by Darwinian means. In that case, however, there would have been no question even raised by investigators about the reversibility of that evolutionary path, because the "path" would not exist—it would have been blocked at the start, in the forward direction. Questions do not arise about hypothetical pathways that would have to pass through brick walls.

The old, organismal, time-asymmetric Dollo's law supposedly blocked off just the past to Darwinian processes, for arbitrary reasons. A Dollo's law in the molecular sense of Bridgham et al., however, is time-symmetric. A time-symmetric law will substantially block both the past and the future, for well-understood reasons: Natural selection fits a protein to a current, not any future (nor any previous), task; thus, it tends strongly to restrict other potential structures/functions. The very same considerations ("shifts in protein structure-function relations," "epistatically interacting substitutions," and so on) that frustrate the reacquisition of complex molecular features will tend strongly to stymie their acquisition in the first place, because no potential protein component would ever be without a prior history of selection. A time-symmetric Dollo's law turns the notion of "pre-adaptation" on its head. The law instead predicts something like "pre-sequestration," where proteins that are currently being used for one complex purpose are very unlikely to be available for either reversion to past functions or future alternative uses.

Yet here we are, with complex life all around us and in us. If a time-symmetric Dollo's law were really such a big roadblock, how did life come to be? Here is where their Darwinian framework most seriously blinkers their vision. Bridgham et al. aimed to test "the importance of contingency and determinism in evolution." But chance and necessity are not the only things that exist. There are also mind and plan. In fact, in their own work the authors themselves reconstructed the ancestral protein from the descendant protein, easily overcoming the hurdle that they realized would block a Darwinian process. Their own minds directed events that chance and necessity never could.

64. RESPONSE TO CARL ZIMMER AND JOSEPH THORNTON

At the time, my last three essays attracted the attention of the *New York Times* science writer Carl Zimmer, who quickly got in touch with Joseph Thornton to raise the alarm. In high dudgeon, Thornton wrote a long reply, which Zimmer posted on his blog. That was my dream come true—there's nothing that shows up the limpness of Darwinism better than the spectacle of really smart people defending it so weakly.

"Piddling Pebbles and Empty Promises: Response to Carl Zimmer and Joseph Thornton," Evolution News and Views, Discovery Institute, October 26, 2009.)

THE SCIENCE WRITER CARL ZIMMER POSTED AN INVITED REPLY[14] on his blog from Joseph Thornton of the University of Oregon to my recent comments about Thornton's work. This is the first of several essays addressing it.

I must say, it never ceases to amaze me how otherwise-very-smart folks like Zimmer and Thornton fail to grasp pretty simple points when it comes to problems for Darwinian mechanisms. Let me start slowly with a petty complaint in Carl Zimmer's intro to the post. Zimmer is annoyed that I think Thornton's latest work is "great," yet I thought his previous work published a few years ago was "piddling." "Why the change of heart?" wonders Zimmer.

It's really not that hard to understand. Here's a little analogy to illustrate. Suppose some company claimed they could build a super-crane (tip of the hat to Daniel Dennett) which could hoist a whole mountain using a novel technology. Though this super-crane had not been tested, the great majority of the relevant engineering community was serenely confident it would work as advertised. In a carefully devised, initial "proof-of-principle" experiment, a laboratory at the University of Oregon demonstrated that the crane technology could lift a smooth pebble. The work was published in *Science*, accompanied by a breathless editorial and a story in the *New York Times*. In a subsequent careful study published several years later in *Nature*, however, the same lab unexpectedly showed that if a pebble were even somewhat rough, the crane technology would not lift it. Since mountains tend to be rough, too, if a super-crane wouldn't move a rough pebble, then it certainly wouldn't lift a mountain.

Of course, the initial work, although technically well-done, can fairly be called "piddling" compared to the promised capacity of the crane. The subsequent work, again technically well-done, was "great" because it revealed formidable difficulties for the technology at a very basic level that no one—not even (ahem...) the few skeptics—had expected. (I hate to be so pedantic, but unfortunately it seems necessary on this topic.)

65. NOT SO MANY PATHWAYS

"Not So Many Pathways: Response to Carl Zimmer and Joseph Thornton,"
Evolution News and Views, Discovery Institute, October 27, 2009.

THE SCIENCE WRITER CARL ZIMMER POSTED AN INVITED REPLY[15] on his blog from Joseph Thornton of the University of Oregon to my recent comments about Thornton's work. This is the second of several essays addressing it.

Now to Professor Thornton's reply. He writes at length but makes just two substantive points: 1) neutral mutations occur and can serendipitously help a protein evolve some function ("[Behe] ignores the key role of genetic drift in evolution"); and 2) just because a protein may not be able to evolve a particular function one way does not mean that it, or some other kind of protein, can't evolve the function another way ("nothing in our results implies that, if selection were to favor the ancestral function again, the protein could not adapt by evolving a different, convergent, underlying basis for the function").

I'll start with the second point since I can just quote myself to answer it. I wrote in one of my previous posts on Thornton's work:

Another point worth driving home in this post concerns the frequently encountered argument that, well, just because one kind of protein can't develop a useful binding site or selectable property easily doesn't mean that some other kind of cellular protein can't. (In keeping with their Darwinian framework, Bridgham et al.[16] seem to allude to this.) After all, there are thousands to tens of thousands of kinds of proteins in a typical cell. If one of them is ruled out, the reasoning goes, many more possibilities remain.

This argument, however, is specious. For any given evolutionary task, the number of proteins in the cell which are candidates for helpful mutations is almost always very limited. For example, as I discussed in *The Edge of Evolution*, out of thousands of malaria proteins, mutations in only a handful are helpful to the parasite in its fight against chloroquine, and only one is really effective—the mutations in the PfCRT protein. Ditto for the human proteins that can mutate to help resist malaria—there's just a handful. In the case of the hormone receptors discussed by Bridgham et al.,[17] one can note that, out of ten thousand vertebrate proteins, the one that gave rise to a new steroid hormone receptor was an already-existing steroid hormone receptor. This should be quite surprising to folks who believe the many-proteins argument, because the steroid receptor was outnumbered 10,000 to 1 by other protein genes, yet it won the race to duplicate and form a new functional receptor. If all things were equal, we should be very surprised by that. But of course, not all things are equal. The reason the receptor duplicated to give rise to a closely-related receptor is that no other protein in the cell is likely to be able to do so in a reasonable amount of evolutionary time.

(Professor Thornton's post gives no indication that he read this; he certainly gave no response to it.) If the most likely candidate protein has difficulty evolving to yield a given function by the most likely candidate route, there may (or may not) be another, less likely route, or a handful of other less likely candidate proteins that could do so, but there is certainly not a huge reservoir of possibilities, as Thornton seems to think. And, for those who believe this all depends on blind luck, most of the time things should not turn out well at all.

Now let's contrast Thornton's blog reply to me with what he wrote in his paper. In the blog he writes that "nothing in our results implies that, if selection were to favor the ancestral function again, the protein could not adapt by evolving a different, convergent, underlying basis for the function." Yet in his paper he wrote:

> There may be other potentially permissive mutations, of unknown number, that could compensate for the restrictive effect of group W and allow the ancestral conformation to be restored. Reversal by such indirect pathways could be driven by selection, however, only if these other mutations, unlike those we studied, could somehow relieve the steric clashes and restore the lost stabilizing interactions... and also independently restore the ancestral function when helix 7 is in its radically different derived conformation. Whether or not mutations that could achieve these dual ends exist, reversal to the ancestral conformation would require a considerably more complex pathway than was necessary before the ratchet effect of W evolved.

Professor Thornton is playing games. The strongly emphasized point of his paper was to show exactly what I discussed in my posts: the extreme improbability (not "impossibility," which is for suckers—one can't prove a negative in science) of re-acquiring the ancestral structure/function, either by direct or indirect reversal.

66. Severe Limits to Darwinian Evolution

"Severe Limits to Darwinian Evolution: Response to Carl Zimmer and Joseph Thornton," Evolution News and Views, Discovery Institute, October 28, 2009.

THE SCIENCE WRITER CARL ᵃZIMMER POSTED AN INVITED REPLY[18] on his blog from Joseph Thornton of the University of Oregon to my recent comments about Thornton's work. This is the third of several essays addressing it.

Now back to Thornton's first point, the role of neutral mutations (which he sometimes labels "permissive" mutations). At several places in his post Thornton implies I'm unaware of the possibilities opened up by genetic drift:

> Behe's discussion of our 2009 paper in *Nature* is a gross misreading because it ignores the importance of neutral pathways in protein evolution.... Behe's first error is to ignore the fact that adaptive combinations of mutations can and do evolve by pathways involving neutral intermediates.... As Fig. 4 in our paper shows, there are several pathways back to the ancestral sequence that pass only through steps that are neutral or beneficial with respect to the protein's functions.

My interest in evolution by neutral mutation, however, is a matter of public record. It is an old idea that if a gene for a protein duplicates,[19] then multiple mutations can accumulate in a neutral fashion in the "spare" gene copy, even if those mutations would be severely deleterious if they occurred in a single-copy gene. Four years ago David Snoke and I wrote a paper entitled "Simulating Evolution by Gene Duplication of Protein Features That Require Multiple Amino Acid Residues,"[20] where we investigated aspects of that scenario. The primary takeaway is that, although by assumption of the model anything is possible, when evolution must pass through multiple neutral steps the wind goes out of Darwinian sails, and a drifting voyage can take a very, very long time indeed. But don't just take my word for it—listen to Professor Thornton:

> To restore the ancestral conformation by reversing group X, the restrictive effect of the substitutions in group W must first be reversed, as must group Y. Reversal to w and y in the absence of x, however, *does nothing to enhance the ancestral function*; in most contexts, reversing these mutations substantially impairs both the ancestral and derived functions. Furthermore, the permissive effect of reversing four of the mutations in group W *requires pairs of substitutions at interacting sites*. Selection for the ancestral function would therefore not be sufficient to drive AncGR2 back to the ancestral states of w and x, because passage through deleterious and/or neutral intermediates would be required; the probability of each required substitution would be low, and the probability of all in combination would be virtually zero. [my emphasis][21]

Let's quote that last sentence again, with emphasis: "Selection for the ancestral function would therefore not be sufficient... *because passage through deleterious and/or neutral intermediates would be required; the probability of each required substitution would be low, and the probability of all in combination would be virtually zero.*" If Thornton himself discounts the power of genetic drift when it suits him, why shouldn't I?

In his blog response to me Professor Thornton wants to emphasize that selection-plus-drift can sometimes lead from one nearby function to another, as it did in his work on the ancestral MR-like to the GR-like receptor transition.[22] But I and virtually everyone else already thought that was true. That's why at the time I called those results (perhaps impolitely, but accurately) "piddling." A surprise it was not. In his 2009 paper investigating the reverse transition, however, Thornton wants to emphasize (because it is unexpected) that in some cases selection-plus-drift cannot lead (with anything like reasonable probability) even to a very similar function. Now that was surprising to me and apparently to many other folks.

The immediate, obvious implication (which he clearly wants to keep far away from) is that the 2009 results render problematic even pretty small changes in structure/function for all proteins—not just the ones he worked on. (Thornton himself is betting on this: "We predict that future investigations, like ours, will support a molecular version of Dollo's law: as evolution proceeds, shifts in protein structure-function relations become increasingly difficult to reverse whenever those shifts have complex architectures.")[23] So how, other than begging the question, are we now to know that even the small differences we see in related protein systems came about by random mutation/selection (and, yes, drift)? Quite simply, we can't. Yet if even small changes are problematic, then larger changes will be prohibitive, and very big changes essentially unattainable. Thanks to Thornton's impressive work, we can now see that the limits to Darwinian evolution are more severe than even I had supposed.

67. PROBABILITY AND CONTROVERSY

"Probability and Controversy: Response to Carl Zimmer and Joseph Thornton," Evolution News and Views, Discovery Institute, October 29, 2009.

THE SCIENCE WRITER CARL ZIMMER POSTED AN INVITED REPLY[24] on his blog from Joseph Thornton of the University of Oregon to my recent comments about Thornton's work. This is the last of four essays addressing it.

At the end of his post Thornton waxes wroth: "Behe's argument has no scientific merit. It is based on a misunderstanding of the fundamental processes of molecular evolution and a failure to appreciate the nature of probability itself. There is no scientific controversy about whether natural processes can drive the evolution of complex proteins. The work of my research group should not be misinterpreted by those who would like to pretend that there is."

Well, now. I'll leave it to readers of my previous replies to Thornton to decide whether they think those replies have scientific merit, and whether it is I or he who misunderstands the disputed facets of molecular evolution. As for "the nature of probability itself" and "no scientific controversy," I will briefly address those here.

To illustrate his own grasp of probability Professor Thornton talks baseball. "[Behe] supposes that if each of a set of specific evolutionary outcomes has a low probability, then none will evolve," Thornton writes. "This is like saying that, because the probability was vanishingly small that the 1996 Yankees would finish 92-70 with 871 runs scored and 787 allowed and then win the World Series in six games over Atlanta, the fact that all this occurred means it must have been willed by God."

Let me first say that, as a devout fan of the Philadelphia Phillies, I would never think that a Yankees title was intended by the deity. (Bought by George Steinbrenner, perhaps, but certainly not "willed by God.") That aside, I don't think Thornton's analogy captures the evolutionary problem. The example he chose posits a fully functioning team for a very specific game, baseball, performing within the parameters it was designed to—hitting the ball, playing defense, winning and losing games. Even the 1962 Mets did all those things (in somewhat different proportions). Yet the problem for the steroid receptor proteins Thornton's lab designed was to work at all. To do so they needed to have the correct tools (the right amino acid residues) oriented in the right directions. So let's change his example a bit. Instead of asking if the Yankees would have won the title with a different number of runs, let's ask if they would have won if their batters lay down on the ground instead of standing when at bat. And let's ask if they would have won if they swung towels instead of wooden bats. And if their pitchers threw the ball in random directions. And if their fielders all huddled together in left field, or ran away from a hit ball instead of towards it. I'll bet even Professor Thornton would be surprised if they won under those circumstances.

Which of those strange behaviors would the imaginary Yankees have to change to win a Series title?—All of them. And how long would it likely take if each season they randomly changed one behavior a bit (say, fielders ran in a direction 173 degrees from a hit ball instead of 180 degrees straight away from it)?—Very, very long. So it would seem that it is Thornton who doesn't understand probability applied to evolutionary possibilities. His set of conceivable examples is severely restricted to ones that simply have to work, or that lead inexorably in the direction he wants them to go, without comprehending that there is no evolutionary law that says *anything* has to work, or that the best current innovation has to lead along a path to something even better. The remarkable thing is that his own admirable laboratory research illustrates this, but he is too enthralled by Darwinian theory to see it.

As for "no scientific controversy," even a brief excursion into the history of science shows many uncontroversial, widely accepted theories that were in fact wrong. There was no scientific controversy in the nineteenth century about the existence of the ether, or the adequacy of Newton's laws. And, if one relies on science journals for her entire perspective, there is no controversy today about whether undirected natural processes can account for the origin of life. Yet neither can any scientist today detail a plausible theory of the origin of life. So the bare question of whether some idea is or is not controversial within the scientific community is itself simply a sociological question, not a scientific one. And when the idea is defended so weakly by someone as intelligent as Professor Thornton, it would seem that sociology is pretty much all the idea has going for it.

68. WHEEL OF FORTUNE

"Wheel of Fortune: New Work by Thornton's Group
Supports Time-Symmetric Dollo's Law," *Evolution News
and Views*, Discovery Institute, October 5, 2011.

IN THE JUNE 2011 ISSUE OF *PLOS* GENETICS THE LABORATORY OF University of Oregon evolutionary biologist Joseph Thornton published "Mechanisms for the Evolution of a Derived Function in the Ancestral Glucocorticoid Receptor," the latest in their series of papers concerning the evolution of proteins that bind steroid hormones.[25] In earlier laboratory work they had concluded that a particular protein, which they argued had descended from an ancestral, duplicated gene, would very likely be unable to evolve back to the original ancestral protein, even if selection favored it.[26] The reason is that the descendant protein had acquired a number of mutations which would have to be reversed, mutations which, the authors deduced, would confer no benefit on the intermediate protein. They used these results to argue for a molecular version of "Dollo's Law," which says roughly that a given forward evolutionary pathway is very unlikely to be exactly reversed.

In my previous comments on this interesting work, I noted that there is nothing time-asymmetric about random mutation/natural selection, so that the problem they saw in reversing the steroid hormone receptor evolution did not have to be in the past—it could just as easily have been in the future. The reason is that natural selection hones a protein to its present job, with regard to neither future use nor past function. Thus, based on Thornton's work, one would not in general expect a protein that had been selected for one function to be easily modified by RM/NS to another function. I have decided to call this the Time-Symmetric Dollo's Law, or "TSDL."

But if there is such a thing as a TSDL, did the forward evolution of the steroid-hormone protein-receptor manage to avoid it? That question had not yet been addressed. Was the protein lucky this time, encountering no obstacles to its evolution from the ancestral state to the modern state? If so, then maybe TSDL is occasionally an obstacle, but not so often as to rule out modest Darwinian evolution of proteins (a type of modest evolutionary innovation I had thought possible before reading Thornton's earlier work).

Well, thanks to the Thornton group's new work, we can now see that there are indeed obstacles to the forward evolution of the ancestral protein. The group was interested in which of the many sequence changes between the ancestral and derived-modern protein were important to its change in activity, which consisted mostly of a considerable weakening of the protein's ability to bind its steroid ligands. They narrowed the candidates down to two amino acid positions, residues 43 and 116. Each of the changes at those sites decreased binding by over a hundredfold. However, when the researchers combined both mutations into a single protein, as occurs in the modern protein, binding was not only decreased—it was for all intents and purposes abolished. Upon further research the group showed that a third mutation, at position 71, was necessary to ameliorate the effects of the combination of the other two mutations, bringing them back to hundreds-fold loss of function rather than essentially complete loss of function.

Carroll et al.[27] conjecture that the mutation at position 71 occurred before the other two mutations, but that it had no effect on the activity of the ancestral protein. So then let us count the ways "fortune" favored the evolution of the modern protein. First, an ancestral gene duplicated, which would usually be considered a neutral event. Thus, it would not have the assistance of natural selection to help it spread in the population. Next, it avoided hundreds of possible mutations which would have rendered the duplicated gene inactive. Third, it acquired a neutral mutation at position 71. Thus, again, this mutation would have to spread by drift, without the aid of natural selection. Once more, the still-neutral gene manages to avoid all of the possible mutations that would have inactivated it. Next, it acquires the correct mutation (either at position 43 or 116) which finally differentiates it from its parent gene—by reducing its activity a hundredfold! Finally, somehow the wimpy, mutated gene (putatively) confers upon the lucky organism some likely-quite-weak selective advantage.

The need to pass through multiple neutral steps while avoiding multiple likely-deleterious steps to produce a protein that has lost 99% of its activity is not a ringing example of the power of Darwinian processes. Rather, as mentioned above, it shows the strength of TSDL. Darwinian selection will fit a protein to its current task as tightly as it can. In the process, it makes it extremely difficult to adapt to a new task or revert to an old task by random mutation plus selection.

Dollo's law holds going forward as well as backward. We can state the experimentally based law simply: "Any evolutionary pathway from one functional state to another is unlikely to be traversed by random mutation and natural selection. The more the functional states differ, the much less likely that a traversable pathway exists."

69. A Blind Man Carrying a Legless Man

"A Blind Man Carrying a Legless Man Can Safely Cross the Street:
Experimentally Confirming the Limits to Darwinian Evolution,"
Evolution News and Views, Discovery Institute, January 11, 2012.

I NEVER THOUGHT IT WOULD HAPPEN BUT, IN MY ESTIMATION, RICHard Lenski has acquired a challenger for the title of "Best Experimental Evolutionary Scientist." Lenski, of course, is the well-known fellow who has been growing E. coli in his lab at Michigan State for 50,000 generations in order to follow its evolutionary progress. His rival is Joseph Thornton of the University of Oregon who, by inferring the sequences of ancient proteins and then constructing (he calls it "resurrecting") their genes in his lab, is able to characterize the properties of the ancestral proteins and discern how they may have evolved into more modern versions with different properties.

I have written appreciatively about both Lenski and Thornton before, whose work indicates clear limits to Darwinian evolution (although they themselves operate within a Darwinian framework). Thornton's latest work is beginning to show a convergence with Lenski's that greatly boosts our confidence that they both are on the right track. In a recent review[28] I pointed out that all characterized advantageous mutations that Richard Lenski has observed in his twenty-year experiment have turned out to be degradative ones—in which a gene or genetic control structure was either destroyed or rendered less effective. Random mutation is superb at degrading genetic material, which sometimes is help-ful to an organism. In his latest work Thornton, too, shows evolution of a system by degradation, although he speculates that the changes were neutral rather than advantageous.

In Finnegan et al., "Evolution of Increased Complexity in a Molecu-lar Machine,"[29] Thornton and colleagues study a ring of six proteins in a molecular machine (that also has many other parts) called a V-ATPase, which can pump protons (acid) across a membrane. The machine exists in all eukaryotes. In most eukaryotic species, however, the hexameric ring consists of five copies of one protein (let's call it protein 1) and one copy of another, related protein (call it protein 2). In fungi, however, the ring consists of four copies of protein 1, one copy of protein 2, and one copy of protein 3. Protein 3 is very similar in sequence to protein 1, so Finnegan et al. propose that proteins 1 & 3 are related by duplication of an ancestral gene and subsequent modification of the two, originally identical duplicated genes.

How did protein 3 insinuate itself into the ring? The original protein 1, present in five copies in most organisms, already had the ability to bind to itself, plus an ability to bind to one side of protein 2, plus a separate ability to bind to the opposite side of protein 2 (see Finnegan et al.'s Fig-ure 3). Thornton's results are consistent with the idea that, by happen-stance, the gene for protein 1 duplicated and spread in the population. These events apparently were neutral, the authors think, not affecting the organism's fitness.

Eventually one of the duplicates acquired a degradative mutation, losing the ability to bind one side of protein 2. This was not a problem because the second copy of the protein 1 gene was intact, and could bind both sides of protein 2, so a complete ring could still be formed. This also spread by neutral processes. As luck might have it, the second gene copy subsequently acquired its own degradative mutation, so that it could no longer bind the other side of protein 2. Again it's no problem, however, because the first mutant copy of protein 1 could bind to the first side of protein 2, bind a few more copies of itself, then bind a copy of protein 3, which still had the ability to bind the other side of protein 2. So a closed, six-member ring could still be formed. This apparently also spread by neutral processes until it took over the entire kingdom of fungi.

The work of Finnegan et al. strikes me as quite thorough and elegant. I have no reason to doubt that events could have unfolded that way. However, the implications of the work for unguided evolution appear very different to me than they've been spun in media reports. The most glaringly obvious point is that, like the results of Lenski's work, this is evolution by degradation. All of the functional parts of the system were already in place before random mutation began to degrade them. Thus it is of no help to Darwinists, who require a mechanism that will construct new, functional systems. What's more, unlike Lenski's results, the mutated system of Thornton and colleagues is not even advantageous; it is neutral, according to the authors. Perhaps sensing the disappointment for Darwinism in the results, the title of the paper and news reports emphasize that the "complexity" of the system has increased. But increased complexity by itself is no help to life—rather, life requires *functional* complexity. One can say, if one wishes, that a congenitally blind man teaming up with a congenitally legless man to safely move around the environment is an increase in "complexity" over a sighted, ambulatory person. But it certainly is no improvement, nor does it give the slightest clue how vision and locomotion arose.

Finnegan et al.'s work intersects with several other concepts. First, their work is a perfect example of Michael Lynch's idea of "subfunctionalization," where a gene with several functions duplicates, and each duplicate loses a separate function of the original.[30] Again, however, the question of how the multiple functions arose in the first place is begged. Second, it intersects somewhat with the recent paper by Austin Hughes[31] in which he proposes a non-selective mechanism of evolution, abbreviated "PRM" (plasticity-relaxation-mutation), where a "plastic" organism able to survive in many environments settles down in one and loses by degradative mutation and drift the primordial plasticity. But again, where did those primordial functions come from? It seems like some notable workers are converging on the idea that the information for life was all present at the beginning, and life diversifies by losing pieces of that information. That concept is quite compatible with intelligent design. Not so much with Darwinism.

Finally, Thornton and colleagues' latest research points to strong limits on the sort of neutral evolution that they envision. The steps needed for the scenario proposed by Finnegan et al. are few and simple: 1) a gene duplication; 2) a point mutation; 3) a second point mutation. No event is deleterious. Each event spreads in the population by neutral drift. Notice that the two point mutations do not have to happen together. They are independent, and can happen in either order. Nonetheless, this scenario is apparently exceedingly rare. It seems to have happened a total of one (that is, 1) time in the billion years since the divergence of fungi from other eukaryotes. It happened only once in the fungi, and a total of zero times in the other eukaryotic branches of life. If the scenario were in fact as easy to achieve in nature as it is to describe in writing, we should expect it to have happened many times independently in fungi and also to have happened in all other branches of eukaryotes.

It didn't. Thus, it seems a good conclusion that such neutral scenarios are much rarer than some researchers have proposed,[32] and that more complex neutral scenarios are unlikely to happen in the history of life.

70. Strong Experimental Support

"From Thornton's Lab, More Strong Experimental Support
for a Limit to Darwinian Evolution," Evolution News
and Views, Discovery Institute, June 23, 2014.

Joseph Thornton, the University of Chicago biologist whose work on hormone receptor proteins has been followed closely here, has published a new paper in *Nature* ("Historical Contingency and Its Biophysical Basis in Glucocorticoid Receptor Evolution").[33] Ann Gauger wrote about it last week.[34] Although Thornton himself always interprets his results in a standard Darwinian framework, in my view the work strongly confirms that severe problems face even relatively minor Darwinian evolution of proteins.

Here's some background. Vertebrates have two proteins that bind different-yet-similar steroid hormones. Since the proteins themselves are very similar in sequence and structure, the conventional view holds that an ancestral gene coding for one such protein duplicated, and the second copy underwent random mutation plus natural selection to yield the second protein.

While investigating the proteins over the past decade, Professor Thornton's laboratory showed that the more modern hormone receptor protein would be quite unlikely to be able to reverse-evolve into the ancestral form by random processes, since it would have to pass through multiple, neutral unselected changes (that is, mutations which by themselves neither help nor hinder an organism's survival). They subsequently showed, quite unexpectedly, that the ancestral form itself had to accumulate specific, neutral, unselected, improbable mutations to yield the modern protein.

In prior comments on Thornton's work I proposed something I dubbed a "Time-Symmetric Dollo's Law" (TSDL). Briefly that means, because natural selection hones a protein to its present job (not to some putative future or past function), it will be very difficult to change a protein's current function to another one by random mutation plus natural selection.

But there was an unexamined factor that might have complicated Thornton's work and called TSDL into question. What if there were a great many potential neutral mutations that could have led to the second protein? The modern protein that occurs in land vertebrates has very particular neutral changes that allowed it to acquire its present function, but perhaps that was an historical accident. Perhaps any of a large number of evolutionary alterations could have done the same job, and the particular changes that occurred historically weren't all that special.

That's the question Thornton's group examined in their current paper. Using clever experimental techniques, they tested thousands of possible alternative mutations. The bottom line is that *none* of them could take the place of the actual, historical, neutral mutations. The paper's conclusion is that, of the very large number of paths that random evolution could have taken, at best only extremely rare ones could lead to the functional modern protein.

A few thoughts:

+ Thornton's lab does terrific work. His is one of the most careful, thorough, meticulous experimental evolution programs in operation anywhere today.

+ The level at which Thornton's group addresses evolution—at the amino acid residue level, and considering many possible mutations—is the *minimum* level of detail necessary to draw even moderately firm conclusions about the ability of random-processes-plus-selection to explain life. In nature, evolution occurs at the molecular level of specific, individual mutations, so it is there we must look to evaluate possible evolutionary paths. Studies with less detail can say very little on the topic.

+ As good as it is, however, Thornton's work still does not address many important biological factors that would be critical in the history of life, such as the degree of selective pressure or the effect of other genes on the experimental system. So random evolution could well be much *less* effective than he has shown. It can't be more effective, because the biological factors are additional constraints, on top of the molecular properties he studied. (It's like an athlete making it through all the visible challenges in her effort to complete an obstacle course race, but then turning a corner onto a previously hidden stretch of the course. There may well be worse obstacles there.)

+ Thornton's work is the first of its kind. So, since the very first protein studied in sufficient detail is found to encounter severe problems in changing its function by even a modest amount by unguided processes, that strongly suggests proteins in general

will, too—not just the particular one he studied. Which is
exactly what you'd expect from a "Time-Symmetric Dollo's
Law."

+ Thornton's approach holds great promise for helping to
determine a rigorous edge to random evolution. To the extent
that a pre-existing system had to pass through improbable,
unselected, or even detrimental states—unguided by natural
selection or any other unintelligent factor—to reach a rare new
function, then to that extent we can say Darwinian evolution
does not explain life.

Thornton himself—apparently a conventional Darwinist, and certainly no sympathizer with intelligent design—does not attribute the
protein receptor's new function to Darwinian processes. Rather, he ascribes it mostly to "historical contingency." That's another way of saying
"dumb luck."

The edge of evolution lies where reasonably probable, random muta-
tion-selection runs out of steam and "dumb luck" (or—for those willing
to consider it—purposeful design) takes over. Thornton's work shows
that the edge occurs far deeper into life than even I had thought.

PART SIX: DARWINISM SICKENED BY MALARIA

IN *THE EDGE OF EVOLUTION* I ARGUED THAT MALARIA'S UNEXPECT-edly slow development of chloroquine resistance pointed to severe limits to the Darwinian mechanism of random mutation plus natural selection. It took mutations in an astronomical number of the deadly parasites to find the specific, required couple of amino acid residues in a pre-existing malarial protein, apparently because multiple changes were needed before a beneficial effect was seen. Darwin's Achilles' heel—that is, required steps in an evolutionary pathway that are, individually, detrimental or unselected—was on full display for all to see. Darwinian reviewers of the book raised all sorts of objections, including that the apparent need for several changes was balderdash. But new laboratory results have shown that they are indeed required.

71. A KEY INFERENCE CONFIRMED

"A Key Inference of *The Edge of Evolution* Has Now Been Experimentally Confirmed," Evolution News and Views, Discovery Institute, July 14, 2014.

A RECENT PAPER IN *PROCEEDINGS OF THE NATIONAL ACADEMY OF Sciences* confirms a key inference I made in 2007 in *The Edge of Evolution*. Summers et al.[1] conclude that "the minimum requirement for (low) [chloroquine] transport activity... is two mutations." This is the first of three essays on the topic.

Let me start with some background. Darwinian theory proposes that the astoundingly intricate machinery of the cell developed step by excruciatingly tiny step, by natural selection acting on random mutation. I argued against that in 1996 in *Darwin's Black Box*, contending that much cellular machinery was, like a mousetrap, irreducibly complex, could not be made gradually, and required purposeful design. Some Darwinists parried with breezy scenarios, imagining intricate systems forming at the drop of the proverbial hat. Vague as the stories might be, though, they often had a surface plausibility that provided an excuse for the reluctant to not look too deeply. For the case against Darwinism to advance, I thought, it had to move beyond descriptive arguments (which too often are deflected with specious yarns) to quantitative ones (which call for numerical replies that can be tested). So, as far as possible, hard numbers had to be attached to the probabilities of the events Darwinists blithely ask of unaided nature. That was the goal of *The Edge of Evolution*.

A major point of the book was that if evolution has to skip even one baby step to attain a beneficial state (that is, if even one intermediate in a long and relentlessly detailed evolutionary pathway is detrimental or unhelpful), then the probability of reaching that state decreases exponentially. After discussing a medically important example (see below), I argued that the evolution of many multi-protein complexes in the cell were beyond the reach of Darwinian evolution, and that design extended very deeply into life.

However, at the time, the book's chief, concrete example—the need for multiple, specific changes in a particular malarial protein (called Pf-CRT) for the development of resistance to chloroquine—was an inference, not yet an experimentally confirmed fact. It was really an excellent, obvious inference, because resistance to chloroquine arises much, much less frequently than to other drugs. For example, resistance to the antimalarial drug atovaquone develops spontaneously in every third patient, but to chloroquine only in approximately every billionth one. About Pf-CRT I wrote, "Since two particular amino acid changes [out of four to eight total changes] occur in almost all of these cases [of chloroquine resistance in the wild], they may both be required for the primary activity by which the protein confers resistance." If so, then "the likelihood of a particular [malarial] cell having the several necessary changes would be much, much less than the case [for atovaquone] where it needed to change only one amino acid. That factor seems to be the secret of why chloroquine was an effective drug for decades." Still, the deduction hadn't yet been nailed down in the lab.

Now it has, thanks to Summers et al. It took them years to get their results because they had to painstakingly develop a suitable test system where the malarial protein could be both effectively deployed and closely monitored for its relevant activity—the ability to pump chloroquine across a cell membrane, which rids the parasite of the drug. Using clever experimental techniques, they artificially mutated the protein in all the ways that nature has, plus in ways that produced previously unseen intermediates. One of their conclusions is that a minimum of two specific mutations are indeed required for the protein to be able to transport chloroquine.

(Interestingly, one of the two mutations I discussed in *The Edge of Evolution* as possibly required, at position 76 of the protein chain, is in fact one of the two that Summers et al. proved to be needed. But the other one I discussed, at position 220, isn't. Although that change can help, Summers et al. found that the second required mutation can be at either position 75 or position 326. They also showed that, although proteins with just the two required mutations could pump chloroquine past a cell membrane in their test system, the rate was significantly less than for some proteins with additional mutations. What's more, the two required ones weren't necessarily enough to allow malarial parasites to survive better in the presence of chloroquine in the lab. What that means for malaria in the wild is still unclear.)

The need for multiple mutations neatly accounts for why the development of spontaneous resistance to chloroquine is an event of extremely low probability—approximately one in a hundred billion billion (1 in 10^{20}) malarial cell replications—as the distinguished Oxford University malariologist Nicholas White[2] deduced years ago. The bottom line is that the need for an organism to acquire multiple mutations in some situations before a relevant selectable function appears is now an established experimental fact.

72. IT'S TOUGH TO MAKE PREDICTIONS

"It's Tough to Make Predictions, Especially about the Future,"
Evolution News and Views, Discovery Institute, July 16, 2014.

As I NOTED HERE AT ENV ON MONDAY, A RECENT PAPER CONFIRMS a key inference I made in 2007 in *The Edge of Evolution*. Writing in *PNAS*, Summers et al.[3] conclude that "the minimum requirement for (low) [chloroquine] transport activity... is two mutations." This is the second of three essays on the matter.

Actually, with apologies to Yogi Berra, on some topics it's not all that hard to predict the future—like on the need for multiple mutations to get some selectable biological function. Way back in 1970, when thinking about protein sequences was avant-garde, the late, eminent theoretical evolutionary biologist John Maynard Smith worried about how they might evolve. He cast the problem in terms of a little game:

The model of protein evolution I want to discuss is best understood by analogy with a popular word game. The object of the game is to pass from one word to another of the same length by changing one letter at a time, with the requirement that all the intermediate words are meaningful in the same language. Thus word can be converted into gene in the minimum number of steps as follows:

WORD WORE GORE GONE GENE[4]

Change just one letter at a time, and all the while the word must make sense.

Of course, mutations come in many different flavors besides the simple substitutions that Maynard Smith considered, as I discussed in detail in *The Edge of Evolution*. Nonetheless, those are the kind that are relevant to chloroquine resistance. What's more, Smith's general point about the need to proceed one step at a time applies to all mutations. For example, if a selectable effect requires a certain gene duplication plus the deletion of a distinct region, that combination will occur much less frequently than if only one of those two steps were needed.

That having been said, not all words can be reached in this way. Maybe not all proteins either. Later, theoretician Allen Orr also acknowledged the general restriction of evolution to one change at a time: "Given realistically low mutation rates, double mutants will be so rare that adaptation is essentially constrained to surveying—and substituting—one-mutational step neighbors. Thus if a double-mutant sequence is favorable but all single amino acid mutants are deleterious, adaptation will generally not proceed.[5]"

Yes, adaptation generally won't proceed, but in special circumstances it can. For example, one way to get past the one-mutation-at-a-time hurdle would be to increase those low mutation rates, as HIV does. Another way is to have enormous numbers of organisms to increase the chances—an astronomical population size. That's out of the question for larger animals. But for single-cell organisms such as the malarial parasite (*Plasmodium falciparum*), it's doable. The distinguished Oxford University malariologist Nicholas White noted about the world malaria population that "in any 2-day period... ill people would contain [a total of] between 5×10^{16} and 5×10^{17} malaria parasites." And: "Resistance to chloroquine in *P. falciparum* has arisen spontaneously less than ten times in the past fifty years. This suggests that the per-parasite probability of developing resistance de novo is on the order of 1 in 10^{20} parasite multiplications."[6]

If you do the arithmetic, that astounding number of parasites would be produced on Earth over the course of perhaps every few years!

Now, why does it take such an enormous number of organisms to adapt when it takes far, far fewer to counter other malarial drugs? What might be the hurdle to developing chloroquine resistance? A decade ago, Hayton and Su had a good idea: "Based on the mutant *pfcrt* haplotypes known so far, it is likely that simultaneous multipoint changes in *pfcrt* are necessary to confer [chloroquine resistance]."[7] So, from the known mutant sequences, from the math that shows adaptation can't "generally" proceed if multiple mutations are needed, yet *with* the enormous population available to originate resistance, it's just not that hard to predict the future. In fact Hayton and Su did predict it: multiple mutations would be found necessary for malarial parasites to adapt to chloroquine. Yogi Berra would have been impressed.

But some other people weren't pleased at all, not one bit—mostly Darwinist reviewers of *The Edge of Evolution*, where I quoted all the folks above. Not only were they not pleased, they went into denial—either ignoring it, pooh-poohing it, or denying it altogether. So we got the spectacle of Sean Carroll lecturing in *Science*: "Multiple replacements can accumulate when each single amino acid replacement affects performance, however slightly, because selection can act on each replacement individually and the changes can be made sequentially."[8]

Well, that's the theory. But what if a necessary single amino acid replacement doesn't affect performance at all (because it's neutral), or "affects performance" by making it worse? (See my response to Carroll's review in part 3, essay 37 above.) Tell your ideas about "sequential changes" to the malarial parasites dying in droves because they have only one of the two required mutations for chloroquine resistance. Tell it to the people saved over the course of decades because chloroquine was still effective, unlike drugs such as atovaquone that barely work at all because they become ineffective after just one mutation.

Next, Kenneth Miller harrumphed in *Nature*, "Behe waves away evidence suggesting that chloroquine resistance may be the result of sequential, not simultaneous, mutations, boosted by the so-called ARMD (accelerated resistance to multiple drugs) phenotype, which is itself drug induced."[9] But if anyone was obfuscating and waving away evidence, it was Miller. (See my response to Miller's review in part 3, essay 39 above.)

Then we heard from yet another partisan of the rabidly Darwinist National Center for Science Education (Carroll and Miller are also affiliated with it), Nicholas Matzke, writing in *Trends in Ecology and Evolution*: "Behe's two mutations do not always co-occur. As a result, [chloroquine resistance] is both more complex and vastly more probable than Behe thinks."[10]

He's right that the same two mutations that I discussed in my book didn't always occur together. But that didn't mean two mutations weren't required, just that their identities hadn't yet been pinned down with certainty. Now they have, and we see why chloroquine resistance is vastly less probable than Matzke thinks.

And in the *New York Times*, Richard Dawkins himself warned us not to put too much faith in arithmetic: "If correct, Behe's calculations would at a stroke confound generations of mathematical geneticists, who have repeatedly shown that evolutionary rates are not limited by mutation."[11]

But surely something was limiting the rate of development of chloroquine resistance. (As I pointed out in my response, the calculations in *The Edge of Evolution* don't contradict any "mathematical geneticists" at all, simply because no one has ever tried to model the complex systems I described in anything like sufficient detail.) Now what could that be? As Maynard Smith took for granted, as Allen Orr calculated, and as Hayton and Su 2004 easily deduced from the data, if two changes were required, the rate of getting those mutations would be greatly limited. They were all right. Dawkins was all wrong.

The need for multiple mutations for chloroquine resistance in malaria just wasn't that hard to see. So why do you think Darwinist reviewers of *The Edge of Evolution* missed it?

73. THE VIEW FROM 30,000 FEET

"The Edge of Evolution: Why Darwin's Mechanism is Self-Limiting,"
Evolution News and Views, Discovery Institute, July 18, 2014.

A RECENT PAPER CONFIRMS A KEY INFERENCE I MADE IN 2007 IN *The Edge of Evolution*. Summers et al.[12] conclude that "the minimum requirement for (low) [chloroquine] transport activity... is two mutations." This is the last of my three essays on the matter.

Looking down from an airplane at 30,000 feet, the landscape can appear pretty smooth. It can be hard to imagine yourself in the place of pioneers in covered wagons of earlier times, who had to slog over the uncleared ground bump by bump, facing rivers, ridges, and ravines. A lot of thinking about evolution over the years has been like looking down from a plane—imagining that an evolutionary trek from one large feature to another wouldn't be too difficult, that it could even be made while blind-folded and drunk. But in reality life is lived on the ground and, without vision and sober planning, ditches, cliffs, and streams can be impassable.

As science probes ever deeper into the molecular details of life, seri-ous evolutionary thought has been forced to descend from 30,000 feet to ground level, and grave obstacles to undirected evolution have become manifest. Relatively recent, terrific research using the powerful tech-niques available to modern biology shows three general, separate barri-ers to a Darwinian (or, for that matter, to any undirected) evolutionary mechanism.

The first major barrier is random mutation itself. Because genomes code for many sophisticated molecular systems, random changes that have an effect will most frequently break or damage some already-func-tioning system. Nonetheless, breaking or diminishing subsystems of an exceedingly complex entity such as the cell can sometimes be adap-tive—causing the degradation to spread, as Richard Lenski's pioneering Long-Term Evolution Experiment has demonstrated so clearly. Other studies of degradative adaptation in nature strongly reinforce this point. (For example, see recent reports about loss-of-function genetic resistance to diabetes[13] and heart disease[14] in humans, gaitedness[15] in horses, loss of cyanogenesis[16] in clover, and a plethora[17] of helpful broken genes in bacteria.)

The second roadblock is actually natural selection. As Darwin en-visioned, natural selection works relentlessly, honing a selected trait to fit its job more and more closely. The problem is that, the more selection hones a trait, the more specialized it becomes, and the more difficult then to use it for another complex purpose without prohibitively un-likely mutational modification. This has been nicely shown by the work of Joe Thornton's group, where even very modest changes (e.g., the bind-ing of a second, structurally similar steroid hormone to a homologous, structurally similar receptor protein) to a pre-existing system encountered strong, unexpected evolutionary obstacles.

The third obstacle is irreducible complexity, or the need to take multiple steps to reach a selected state. As I discussed in *The Edge of Evolution*, and as Summers et al. have now demonstrated experimentally, some selectable effects require more than one mutation before they kick in. When that's the case, the likelihood of reaching the state drops exponentially with each unselected step. Although special circumstances such as a very high mutation rate or population size can help get over one or a few such steps, those aren't generally available. Even when they are, it doesn't take many such steps to put the state well beyond the reach of random mutation.

This type of barrier is ubiquitous at the molecular level because new protein-protein interactions in general will require multiple mutational steps to attain, many of which will be unselected. I discussed this at length in *The Edge of Evolution* and in later essays (collected in the present volume).

It's important to notice that these three roadblocks are substantially independent of each other. Sequestration of a system to its current function by natural selection is a different problem from the damage done by adaptive-yet-degradative random mutations, both of which are conceptually distinct from the need for multiple, unselected steps to reach some adaptive states. A result of their independence is that they will work synergistically. Undirected evolutionary change faces multiple overpowering restraints.

UC Berkeley law professor emeritus Phillip Johnson once used an analogy for Darwinian evolution that I thought at the time was intriguing but unpersuasive. He pointed out that the same physical mechanism that causes a hot air balloon to rise in the sky prevents it from continuing up indefinitely. The mechanism of flight itself limits the balloon's ascent—it will never get beyond Earth's atmosphere.

Similarly, he offered, the Darwinian mechanism both permits and limits evolutionary change. In light of recent outstanding research,

I've changed my mind—Johnson's image is delightfully apposite. We see clearly at the detailed, molecular, ground level of life that Darwin's mechanism is self-limiting. It can lift an evolutionary balloon only so high and no higher, no matter how much hot air is blown into it.

74. AN OPEN LETTER TO MILLER AND MYERS

The essays above, pointing out the experimental vindication of my argument in *The Edge of Evolution*, annoyed some Darwinists. The atheist blogger/part-time-biologist P. Z. Myers solicited comments from Ken Miller to assure his readers that all was still under control. Miller apparently was busy, so he just sent Myers an excerpt from his review of my book in *Nature*. (My response to that review is collected in the present book.) In the essay below I tried to get them to be specific—to use math, not just words.

"An Open Letter to Kenneth Miller and P. Z. Myers," Evolution News and Views, Discovery Institute, July 21, 2014.

DEAR PROFESSORS MILLER AND MYERS:
Talk is cheap. Let's see your numbers.

In your recent post[18] on and earlier reviews of my book *The Edge of Evolution* you toss out a lot of words, but no calculations. You downplay FRS Nicholas White's straightforward estimate that—considering the number of cells per malaria patient (a trillion), times the number of ill people over the years (billions), divided by the number of independent events (fewer than ten)—the development of chloroquine resistance in malaria is an event of probability about 1 in 10^{20} malaria-cell replications.[19] Okay, if you don't like that, *what's your* estimate? Let's see *your* numbers.

The malaria literature shows strong population genetics evidence for fewer than ten independent origins of resistance to chloroquine.[20] The riddle is, why so few? Show us *your* calculation for that. Here's a number to keep in mind—10^{12}. That's roughly the number of malarial cells in one sick person.[21] Here's another—10^{-8}. That's a generous rounding-up of the mutation rate for malaria.[22] (Multicellular eukaryotes are an order of magnitude less.) That means that on average *ten thousand copies of each and every point mutation* of the malarial genome will be present in *every person* being treated with chloroquine.

Here's another—3. That's the number of patients it takes for spontaneous resistance to atovaquone to appear.[23] That makes a lot of sense, since resistance to atovaquone needs only one point mutation. If atovaquone were used as widely as chloroquine, we'd expect about a billion or more origins of resistance to it by now, in contrast to the measly handful we find with chloroquine. So how do you quantitatively account for that difference—give or take an order of magnitude?

Your chief complaint against my ideas seems to be this:

That the malaria parasite needs two mutations was never a point of contention, nor was it particularly worrisome. What was wrong with Behe's work is that he naïvely claimed that the two mutations had to occur simultaneously in the same individual organism, so that the probability that could happen was the product of multiplying the two individual probabilities. That's ridiculous.[24]

What's puzzling to me is your thinking the exact route to resistance matters much when the bottom line is that it's an event of probability 1 in 10^{20}. From the sequence and laboratory evidence it's utterly parsimonious and consistent with all the data—especially including the extreme rarity of the origin of chloroquine resistance—to think that a first, required mutation to PfCRT is strongly deleterious[25] while the second may partially rescue the normal, required function of the protein, plus confer low chloroquine transport activity. Those two required mutations—including an individually deleterious one which would not be expected to segregate in the population at a significant frequency—by themselves go a long way (on a log scale, of course) to accounting for the figure of 1 in 10^{20}, perhaps 1 in 10^{15} to 10^{16} of it (roughly from the square of the point mutation rate up to an order of magnitude more than it). So how do your calculations account for it?

It's also entirely reasonable shorthand to characterize such a situation as needing "simultaneous" or "concurrent" mutations, as has been done by others in the malaria literature,[26] even if the second mutation actually occurs separately in the recent progeny of some sickly, rare cell that had already suffered the first, harmful mutation. Guys, please don't hide behind some dictionary or Einsteinian definition of "simultaneous." It matters not a whit to the practical bottom line. If you think it does, don't just wave your hands, show us your calculations.

From the recent work of Summers et al.[27] it's possible that a third mutation in PfCRT may also be needed (perhaps already segregating in the population as a nearly neutral or marginally deleterious mutation) to allow the parasite to survive at therapeutic levels of chloroquine. That may contribute another factor of 1 in 10^3 to 10^4 or so to the probability, to reach an aggregate factor of approximately 1 in 10^{20}. After that minimally functioning foundation is established, further mutations could rapidly be added individually and cumulatively—the way Darwinists like—to help balance the complex demands on PfCRT for its native activity plus chloroquine transporting, as Summers et al. discuss.

If you think that direct, parsimonious, rather obvious route to 1 in 10^{20} isn't reasonable, go ahead, calculate a different one, then tell us how much it matters, quantitatively. Posit whatever favorable or neutral mutations you want. Just make sure they're consistent with the evidence in the literature (especially the rarity of resistance, the total number of cells available, and the demonstration by Summers et al. that a minimum of two specific mutations in PfCRT is needed for chloroquine transport). Tell us about the effects of other genes, or population structures, if you think they matter much, or let us know if you disagree for some reason with a reported literature result.

Or, Ken, tell us how that ARMD[28] phenotype you like to mention affects the math. Just make sure it all works out to around 1 in 10^{20}, or let us know why not.

Everyone is looking forward to seeing your calculations. Please keep the rhetoric to a minimum.

With all best wishes (especially to Professor Myers for a speedy recovery),

Mike Behe

75. Sandwalk Evolves Chloroquine Resistance

"Laurence Moran's Sandwalk Evolves Chloroquine Resistance,"
Evolution News and Views, Discovery Institute, August 13, 2014.

First, a bit of background. As I discussed previously, in a new paper Summers et al. show that a minimum of two mutations to the malarial protein PfCRT are needed to confer an ability to pump the antibiotic chloroquine (which is necessary but may not be sufficient for chloroquine resistance in the wild).[29]

That result agrees with my discussion in *The Edge of Evolution* and goes a long way toward quantitatively explaining the rarity of the development of resistance. Over at P. Z. Myers's blog Pharyngula, he and Kenneth Miller disagreed strongly with me in words, but cited no numbers. I then invited them, since they don't like *mine*, to show us *their* calculations for how frequently chloroquine resistance should arise in the malarial parasite.

The bad news is that so far neither has responded. The good news is that Laurence Moran, professor of biochemistry at the University of Toronto, has done so.[30] Professor Moran is an intelligent, informed, direct, and generally civil critic of intelligent design who maintains a popular blog, Sandwalk, on evolution-related matters. So his response gives us a great opportunity to see what the best alternative explanations might be.

Moran begins his own calculation by assuming that the first required mutation is strictly neutral and spreads in the growing population before the second one arises. His straightforward computation leads him to conclude, "What this means is that if you start with an infection by a cell that has none of the required mutations then you will only get the right pair of mutations once in one million infected people."

Once in one million infected people? Since there are a trillion malarial cells in one sick person, then according to Moran's own calculation there are a million times a trillion malaria cells needed for resistance to arise, which in scientific notation is 10^{18}. On a log scale that's a stone's throw from Nicholas White's[31] estimate of 10^{20} cells per origin of resistance that I have been citing, literally an astronomically large number (there are only a paltry hundred billion, 10^{11}, stars in our galaxy). So let me just say thank you and welcome aboard to Professor Moran. Unfortunately, he seems not to have realized the import of his calculation at the time, and has shown no enthusiasm for exploring it much after it was brought to his attention by a commenter.

Right after his calculation Moran writes, "We know that the right pair of mutations... is not sufficient to confer resistance to chloroquine so the actual frequency of chloroquine resistance is far less." Far less? Far less than 1 in 10^{18}? Now, it's true that at least four mutations have been found in all known resistant strains of malaria. And it's true that, although Summers et al. showed two mutations are necessary for pumping chloroquine at a low level, they might not be sufficient for chloroquine resistance in the wild. Nonetheless, a need for further mutations would only make the problem for Darwinism much worse. It wouldn't make it better. Let me emphasize: Professor Moran's own reasoning would make the problem much more severe than I myself have ever argued. Yet he doesn't take any time on his blog to explore the ramifications of his own reckoning. Why doesn't he think that's an interesting result? Why not ponder it a bit?

Moran doesn't seem to actually have much confidence in his own numbers. He asks the readers of his blog to help him correct his calculations—which is a commendable attitude but makes one wonder, if he's so unsure of the likelihood of helpful combinations of mutations, whence his trust in mutation/selection? In response to the commenter who alerted him to the huge number of parasites in a million people he writes, "This is why meeting the Behe challenge is so difficult. There are too many variables and too many unknowns. You can't calculate the probability because real evolution is much more complicated than Behe imagines." But, again, if he thinks everything is so darn complicated and incalculable, on what basis does he suppose he's right?

That's the reason I issued the challenge in the first place. In my experience almost all Darwinists and fellow travelers (Professor Moran doesn't consider himself a Darwinist[32]) simply don't think quantitatively about what their theory asks of nature in the way of probability. When prodded to do so, they quickly encounter numbers that are, to say the least, bleak. They then seem to lose all interest in the problem and wander away. The conclusion that an unbiased observer should draw is that Darwinian claims simply don't stand up to even the most cursory calculations.

Another commenter at Sandwalk didn't like Moran's calculation, so came up with his own. Great! The more the merrier! He also assumed the first mutation to be neutral, but kept a more careful accounting of its accumulation through the generations and ended up with a result of one necessary double-mutation per 420 patients. That actually strikes me as a more realistic value for a neutral mutation than Professor Moran's. Now, at first blush 420 may seem much smaller than Moran's number of a million patients, but that's only because we haven't yet considered the factor of a trillion parasites per patient. When we multiply by 10^{12} to get the total number of parasites per double mutation, the commenter's odds turn out to be 1 in $10^{14.6}$ versus Moran's 1 in 10^{18}, again not all that far on a log scale. Either or both of these values can easily be reconciled to White's calculation of 1 in 10^{20} by tweaking selection coefficients or by inferring that a further mutation is needed for effective chloroquine resistance in the wild, as Professor Moran noted.

What if the first necessary mutation isn't neutral? What if—as seems very likely from the failure of malaria cells with one required mutation (K76T) to thrive[33] in the lab—the first mutation is rather deleterious? The commenter estimated that, too (and also added another consideration, a selection coefficient), and came up with a value of one new double mutant per 818,500 patients. Let's relax the admirable precision a bit and round the number up to a million. That's the same count Professor Moran got in his (supposedly neutral) calculation, which we saw means there is one new origin per 10^{18} malarial parasites—not far at all on a log scale from White's number that I cited.

The bottom line is that numbers can be tweaked and a few different scenarios can be floated, but there's no escaping the horrendous improbability of developing chloroquine resistance in particular, or of getting two required mutations for any biological feature in general, most especially if intermediate mutations are disadvantageous. If a needed step does not confer an advantage, the wind goes out of Darwin's sails.

76. HOW MANY WAYS TO WIN AT SANDWALK?

"How Many Ways Are There to Win at Sandwalk?," Evolution
News and Views, Discovery Institute, August 15, 2014.

AT UNIVERSITY OF TORONTO PROFESSOR LAURENCE MORAN'S blog Sandwalk, named for Darwin's famous "thinking path," I've followed a discussion of the evolution of de novo chloroquine resistance by malaria.[34] The exchange has touched on a few issues that seem to confuse people easily.

One is how we should view the probability of winning something. In questioning my malaria numbers, a commenter remarked that it's misleading to focus retrospectively on a single event, such as winning a familiar game of cards, to calculate the odds of that exact arrangement of cards and declare it to be the likelihood of winning at the game. After all, there may be very many other ways to win, too. In order to correctly calculate the odds, he explained, one would have to take into account all of the ways to win, not just a single hand.

I agree completely. Fortunately, in the huge number of malaria cells exposed to chloroquine, all the proverbial hands have already been dealt many times over, so we can confidently calculate the odds from the statistics.

Here's an analogy. Suppose we observe a hall where around a thousand people are each dealt ten cards—but not our normal playing cards. Instead there's a variety of strange symbols on the cards, in different colors and sizes, which we assume are distributed randomly to the players. We don't know the game they are playing or any of the rules, but we see that nobody in the group wins. That group shuffles out of the hall and a fresh group of a thousand people takes its place, is dealt ten cards, and again no one wins. This goes on until the forty-third group, where one person jumps up smiling and is declared to be a winner. Another sixty-one groups follow before there is another winner. After watching for a long time we record that on average the size of each crowd is a thousand people and somebody wins once every fifty crowds.

So what are the odds of winning that game? Of course it's about 1 in 50,000—the statistical average number of people it takes to get a winner. Since we don't know the rules, there may be just one way to win the game, or many different ways. There may be one rare card that is needed, or multiple different specific combinations of cards. When we eventually learn more about the game we might be able to figure out the rules and understand why the odds are what they are. But that doesn't matter for this. The odds of winning themselves won't change outside the range of our statistical uncertainty. They'll remain approximately 1 in 50,000.

We can deduce another pertinent lesson. There may be card combinations other than what have so far been dealt that win the game. But if there are, the probability of their occurrence is *less than or equal to* that of other ways to win that have already happened. The reason of course is that card combinations with significantly higher probabilities would already have been dealt in the large number of hands. The lesson, then, is that once we have good statistics, the probability of winning is fixed. It already implicitly includes any and all of the ways there are to win.

So, too, with chloroquine resistance in *Plasmodium falciparum*. The best current statistical estimate of the frequency of de novo resistance is Nicholas White's value of 1 in 10^{20} parasites.[35] That number is now essentially fixed—*no pathway to resistance will be found that is substantially more probable* than that. Although with more data the value may be refined up or down by even as much as one or two orders of magnitude (to between 1 in 10^{18} and 1 in 10^{22}), it's not going very far on a log scale. Not nearly far enough to lift the shadow from Darwinism.

What's more, we can also conclude that the mutations that have already been found are the most effective available of any combination of mutations whose joint probability is greater than 1 in 10^{20}, since more effective alternatives would already have occurred and been selected if they were available. That's a point of great public health consequence.

Before investigating what it takes at the molecular level to confer chloroquine resistance, we might have conjectured that there was one exceedingly rare, necessary mutation, or a combination of several mutations, or a dozen different paths each with several required mutations. We would nonetheless expect that when we did uncover the pathway, we would be able to reconcile the likelihood of each of its steps with the statistical evidence. Although it was pretty easy to predict from the sequence evidence even as early as ten years ago, that is what Summers et al.'s recent work[36] allows us to do now with great confidence. The fact that several point mutations are required before low chloroquine-pumping activity is observed for PfCRT, coupled with the known mutation rate, easily gets us very close on a log scale to Nicholas White's statistic, 1 in 10^{20}. There is no particular reason to grasp for other explanations.

Nor would it do any good. There is a lot of chatter at Sandwalk deriding the idea of "simultaneous" mutations (which was not intended in my book *The Edge of Evolution* in the sense it is being taken there, and which at this point I would gladly replace with other words simply to avoid the distraction). Yet it matters not a whit for the prospects of Darwinian theory whether the pathway consists of two required mutations that are individually lethal to a cell and must occur strictly simultaneously (that is, in the exact same replication cycle), or whether it consists of several mutations each with moderately negative selection coefficients, or consists of, say, five required mutations that are individually neutral and segregating at some appreciable frequency in the population, or some other scenario or combination thereof. The bottom line for all of them is that the acquisition of chloroquine resistance is an event of statistical probability 1 in 10^{20}.

It is the outlandish improbability of the pathway—not its particular features—that is the crux. It puts strong limits on what we can expect from Darwinian processes. And that is an important point for any biologist—whether in a medical field or not—to appreciate.

One other interesting point was raised in the comments at Sandwalk, a point that sounds in both science and philosophy of science. It is similar to, yet much broader than, the one dealt with above. Instead of asking merely whether we have counted all the ways to win a particular game, it essentially asks whether we have considered all the games that could have been played. The commenter writes:

> It is one thing to calculate the probability of a specific change... from a specific, randomly chosen starting point..., and it is a TOTALLY DIFFERENT THING to calculate the probability of SOMETHING interesting happening anywhere amongst thousands of genes and thousands of species over millions of years. When the modern scientist discovers that interesting evolutionary change, it is totally illegitimate to forget about all of the things that didn't happen.

I agree with this, too. Happily, the tacit hypothesis—that very many possible-but-unrealized complex biochemical systems could have been made by random mutation/selection—can actually be tested to a degree. To the extent it can't be tested, it is unfalsifiable. To the extent that it can be tested, it has, in my view, already been falsified.

Let's start with the untestable part. Could a plethora of very complex biological systems, other than the ones that in fact exist, have arisen by Darwinian processes over the course of life on Earth? That question strikes me as reminiscent of the multiverse hypothesis in physics—the postulation of copious unseen and probably unknowable systems to account for the existence of apparently very improbable known ones. So, like the multiverse, to the extent that this proposal is untestable I regard it as not scientific.

How can the claim be at least partially tested? One way is to examine huge numbers of organisms under great selective pressure and see whether they do in fact evolve anything that looks like it might at least be on the way to becoming one of those postulated, frequent, hidden, new complex systems. Now, what data do we have that might be relevant?

Data on malaria, of course. And what we see there is that in a gargantuan number of organisms under relentless selective pressure from chloroquine, no new complex system evolved, only a few crummy point mutations in a pre-existing protein.

Let's consider the flip side of the malaria question. What mutations in more than ten thousand available genes have been selected over the past ten thousand years in billions upon billions of humans as a result of malaria exposure?—A handful of point mutations and broken genes, none of which appears to be leading to anything like a new complex system.

Other examples of Darwinian futility can be found in *The Edge of Evolution*. For the similarly very modest results of many laboratory evolution experiments, see my 2010 paper in *The Quarterly Review of Biology*.[37]

If nature were thick with possible complex biochemical systems that could be found by Darwinian processes, we should expect to in fact find some when we go looking. We don't.

77. A QUICK REPRISE OF THE EDGE OF EVOLUTION

"Guide of the Perplexed: A Quick Reprise of *The Edge of Evolution*," Evolution News and Views, Discovery Institute, August 20, 2014.

ON HIS BLOG, SANDWALK, UNIVERSITY OF TORONTO BIOCHEMIStry professor Laurence Moran expressed uncertainty[38] concerning the basic argument of *The Edge of Evolution: The Search for the Limits of Darwinism*. So for anyone who wants a quick reprise of the book's reasoning, below is a list of annotated bullet points plus some commentary summarizing it.

+ If the development of some particular adaptive biochemical feature requires more than one specific mutation to an organism's genome, and if the intermediate mutations are deleterious (and to a lesser extent even if they are neutral), then the probability of the multiple mutations randomly arising in a population and co-existing in a single individual so as to confer the adaptation will be many orders of magnitude less than for cases in which only a single mutation is required.

+ The decreased probability means either that a much larger population size of organisms would be required on average to produce the multiple mutations in the same amount of time as needed for a single mutation, or that for the same population size a multiple-mutation feature would be expected to require many more generations to appear than a single-mutation one.

As a matter of simple population genetics theory, the two points above should be uncontroversial. Now let's look at some empirical data.

+ In *The Edge of Evolution* I cited the development of chloroquine resistance in the malaria parasite *Plasmodium falciparum* as a very likely real-life example of this phenomenon. The recent paper by Summers et al.[39] confirms that two specific mutations are required to confer upon the protein PfCRT the ability to pump chloroquine, which is necessary but may not be sufficient for resistance in the wild.

+ The best estimate of the per-parasite occurrence of de novo resistance is Nicholas White's[40] value of 1 in 10^{20}. This number is surely made up of several components, including: 1) the probability of the two required mutations identified by Summers et al. coexisting in a single *pfcrt* gene; 2) the value of the selection coefficient (which can be thought of as the likelihood that the de novo mutant will successfully recrudesce in a person treated by chloroquine and be transmitted to another person); and 3) the probability of any possible further PfCRT mutation needed to confer chloroquine resistance in the wild coexisting in the same gene with the other mutations.

+ The known point mutation rate of *P. falciparum*, combined with the apparent deleterious effect of the required mutations occurring singly, suggests that component 1 from the previous bullet point will account for the lion's share of White's estimate, probably at least a factor of 1 in 10^{15}–10^{16} of it. The other factors would then account for 1 in 10^4–10^5. These values are somewhat flexible, accommodating the uncertainty in our knowledge of the exact values in the wild. In other words, a decrease in our best estimate of the value of one factor can be conceptually offset relatively easily without affecting the argument, offset by supposing another factor is larger and thereby again matching the empirically derived 1 in 10^{20}.

These last three points, driven by inferences from empirical data rather than by pure theory, should also be pretty uncontroversial. Now let's pass on to the dicier stuff.

+ Any particular adaptive biochemical feature requiring the same mutational complexity as that needed for chloroquine resistance in malaria is forbiddingly unlikely to have arisen by Darwinian processes and fixed in the population of any class of large animals (such as, say, mammals), because of the much lower population sizes and longer generation times compared to that of malaria. (By "the same mutational complexity" I mean requiring two to three point mutations where at least one step consists of intermediates that are deleterious, plus a modest selection coefficient of, say, 1 in 10^3 to 1 in 10^4. Those factors will get you in the neighborhood of 1 in 10^{20}.)

+ Any adaptive biological feature requiring a mutational pathway of twice that complexity (that is, four to six mutations with the intermediate steps being deleterious) is unlikely to have arisen by Darwinian processes during the history of life on Earth.

In the book I then go on to make a general argument that Darwinian processes could not have constructed the molecular foundation of life, but let's leave that aside for now. Let's just concentrate on the last two bullet points here.

Considered in the calmer context of the development of resistance to particular antibiotics (such as, say, a combination of chloroquine plus a second drug that is as difficult to evolve resistance to and works by an independent mechanism)—rather than in the highly charged context of intelligent design—even these two statements should seem reasonable to critics of ID. After all, many medical professionals searching for treatments for malaria are trying to do exactly that—to combine two very improbable mutational steps into an insuperable mutational pathway. If there were a second drug with the efficacy of chloroquine which had always been administered in combination with it (but worked by a different mechanism), resistance to the combination would be expected to arise with a frequency in the neighborhood of 1 in 10^{40}—a medical triumph.

Where a critic might demur is on the question of how many ways exist to solve an evolutionary problem of that mutational complexity. I think that's due to a confusion about the need for particular mutations versus nonspecific mutations. While comparing the math of chloroquine resistance to mutations that have occurred in the primate line leading to humans, Professor Moran wrote, "Does he really mean that there can't be any examples of two mutations occurring in the same gene since humans and chimps diverged?" No, of course not. That overlooks the requirement for the great specificity needed to build biochemical systems. For example, to achieve chloroquine resistance malaria must at least acquire the mutations K76T plus either N75E or N326D in PfCRT—two very particular amino acid positions in a very particular protein—not just any two amino acid codons in any gene. That of course makes a huge difference to the probability.

Moran also writes, "He seems to think that whenever we see such mutations they must have been the only possible way to evolve some new function or feature." Well, no, not the "only possible" way, but, yes, one of a very limited number of possibilities. (I wrote about this in my last article,[41] too.)

In fact the number is limited enough that we can conclude with confidence that it won't affect my argument summarized above. For example, suppose there were ten, or a hundred different ways to address a particular biochemical challenge. That would barely move the dial on a log scale that's pointing at 1 in 10^{20}.

What's more, Nicholas White's factor of 1 in 10^{20} already has built into it all the ways to evolve chloroquine resistance in *P. falciparum*. In the many malarial cells exposed to chloroquine there have surely occurred all possible single mutations and probably all possible double mutations—in every malarial gene—yet only a few mutational combinations in *pfcrt* are effective. In other words, mutation and selection have already searched all possible solutions of the entire genome whose probability is greater than 1 in 10^{20}, including mutations to other genes. The observational evidence demonstrates that only a handful are effective. There is no justification for arbitrarily inflating probabilistic resources by citing imaginary alternative evolutionary routes.

To summarize, my argument concerns the evolutionary construction of biochemical features of specificity similar to that of malarial chloroquine resistance. The little-appreciated point I wanted to emphasize is that the likelihood of success decreases enormously if even a single mutational step of a pathway is disfavored. With more such steps, its improbability becomes prohibitive.

78. DRAWING TO A CLOSE WITH MORAN

"Drawing My Discussion with Laurence Moran to a Close," Evolution
News and Views, Discovery Institute, August 26, 2014.

EVER HAVE A CONVERSATION WITH SOMEONE WHERE, TRY AS YOU
might, you just can't seem to get through? Over the past few weeks
University of Toronto professor of biochemistry Laurence Moran and
I have been responding to each other's posts concerning my book *The
Edge of Evolution* and a recent paper by Summers et al.[42] on chloroquine
resistance in malaria (a major topic of the book). Unfortunately, I think
we have reached the point of diminishing returns, where arguments are
repeated to little effect. So I will address his latest (entitled, ironically it
seems to me, "Understanding Michael Behe"[43]) and let him have the last
word if he wishes. Readers can decide for themselves who had the better
of the exchange.

He starts off well enough, agreeing with me on two simple points. To
my contention that an effect that requires two mutations will be much
rarer than an effect that requires one Moran writes, "This is correct."
And to my statement that that means a species would either require a
much larger population size or a much longer time to acquire the double
mutation he also responds, "This is correct." Great! Progress.

But then he goes off the rails. To make my response as clear as possible, I'm going to go through his post in order and in some detail.

I wrote in my last essay (see essay 77 above) that I had cited chloroquine resistance in *Edge* as a likely example of the two-mutation phenomenon, and that Summers et al. recently "confirmed" that it did need
two mutations to pump chloroquine. Moran responds, "This is a little bit
misleading and possibly a little bit disingenuous. Everyone understood
that chloroquine resistance was rare and that it almost certainly required
multiple mutations."

I'm afraid it is he who is playing the ingenue. There's a big difference
between simply requiring serial additive mutations for some maximal effect and requiring multiple mutations before you get an effect at all. The

first is a run-of-the-mill, gradualist Darwinian scenario: one mutation comes along, helps a bit, spreads in the population by selection, which increases the base from which the second mutation may arise; the second appears, helps a bit more, spreads, and so forth. Lather, rinse, repeat.

But if the first required mutation (or second, or third) doesn't help, or actually hurts, then the gradualist scenario is interrupted. The first mutation does not spread in the population (in fact if it is detrimental it's actively kept in check by negative selection), so the number of organisms with the mutation does not increase and can't provide a larger base within which the second mutation can arise. The Darwinian magic is turned off.

How much does that hurt? Enormously.

For most species, missing even one such baby step increases the required population size/waiting time by a factor of millions to billions. If even one step in a long and relentlessly detailed evolutionary pathway is deleterious, then a Darwinian process is woefully impaired. If several steps in a row are deleterious, you can kiss the Darwinian explanation goodbye.

From here on I will first quote Moran, then respond.

Moran: "Everyone understood that chloroquine resistance was rare and that it almost certainly required multiple mutations."

Me: That's revisionist history. Other than the malariologists that I cited in my book (Hayton and Su[44] and Nicholas White[45]), no one I've come across who wrote before Summers et al. was published thought multiple mutations would be needed before any chloroquine resistance appeared. For example, the 2002 *Science* article "A Requiem for Chloroquine"[46] specifically proposed a one-beneficial-mutation-at-a-time scenario.

In *The Edge of Evolution* I wrote:

Suppose that *P. falciparum* needed several separate mutations just to deal with one antimalarial drug. Suppose that changing one amino acid wasn't enough. Suppose that two different amino acids had to be changed before a beneficial effect for the parasite showed up. In that case, we would have a situation very much like a combination-drug cocktail, but with just one drug.

Reviewers of the book writing in major periodicals frothed at the mouth over that idea. As I noted earlier:

But some other people weren't pleased at all, not one bit—mostly Darwinist reviewers of *The Edge of Evolution*.... So we got the spectacle of Sean Carroll lecturing in *Science*: "Multiple replacements can accumulate when each single amino acid replacement affects performance, however slightly, because selection can act on each replacement individually and the changes can be made sequentially."

Well, that's the theory. But what if a necessary single amino acid replacement doesn't affect performance at all (because it's neutral), or "affects performance" by making it worse?

As I show in the post, Kenneth Miller writing in *Nature*, Nicholas Matzke in *TREE*, and Richard Dawkins in the *New York Times* all say similar things. No, "everyone" most definitely did not understand "that it almost certainly required multiple mutations" that had to be present before the effect kicks in. On the contrary, reviewers actively denied it.

Moran: "Behe developed an explanation based on the idea that two mutations were required and one of them, by itself, had to be very deleterious."

Me: And that turned out to be exactly correct. Summers et al. show that at least two mutations are indeed required for chloroquine-pumping activity, and Lakshmanan et al.[47] showed that parasites with the single K76T mutation languish in the lab. So what's the problem?

Moran: "This is because he used an incorrect value for the mutation rate that was several orders of magnitude too low."

Me: No. I worked with a mutation rate of 10^{-8}, and the best literature value for malaria is 2.5×10^{-9}. That's a difference of only a factor of four from the rate I used, not "several orders of magnitude." What's more, it's a factor of four *greater* than the literature value, not "too low"—the opposite direction from what Moran thought. (Those negative exponentials can be confusing.) I prefer rounding the value up to 10^{-8} because using several significant figures gives a misleading impression of undue precision.

Moran: "He based his calculations on the assumption that the two mutations had to arise in a single infected patient."

Me: Those are Moran's words, not mine, but it's probably correct anyway and certainly makes little difference to the math. The two required mutations do have to be present, not only in a single patient but in the very same *pfcrt* gene of a single malaria cell. That's why they're called "required."

Moran: "The recent paper by Summers et al. (2014) shows that seven of the chloroquine resistant strains that have been observed have at least four mutations and some of them are relatively neutral. This refutes and discredits the scenario that Michael Behe put forth in his book."

Me: It's hard to see why Professor Moran thinks the second sentence in this quote belongs after the first. Close your eyes and envision a pathway to a malaria parasite that has four mutations. The first mutation is deleterious, the second rescues the first and makes the parasite marginally chloroquine resistant. Subsequent steps are all beneficial by dint of either improving chloroquine resistance or of stabilizing the structure of the mutated PfCRT, which is required for malaria survival. Once a parasite can survive at least marginally in the presence of chloroquine, further mutations can be added one at a time (no longer two at a time) in each cycle of infection because the population size (10^{12}) greatly exceeds the inverse of the mutation rate.

In the argot of chemical kinetics, getting beyond the deleterious mutation is the "rate-limiting step." After that hurdle is passed further mutations can be added singly—the way Darwinists like—and comparatively rapidly. Since they would be added rapidly, they would be difficult to detect in the wild. Hence the pattern described by Summers et al. fits the scenario I described perfectly.

Moran: "His sycophants are promoting the idea."

Me: Name-calling is puerile.

Moran: "As we all know by now, the "guesstimate" by Nicholas White refers to the possible frequency that a chloroquine resistant strain will be detected someplace in the world. This is a far cry from the probability that the correct mutations will actually occur."

Me: It takes chutzpah to deride a value calculated by one of the world's foremost malariologists—a Fellow of the Royal Society, who lives and works in the middle of malarious regions of the world, who reasons deeply and quantitatively about the development of drug resistance—as a "guesstimate." White's calculation was based on intimate knowledge of many details of the biology and epidemiology of the parasite.

Malaria is a horrendous scourge, and so is followed closely by international health organizations. In this age of easy and rapid sequencing, one would expect independent strains of malaria to be detected quickly by the many researchers in the area. If Professor Moran has any actual evidence that a large number of de novo origins of malaria have gone undetected, we would all love to see it. Otherwise it's reasonable to assume that the number so far identified is close to the number that have in fact arisen.

Moran: "But, according to Michael Behe, there's a 10,000 (10^4) fold difference between the probability of the mutations occurring and their actual frequency. This is important because he attributes that difference to the fact that both of the two mutations are deleterious on their own."

Me: For brevity, I didn't quote Moran's preceding paragraph, which refers to this. Professor Moran seems terribly confused here. The mutation rate he himself favors, about 10^{-10}, would also require a mutational step to be deleterious to get to the observed frequency of 1 in 10^{20}. Otherwise, accumulation of neutral mutations would raise it. I myself do not "attribute that difference" of a factor of 10^4 between the (very reasonable) rate of genetic production I use and Nicholas White's number of 1 in 10^{20} to the mutations being deleterious. Rather, the factor of 10^4 or so is simply the factor remaining of White's empirical statistic after accounting for the double mutation rate.

What's more, the factor of 10^4 accommodates an expected fractional selection coefficient and leaves room for the possibility of an additional required mutation (a third and/or fourth one). Moran's number does not. Thus, he is in the awkward position of advocating a scenario that would seem to predict an even less frequent occurrence of de novo chloroquine resistance than White estimates, all the while arguing (above) that many origins have gone undetected, which would make it more frequent. He seems to really believe the old saw that consistency is the hobgoblin of little minds.

Moran: "Summer et al. (2014) showed conclusively that this calculation is wrong. They showed that among the seven strains analyzed there were multiple pathways to resistance and that a minimum of four mutations were required for effective resistance. They showed that one mutation (K76T) was essential in all strains. All strains had one of two additional mutations that seemed to be required early on, N75E or N326D."

Me: This is referring to a quotation of mine, not cited here for brevity's sake. I have no idea why Professor Moran thinks any of this shows my "calculation is wrong."

Moran: "The recent data by Summer et al. refutes three of Behe's assumptions: (a) that only two mutations are required to account for the appearance of resistant strains, (b) that a resistant parasite had to arise from wild type within a single infected individual, and (c) that one or more of the individual mutations are deleterious on their own. Other than that, his calculations and his "predictions" are perfect!!!!"

Me: Summers et al.'s paper does not even address Moran's (a), (b), or (c), let alone resolve matters the way he claims. The paper measured the ability of a number of mutant PfCRT proteins to pump chloroquine across a membrane in an artificial test system consisting of frog eggs. It also examined the ability of two mutant malaria constructs carrying artificial genes on a transfected plasmid to survive in a medium containing a certain level of chloroquine that usually kills half of sensitive malaria cells. That's it!

They did not determine that more than "two mutations are required" for resistant strains. They did show that several strains with three mutations that could pump chloroquine nonetheless did not have higher resistance against a certain level of chloroquine in test tubes in the lab. However, this does not show that the strains could not survive at some levels of chloroquine in some humans in the wild. That remains an unanswered question.

Summers et al. did absolutely nothing to determine whether or not "a resistant parasite had to arise from wild type in a single infected individual." Nor did they even try to investigate whether "one or more of the individual mutations are deleterious on their own." Rather, mutant PfCRT proteins were examined in an elegant test system of frog eggs, which can't tell you what the effect of the protein will be in a malarial cell within a human. The several mutant proteins they did test in malaria cells were coded by genes carried on plasmids—meaning that the malaria cells also expressed their own, wild-type, normal, genomic, unmutated gene, which of course could compensate for any missing required activity of a nonfunctional mutant plasmid gene, so they couldn't detect if a mutant protein were deleterious by itself. On the other hand, it was shown years ago that malaria cells constrained to express only mutant K76T PfCRT would not grow in the lab, demonstrating the mutation is indeed deleterious.

Other than that, Moran's comments are perfect.

Moran: "It is true that any specific set of 4–6 mutations is extremely unlikely. Nobody disputes that any more than they dispute the probability of a specific hand of bridge being dealt in the next deal. That's only going to happen once in 1028 tries. It's impossible that you will see it in your lifetime. But you still get to play bridge."

Me: The bridge analogy is inapt (see my previous post). *P. falciparum* dies without the needed mutations. It will never play bridge again.

Moran: "For example, let's say that the difference in needle clusters between red pine and white pine are determined by a single gene. Let's say that three specific mutations are required to change from a cluster of two needles to a cluster of five needles.... When such a triple mutation arises we recognize that it was only one of millions and millions of possible evolutionary outcomes. There was no a priori requirement that Earth contain red pines and white pines just as there's no a priori requirement that you get a specific hand in the next deal."

Me: Were there "millions and millions of possible evolutionary outcomes" that could make malaria resistant to chloroquine? I can think of a few possible alternative scenarios. Instead of a chloroquine pump appearing, maybe a chloroquine-degrading enzyme could have arisen, or maybe malaria's membrane could somehow be altered to stop chloroquine from entering the cell, or maybe the cell could become dormant until the chloroquine passed.

But none of those scenarios happened. Why not? Because any evolutionary pathway leading to those outcomes was even less probable than the pathway that occurred. The odds of 1 in 10^{20} are not just the probability of PfCRT acquiring the right mutations—they are the minimum odds of any chloroquine-resistance mechanism arising in *P. falciparum*. It may very well be the case that there are no other resistance mechanisms that can be reached by malaria.

And just as those alternative chloroquine resistance pathways are imaginary, Professor Moran's "millions and millions of possible evolutionary outcomes" are imaginary. In the absence of actual evidence that a huge number of relevant unrealized biochemical features could have been built by Darwinian processes—and indeed, in the face of robust contrary empirical evidence—it is illegitimate to arbitrarily multiply probabilistic resources.

Moran is right that there is "no a priori requirement that Earth contain red pines and white pines." But he doesn't follow his own logic far enough. In fact there's no a priori requirement that Earth contain any life at all, or that any life that does exist be able to successfully traverse a mutational pathway by Darwinian means to give rise to a form of life significantly different from itself.

In the absence of an a priori requirement, science is obliged to investigate whether or not such pathways exist. Right now the evidence we have in hand militates strongly against it.

Moran: "This is hard for IDiots to understand."

Me: See my remark above about name-calling.

79. KENNETH MILLER RESISTS CHLOROQUINE RESISTANCE

P. Z. Myers never did respond, but five months after I posted my letter challenging him and Ken Miller to show us their calculations for the development of malaria chloroquine resistance, Miller finally posted a reply on his blog. I was delighted. Again, nothing shows the emptiness of Darwinism better than when very smart, dedicated proponents defend it so poorly.

"Kenneth Miller Resists Chloroquine Resistance," Evolution
News and Views, Discovery Institute, January 14, 2015.

BROWN UNIVERSITY BIOLOGIST KENNETH R. MILLER HAS POSTED a reply[48] to my challenge to him[49] to give a quantitative account for the extreme rarity of the origin of chloroquine resistance in malaria. I'm grateful to him for doing so. Although I strongly disagree with nearly everything he wrote, his essay gives the public a chance to see directly how one informed Darwinist reacts to a basic empirical challenge to the theory.

Last April, a paper by Summers et al.[50] appeared in the *Proceedings of the National Academy of Sciences*, confirming that at least two mutations to the protein PfCRT are required before chloroquine resistance appears in the malaria parasite, as I had surmised in *The Edge of Evolution*. I wrote a series of posts analyzing the result, and another one that challenged Miller and P. Z. Myers (who had downplayed the unsettling implications of the result for Darwinism) to provide their own quantitative—not verbal—account for the extreme rarity of chloroquine resistance. Last month Miller posted an eleven-page reply.

The first two and a half pages of the PDF version of Miller's essay consist of stage-setting and throat-clearing. The last six pages are a reprise of his review of *The Edge of Evolution* and a defense of the evolutionary musings of University of Chicago biologist Joseph Thornton from my criticism. I'll deal with those later. (See essay 82 below.) Miller's only response to my take on the importance of Summers et al. is in the section "Parasites and Drugs."[51] Although the section is less than three pages (including several large figures), as we shall see it includes a number of serious mistakes.

Unfortunately, Miller dodges my challenge to provide a quantitative account of the rarity of the origin of chloroquine resistance. I had asked him to "please keep the rhetoric to a minimum." Alas, to no avail. He cites no relevant numbers, makes no calculations—just words.

Miller begins the section by questioning whether the mutation K76T (which replaces a lysine, "K," at position 76 with a threonine, "T") in the protein PfCRT is important to its ability to transport chloroquine: "There is indeed one required mutation in the PfCRT protein, which is a change of an amino acid at position number 76 from lysine to threonine.... But Behe was dead wrong about it being 'strongly deleterious.' In fact, it seems to have no effect on transport activity at all."[52]

It's nothing short of incomprehensible to claim that the K76T mutation has "no effect on transport activity." Figure 2 of Summers et al.—the very paper to which Miller is referring—shows that one variant of PfCRT (dubbed "D39") with a particular mutation (N75E) has no chloroquine transport activity. When the K76T mutation is added to it to make a double mutant (variant D32), it gains such activity. So it had no effect? Variant E1 has three mutations (none are K76T) but no activity. When K76T is added to those mutations to make the Ecu variant, activity appears. The only difference between the non-transporting and transporting variants is the addition of K76T, but Miller maintains it has no effect on transport activity?

Summers et al. did show in their Figure 4 that, on the background of the mature malaria resistant variant Dd2, which has a total of eight mutations, other amino acid residues could replace T at position 76 and retain activity. But that has nothing to do with the claim that a K76T mutation would have "no effect on transport activity" in a strain that is newly developing chloroquine resistance.

Miller writes, "Quite frankly, [Be]he must be secretly hoping that nobody actually looks at the details in the PNAS paper."[53] Actually, I'm very publicly encouraging everyone to read it with a lot more attention to detail than he did. Anyone who relies on Miller's characterization of the paper will be badly misled.

Miller's claim that the paper shows K76T isn't deleterious is equally unsupported. Summers et al. did not even try to test whether the K76T mutation is deleterious. The word doesn't even appear in their paper. Rather, the researchers were interested mostly in testing what mutations were required just for chloroquine transport activity.

To test if a certain mutation was itself "strongly deleterious" takes particular conditions. That mutation would at least have to be examined: 1) alone on the background of the wild-type sequence (that is, with no other mutations present; more about this later); and 2) in the relevant organism. Yet most of the work described in the Summers et al. paper used frog eggs (*X. laevis*) as a test system, not malaria (*P. falciparum*). Whether a mutation has "no effect on transport activity" in frog eggs says nothing at all about whether it would be deleterious to malaria.

Summers et al. did test a PfCRT variant ("D38") that had only the K76T mutation. It did not transport chloroquine, showing that multiple mutations are needed—which is by far the most ominous result for Darwinism. But since the test system was frog eggs, that of course couldn't determine whether K76T might have any deleterious effect in malaria. The authors did also test five PfCRT variants (encoded on plasmids) in malaria cells in the lab to see how they would affect the cells' survival in the presence of chloroquine. But all had multiple mutations—not K76T alone—so that couldn't determine possible deleterious effects of the single mutation. What's worse, the malaria cells also retained their own, genomic, functional, wild-type PfCRT, which would likely mask any deleterious effects of a nonfunctional plasmid-encoded mutant protein.

One of the variants ("E2") had three mutations, including K76T. The protein could transport chloroquine modestly well in frog eggs, but in malaria cells in the lab it didn't help increase survival rate much above background. That result might indicate that more than those three mutations are needed for net beneficial activity in the wild. But it says nothing about whether a single K76T mutation would be deleterious to malaria cells.

Miller's reading of Summers et al. is seriously mistaken. Sadly, a person who can't accurately report the results of a paper makes for an unreliable guide. I urge everyone who has sufficient background to read at least the disputed parts of Summers et al. Determine for yourself which account is correct.

80. THE VERY NEUTRAL KENNETH MILLER

"The Very Neutral Kenneth Miller," Evolution News and
Views, Discovery Institute, January 15, 2015.

ABOVE, I SHOWED MILLER'S CLAIM THAT THE K76T MUTATION IN the chloroquine-resistance protein factor PfCRT had "no effect on transport activity" was simply wrong. Also wrong was his claim that Summers et al.[54] had shown the single mutation not to be deleterious. Here I rebut several other statements of Miller's concerning whether the mutation is selectively neutral.

He writes:

In fact, a 2003 study recommended against using the K76T mutation to test for chloroquine resistance since that same mutation was also found in 96% of patients who responded well to chloroquine. Clearly, K76T wouldn't have become so widespread if it were indeed "strongly deleterious," as Behe states it must be. This is a critical point, since Behe's probability arguments depend on this incorrect claim.[55]

Wrong again. A mutation could easily be both deleterious by itself yet widespread in a population if an organism has acquired other, compensating mutations that block its injurious effects. In the case of malaria this could happen in the following highly plausible scenario. First, as chloroquine is deployed in an afflicted country, the initial ultra-rare two required mutations eventually appear together in one cell (the deleterious-by-itself K76T plus a second mutation that confers transporting activity) and are selected to survive in the presence of chloroquine.

After those first two required mutations get their proverbial foot in the door and allow the altered gene to increase in the population, subsequent additional helpful mutations could do two things: 1) improve chloroquine-transporting, and 2) compensate for the destabilizing effect of K76T and/or other mutations on the normal, required function of PfCRT. (Malaria cells cannot survive without that protein, even in the absence of chloroquine.) The next milestone is that, once that multi-mutated protein becomes widespread, chloroquine is rendered medically useless and is prescribed much less frequently in the geographic region. With diminished pressure from chloroquine, the protein can then back-mutate to improve its necessary native function and spread in the population, without regard to maintaining the now-unnecessary transporting ability.

Consistent with this scenario are all of the following: The study Miller cites[56] did not check for other, possibly compensating mutations in PfCRT, only for K76T. Chloroquine had been the drug of choice in India, the locale of the study, but was discontinued as the first-line drug way back in 1973—thirty years before the study Miller cites—when chloroquine-resistant malaria became endemic.[57] Work in Malawi[58] and China[59] has shown that chloroquine-resistant malaria are replaced rapidly by sensitive malaria once the drug is discontinued in a region, likely reflecting selective pressure to alter the PfCRT protein in the absence of chloroquine. Although the Malawi work showed simple replacement of the resistant strain by the original unmutated one, other reversion pathways may well be possible. In doing so the PfCRT from the study Miller cites could easily have retained K76T if it also retained compensating mutations.

The result—widespread, chloroquine-sensitive K76T PfCRT in a population—is medically and epidemiologically interesting, but would have nothing to do with how resistance originally arose. It shouldn't need pointing out to professional biologists that the major unsolved problem for Darwinian evolution is to explain how a new beneficial property first appears under a new selective pressure—in this case, how chloroquine resistance arises from the wild-type protein—not how a feature decays once selective pressure is removed.

Here's a homey analogy to help explain. Suppose you wanted to replace the support columns of an old porch. If you simply pulled one away, the porch roof might collapse. But if you first braced the roof with strong poles, the columns could safely be replaced. Ken Miller is in the position of someone who points to a braced porch under repair, and claims that it shows that removing a column from a normal, unbraced structure would not be harmful. He writes: "A neutral mutation like this can easily propagate through a population, and field studies of the parasite confirm that is exactly what has happened."[60]

In my last post I showed Miller's claim that Summers et al. found K76T to be selectively neutral was incorrect. Above I showed that the study he cites, showing the mutation is found in chloroquine-sensitive malaria, is easily compatible with its being deleterious on its own. But is there any direct, positive, experimental evidence indicating whether a single, uncompensated K76T mutation is deleterious or neutral?

Yes, there is. As I wrote earlier, to see if a mutation is harmful by itself, at the very least you have to test it without any other mutations present in the relevant organism. Lakshmanan et al. did this carefully in the lab in 2005:

> To test whether K76T might itself be sufficient to confer VP-reversible [chloroquine resistance] in vitro... we employed allelic exchange to in-

troduce solely this mutation into wild-type *pfcrt* (in GC03). From multiple episomally transfected lines, one showed evidence of K76T substitution in the recombinant, full-length *pfcrt* locus (data not shown). However, these mutant parasites failed to expand in the bulk culture and could not be cloned, despite numerous attempts. *These results suggest reduced parasite viability resulting from K76T in the absence of other* pfcrt *mutations.* This situation is not reciprocal however, in that parasites harboring all the other mutations except for K76T (illustrated by our back-mutants) show no signs of reduced viability in culture. [emphasis added][61]

Frankly, it was never a good bet that the K76T mutation—a nonconservative change in a required protein that's likely to be near a binding site—would be selectively neutral. The best relevant experimental evidence indicates that K76T is indeed strongly deleterious by itself in the wild-type protein, but not if compensatory mutations are present. Miller's claim that the individual mutation is neutral is wrong.

81. THE MANY PATHS OF KENNETH MILLER

"The Many Paths of Kenneth Miller," Evolution News and Views, Discovery Institute, January 16, 2015.

IN MY LAST TWO ESSAYS I SHOWED MILLER'S CLAIM THAT THE K76T mutation in the chloroquine-resistance protein factor PfCRT had "no effect on transport activity" was simply incorrect, and that his argument that the mutation is selectively neutral strongly conflicts with the best relevant experimental evidence. Here I deal with his discussion that the existence of several pathways leading from partial to full chloroquine resistance somehow mitigates the improbability of its origin from zero resistance, and the baleful implications for Darwinism.

Miller writes, "Directly contradicting Behe's central thesis, the *PNAS* study also showed that once the K76T mutation appears, there are *multiple* mutational pathways to drug resistance."[62]

And once you have jumped over the Grand Canyon, there are *multiple* pretty trails you can explore....

Here Miller's mistake helps us to see the bad effects of the lack of quantitative rigor. He exclaims that there is not just one pathway from partial to full chloroquine resistance; there are *"several mutational routes."* His italics indicate that he thinks this is very important. But a moment's quantitative thought shows that the number of pathways makes precious little practical difference. Suppose you were given a choice of a billion trillion roads to travel, but were told that only an unknown one of them led to safety; the others all led to certain death. You would likely feel pretty pessimistic about your chances. But suppose someone came along to tell you that there are actually several paths—in fact, five of them! So your odds of finding a safe path home have jumped from one in a billion trillion to five in a billion trillion. Feeling better? I didn't think so.

If the odds of finding something by one path are an astronomically unlikely 1 in 10^{20}, then the odds of finding it by either of two equally probable ones are 2 in 10^{20}—still astronomically unlikely. Even with a hundred paths the odds would be a profoundly prohibitive 1 in 10^{18}. So it turns out that Miller's strongly emphasized point has little practical importance. To avoid being misled by our imprecise intuitions, it is necessary to be as quantitative as the data allow.

The problem is actually somewhat worse than the above discussion indicates. Here's how. Suppose there were an enormous number of routes a traveler could take. Almost all lead to oblivion, but one or more (it's unknown how many) lead to a particular safe destination. After counting many, many travelers embarking and arriving by whatever route they happened upon, we reliably determined that approximately one in a billion trillion arrived at the safe port. That means the odds of finding *any* safe route to the destination is one in a billion trillion. It does not matter if in reality there are a thousand routes or just one, the odds of finding one of them remain one in a billion trillion. That's because our numbers are derived from statistics, not from some preconceived, theoretical way of arriving at the destination.

The same goes for chloroquine resistance. The number of 1 in 10^{20} against developing chloroquine resistance comes from estimating the number of malaria cells without resistance that it takes to produce and select one with resistance, no matter what genetic route is taken. So the number of routes that Miller emphasizes turns out to have no effect at all on the statistical likelihood of developing chloroquine resistance. Each route itself is actually less likely than the cumulative probability. All of the routes together add up to only 1 in 10^{20}.

Miller writes:

> In most of these [pathways], each additional mutation is either neutral or beneficial to the parasite, allowing cumulative natural selection to gradually refine and improve the parasite's ability to tolerate chloroquine. One of those routes involves a total of seven mutations, three neutral and four beneficial, to produce a high level of resistance to the drug. Figure 4, taken from the Summers et al. paper, makes this point in graphic fashion, showing the multiple mutational routes to high levels of transport, which confer resistance to chloroquine.[63]

A chain is only as strong as its weakest link. And a Darwinian pathway is only as likely as its most improbable step. It matters not a whit whether later helpful mutations are easily acquired. It matters only that some steps are exceedingly unlikely. *All* of the pathways in Miller's Figure 4 (which is Figure 3 from Summers et al.[64]) require K76T—the most difficult, daunting change—as the first or second step. The pleasantly colored crisscrossing arrows of the figure might distract a person's gaze, but they are not intended to quantitatively represent the improbabilities of the transitions.

Here's an analogy. Suppose to win a prize you had to match seven numbers. The last five can be any number between one and three. The first two numbers, however, might be anything between one and a billion. Of course it doesn't matter that the last five are pretty easy to guess, or that you might be permitted to guess them in any order. The steps that overwhelmingly control your odds of winning are the first two, the most improbable ones.

Here's another one. Suppose there were a nice, pretty, level meadow where a person could easily walk around, smelling the flowers—but the meadow was situated on the top of a sheer-cliffed butte hundreds of feet high. If a travel agent told you how easy it was to walk around the pretty meadow without mentioning the brutal climb it would take to get there you would rightly conclude that, whatever other admirable qualities he might have, he was an unreliable guide.

Let me also emphasize, if any of the pathways or other factors Miller discusses made much difference, then the odds of malaria developing chloroquine resistance would be better than they are known to be. Miller's argument has both a quantitative and a conceptual problem. He agrees that the development of chloroquine resistance is an extremely rare 1 in 10^{20}, but he doesn't know why. He seems to really want one beneficial mutation to be available at a time, but the mutation rate in malaria is about a trillion times greater than the origin of chloroquine resistance. So why is the origin of resistance so rare? Resistance to another antimalarial drug (atovaquone) arises de novo in nearly every infected person it's given to. So why is de novo chloroquine resistance much, much, much less frequent?

That question is the bane of Miller's perspective. Figure 3 of Summers et al. (Miller's "Figure 4") shows that it takes a minimum of two mutations for chloroquine transport function to appear, that before both of them appear there is zero activity. That is the big problem for the evolution of resistance. And that is the reason why de novo chloroquine resistance appears so much less frequently than resistance to other antimalarial drugs that require only one mutation.

Miller: "There are indeed *several mutational routes* to drug resistance, and they are indeed the result of sequential, not simultaneous mutations."

Sequential versus simultaneous is the wrong distinction. The only question relevant to Darwinian evolution is whether the helpful, selectable activity appears incrementally, with each additional mutation. Summers et al. shows that it doesn't. There is zero chloroquine-transport activity until two mutations have occurred to the wild-type sequence. The relevant activity appears discontinuously, not incrementally.

82. MILLER STEPS ON DARWIN'S ACHILLES' HEEL

In the previous three essays, I showed that Miller's claims concerning the evolution of the chloroquine-transporting protein PfCRT of malaria were at best contrary to strong experimental evidence and at worst simply wrong. That section occupied less than a third of his essay. Now I'll move on to the remaining two-thirds. Mercifully, that can be dealt with more briefly here, because in previous writings I have already answered almost all the objections he raises. For most of that final portion Miller simply calls on the prominent biologists Sean Carroll and Joseph Thornton for help. My answers to Carroll and Thornton can be found elsewhere in this book.

"Kenneth Miller Steps on Darwin's Achilles' Heel," Evolution
News and Views, Discovery Institute, January 17, 2015.

JUST ONE ISSUE REMAINS. IN A SECTION TITLED "RIGGING THE Odds"[65] (the hyperlinked title to that section in the HTML document is more bluntly called "Fabricating the Odds"), Miller objects to my saying that if two new protein binding sites were required to evolve for some new, useful, selectable function, the likelihood would be about the square of the odds of one new selectable protein binding site evolving. I put the latter odds at about 1 in 10^{20}, about the same as the odds of malaria developing chloroquine resistance (which I dubbed a "Chloroquine Complexity Cluster," or CCC). The odds of two required sites evolving in my model would then be around 1 in 10^{40}—a very large number indeed, and what I argued was the "edge," the limit, of Darwinian evolution. That would be a big problem for the theory, since most proteins occur in cells as complexes of six or more.

Miller grants the value of 1 in 10^{20} for purposes of discussion ("Let's accept Behe's number of 1 in 10^{20} for the evolution of a complex mutation like his CCC"). But he balks at multiplying the odds for the development of two required sites to get the ultra-huge value of 10^{40}, calling it a "breathtaking abuse of statistical genetics." It's more likely he lost his breath from the speed at which he switched models (see below).

He points to what might happen with chloroquine and a second anti-malarial drug of the same efficacy:

> Chloroquine resistance arose in just a decade and a half, and is now common in the gene pool of this widespread parasite. Introduce a new drug for which the odds of evolving resistance are also 1 in 10^{20}, and we can expect that it will take just about as long, 15 years, to evolve resistance to the second drug. Once you get that first CCC established in a population, the odds of developing a second one are not CCC squared. Rather, they are still 1 in 10^{20}. Behe gets his super-long odds by pretending that both CCCs have to arise at once, in the same cell, purely by chance. They don't.[66]

Miller shows here that he has simply misunderstood the central argument of *The Edge of Evolution.* In my book I stipulated that the two sites in my proposed scenario were interdependent, linked: "Now suppose that, in order to acquire some new, useful property, not just one but two new protein-binding sites had to develop." The sites are defined as belonging to the subset of beneficial traits named "Only Selectable When Partner Site Has Also Evolved." In my thought experiment, there is no selectable property in the presence of only one such site. I postulated that only when the two have developed do we get such an effect. (For example, suppose in Miller's inapt illustration above the two drugs were chemically tied together—perhaps they were simply different regions of the same molecule. In that case resistance would have to be developed against both at once to do any good. And the likelihood of that would indeed be the multiple of the odds of developing resistance against each one separately.)

Miller's incongruous response essentially is to say: "No, I have decided to change your own model. I will switch the premise to one in which each protein binding site will necessarily be beneficial by itself." Much worse, he doesn't tell his readers that's what he's doing. His writing leads them to believe he is describing the same situation as I did. Let me be clear: if Miller had simply said that he thought there would be no actual situations in nature like I modeled—that the subset was empty; that never in reality would two new protein binding sites be needed before any new selectable property resulted—then that would have been fine. We could have argued amicably about whether that was true. But he didn't. Instead he conjured an entirely separate scenario, and then claimed it was I—not he—who was "fabricating," trying to deceive readers with a "statistical trick."[67]

Miller's efforts to divert readers' attention from features that require multiple mutations follows inexorably from Darwinism's profound Achilles' heel. Let's play off Miller's two-antimalaria-drugs example to help see what it is. He wrote that, sure, the odds of malaria developing resistance to chloroquine are about 1 in 10^{20}. But if a second drug came along the odds would still be 1 in 10^{20}. They wouldn't be multiplied, he said.

Well, we can note that the odds of developing resistance to the malaria drug atovaquone are only about 1 in 10^{12}.[68] We can then ask, why is the probability so much better for atovaquone than for chloroquine? And following Miller's lead we can ask, if malaria developed resistance to atovaquone at a frequency of 1 in 10^{12}, shouldn't it subsequently develop resistance to chloroquine at 1 in 10^{12}? Why not just another round of 1 in 10^{12}? Why the jump to 1 in 10^{20}?

Enter Achilles and his heel. It turns out that the odds are much better for atovaquone resistance because only one particular malaria mutation is required for resistance. The odds are astronomical for chloroquine because a minimum of *two* particular malaria mutations are required for resistance. Just one mutation won't do it, not even a little bit. For Darwinism, that is the troublesome significance of Summers et al.: "The findings presented here reveal that the minimum requirement for (low) CQ transport activity... is two mutations."[69]

Darwinism is hounded relentlessly by an unshakeable limitation: if it has to skip even a single tiny step—that is, if an evolutionary pathway includes a deleterious or even neutral mutation—then the probability of finding the pathway by random mutation decreases exponentially. If even a few more unselected mutations are needed, the likelihood rapidly fades away.

Without telling his readers, Miller switches from my model to one with the tendentious assumption that new protein binding sites would necessarily always be helpful on their own. (New protein binding—that sounds so nice. What could possibly go wrong?[70]) Yet one mutation in the chloroquine-resistance protein isn't helpful at all. In fact the best evidence indicates it is harmful on its own. Two mutations are needed before it's helpful. So why should we think that just one binding site must always be helpful? Who made up that rule? The answer is that we have no particular reason to think it, and good reason to disbelieve it.

So what should we conclude from all this? Miller grants for purposes of discussion that the likelihood of developing a new protein binding site is 1 in 10^{20}. Now, suppose that, in order to acquire some new, useful property, not just one but two new protein-binding sites had to develop. In that case the odds would be the multiple of the two separate events— about 1 in 10^{40}, which is somewhat more than the number of cells that have existed on earth in the history of life. That seems like a reasonable place to set the likely limit to Darwinism, to draw the edge of evolution.

PART SEVEN: DARWIN DEVOLVES

ONE OF MY FAVORITE SINGERS IN THE 1970S WAS NEIL DIAMOND. On his album *Hot August Night* he recorded the funny ode to country lovin', "You're So Sweet, Horseflies Keep Hangin' 'Round Your Face." The song had the memorable lyric about his sweetheart: "Front teeth missin', that's fine for kissin'."

It turns out that, in love, improvements can often be made by losing things that were already there—lose a few front teeth, lose some weight, lose a bad habit. So, too, with machinery. One quick way to improve the gas mileage of your car might be to break off the side mirrors. Without them, there's less wind resistance. Of course the mirrors are then gone, but maybe you're in an unusual situation where you don't need them as much as the slight uptick in gas mileage.

Living organisms contain many molecular machines. In recent years a surprising discovery is that a very great deal of beneficial, helpful evolutionary change occurs by degradatory mutations—by breaking or throwing away things that were already there. Understanding that process can tell us how organisms may adapt, but it's a huge difficulty for the idea that Darwinian mechanisms built life.

In 1859 Darwin proposed that evolution worked mainly by natural selection culling random variation. If a variation occurred that was helpful to an organism's survival, the organism would tend to live longer and produce more offspring. In turn, the offspring that inherited the variation would also prosper. Over time the variation would become the norm as it spread throughout the population. Darwin and his contemporaries, however, had no idea what caused biological variation.

Almost a century later, the structure of DNA was determined and the genetic code broken. Biological variation, modern science discovered, was based on mutations, i.e., changes in the sequence of a species' DNA. Since Darwin's mechanism depends on molecular changes, helpful mutations have to be tracked down at the molecular level in order to properly determine the scope of his theory. Only in the past few decades has science—through the development of new laboratory techniques—acquired the ability to do so. As I showed in 2019 in *Darwin Devolves*, it can now be seen that random mutation degrades genes, and sometimes that can help a species survive.

To put a point on it, Darwin's mechanism is actually powerfully devolutionary—it squanders information for short-term gain. Needless to say, that is not a process that would build complex biochemical systems in the first place.

83. EXCERPT FROM "EXPERIMENTAL EVOLUTION"

Excerpt from "Experimental Evolution, Loss-of-Function Mutations, and 'The First Rule of Adaptive Evolution,'" *Quarterly Review of Biology*, 2010.

IN *THE ORIGIN OF SPECIES* DARWIN EMPHASIZED THE RELENTLESS-ness of natural selection: "Natural selection is daily and hourly scrutinising, throughout the world, every variation, even the slightest; rejecting that which is bad, preserving and adding up all that is good; silently and insensibly working, whenever and wherever opportunity offers, at the improvement of each organic being in relation to its organic and inorganic conditions of life."[1]

Yet he realized that the changes that were selected to adapt an organism to its environment did not have to be ones that conferred upon it a new ability such as sight or flight. In some barnacles Darwin discovered unexpected cases of the gross simplification of an organism: "The male is as transparent as glass... In the lower part we have an eye, & great testis & vesicula seminalis: in the capitulum we have nothing but a tremendously long penis coiled up & which can be exserted. There is no mouth no stomach no cirri, no proper thorax! The whole animal is reduced to an envelope... containing the testes, vesicula, and penis."[2]

Adaptive evolution can as easily lead to the loss of a functional feature in a species lineage as to gain of one. To become better adapted to their environments, over the course of evolutionary history snakes have lost legs, cavefish have lost vision, and the parasitic bacterium *Mycoplasma genitalium* has lost its ability to live independently in the wild. Whatever variation sufficiently aids a particular species at a particular moment in a particular environment—be it the gain or loss of a feature, or the simple modification of one—may be selected. Since species can evolve to gain, lose, or modify functional features, it is of basic interest to determine whether any of these tends to dominate adaptations whose underlying molecular bases we can ascertain. Here I survey the results of evolutionary laboratory experiments on microbes that have been conducted over the past four decades. Such experiments exercise the greatest control over environmental variables and they yield our most extensively characterized results at the molecular level....

The First Rule of Adaptive Evolution

As seen in Tables 2 through 4, the large majority of experimental adaptive mutations are loss-of-FCT [a "Functional Coded elemenT such as a gene, protein, or control region] or modification-of-function ones.[3] In fact, leaving out those experiments with viruses in which specific genetic elements were intentionally deleted and then restored by subsequent evolution, only two gain-of-FCT events have been reported—the development of the ability of a fucose regulatory protein to also respond to d-arabinose[4] and antibiotic gene capture by f1.[5] Why is this the case? One important factor is undoubtedly that the rate of appearance of loss-of-FCT mutations is much greater than the rate of construction of new functional coded elements. Suppose an adaptive effect could be secured by diminishing or removing the activity of a certain protein. If the gene for the protein were, say, 1,000 nucleotides in length, then there would be numerous targets of opportunity for a loss-of-FCT mutation. The deletion of any single nucleotide in the coding sequence would alter the reading frame and likely destroy or greatly diminish protein activity. Insertion of a nucleotide anywhere in the coding sequence would do the same. Longer insertions or deletions would commonly have the same effect, as would alteration of a codon from sense to nonsense. All these would fall into the category of loss-of-FCT mutations.

Nucleotide substitutions resulting in missense mutations, although they would likely not completely eliminate protein activity, are very likely to diminish activity to a greater or lesser extent since in multiple experiments the majority of amino acid substitutions have been found to decrease a protein's activity.[6] Although, if residual protein activity remained, these would be categorized as modification-of-function mutations under the accounting system used here, the partial diminishment of the function provides the adaptive effect. (A caveat: in the particular case of microbes that were first allowed to accumulate deleterious mutations and then recover, such as Bull et al.,[7] it is likely that many adaptive point mutations are compensatory for initial deleterious mutations and therefore increase (mutated) protein function. Biochemical analysis would be needed to prove that protein function increased or decreased and that was the basis of the adaptive effect.) If the basic point mutation rate per nucleotide per generation were 10^{-9}, then, because of the many ways to decrease the activity of a protein, the rate of appearance of an adaptive loss-of-FCT mutation would likely be on the order of 10^{-6}. Indeed, an adaptive mutation rate of $\sim 10^{-5}$ was recently measured in E. coli.[8]

Contrast this with the situation in which, to gain an adaptive effect, a particular nucleotide in a gene for a certain protein had to be mutated. (The particular mutation can be supposed to help code for a new binding site in the protein, construct a new genetic control element from a sequence that was already a near-match, or other possibilities. These would be gain-of-FCT mutations.) If the basic point mutation rate per nucleotide were 10^{-9}, then that would also be the rate of appearance of the beneficial mutation. Even if there were several possible pathways to construct a gain-of-FCT mutation, or several possible kinds of adaptive gain-of-FCT features, the rate of appearance of an adaptive mutation that arises from the diminishment or elimination of the activity of a protein is expected to be 100–1,000 times the rate of appearance of an adaptive mutation that requires specific changes to a gene.

This reasoning can be concisely stated as what I call "The First Rule of Adaptive Evolution": *Break or blunt any functional coded element whose loss would yield a net fitness gain.*

It is called a "rule" in the sense of being a rule of thumb. It is a heuristic, a useful generalization, rather than a strict law: other circumstances being equal, this is what is usually to be expected in adaptive evolution. Since the rule depends on very general features of genetic systems (that is, the mutation rate and the probability of a loss-of-FCT versus a gain-of-FCT mutation) it is expected to hold for organisms as diverse as viruses, prokaryotes, and multicellular eukaryotes. It is called the "first" rule because the rate of mutations diminishing the function of a feature is expected to be much higher than the rate of appearance of a new feature, so adaptive loss-of-FCT or modification-of-function mutations that decrease activity are expected to appear first by far in a population under selective pressure.

ILLUSTRATIONS OF THE FIRST RULE

THE FIRST rule, gleaned from laboratory evolution experiments, can be used to interpret data from evolution in nature, including human genetic mutations in response to selective pressure by malaria (Table 1). Hundreds of distinct mutations are known that diminish the activities of G6PD or the α- or β- chains of hemoglobin (leading to thalassemia). Yet it is estimated that the gain-of-FCT mutation leading to sickle hemoglobin has arisen independently only a few times, or perhaps just once, within the past 10,000 years.[9] So loss-of-FCT adaptive mutations in this situation appeared several orders of magnitude more frequently than did a gain-of-FCT mutation. Nonetheless, the sickle hemoglobin mutation did arise and spread in a regional population. Thus, if a gain-of-FCT mutation such as the sickle gene has a sufficiently large selection coefficient then, even though adaptive loss-of-FCT mutations arrive more rapidly and in greater numbers, it is possible for the gain-of-FCT mutation to outcompete them.

Another illustration from nature of the first rule can be seen in the adaptive evolution of the plague bacterium *Yersinia pestis* in the past 1,500–20,000 years.[10] A likely evolutionary scenario for its great virulence is that the plague bacterium serially acquired several plasmids that conferred upon it the ability to be transferred by flea bite.[11] The 101 kb pFra plasmid carries the *yplD* gene, which codes for a phospholipase D that is necessary for the survival of the bacterium in the flea proventriculus. The 9.6 kb pPla plasmid codes for a plasminogen activator, which allows the bacterium to move in its host unhindered by blood clotting. The acquisitions of these genes are gain-of-FCT events. *Y. pestis* has also subsequently lost a large number of chromosomal genes. A general estimate is 150 genes lost.[12] A number of discarded genes have activities in other *Yersinia* species that allow pathogen-host adhesion. The plague bacterium has also acquired hundreds of missense mutations.[13] The discarded genes are of course loss-of-FCT mutations, and many missense mutations are likely to diminish activity. Thus, the organism adapted relatively quickly to its new lifestyle—first made possible by several gain-of-FCT events—through much more numerous loss-of-FCT and modification-of-function mutations.

Two recent laboratory studies also illustrate the working of the first rule. Ferenci[14] discusses the adaptive value of mutations to the gene for a specialized *E. coli* RNA polymerase σ factor that contributes to the general stress response. The author observed that "*rpoS* mutations occurred, and indeed spread at rapid rates within a few generations of establishing glucose-limited chemostats," and also that "the majority of *rpoS* mutations accumulating in glucose-limited cultures are loss-of-function mutations with little or no residual RpoS protein.... The mutations include stop codons, deletions, insertions as well as point mutations." A second study showed that the loss of mating genes during asexual growth in

Saccharomyces cerevisiae provided a 2% per-generation growth-rate advantage.[15] The authors noted, "In bacteria, gratuitous gene expression reduces growth rate.... We suspect that the cost of gene expression is not specific to bacterial enzymes or genes in the yeast mating pathway, but rather reflects a universal cost of gene expression and that this cost must be borne in all environments where the gene is expressed." Thus, in any environment in which a gene becomes superfluous or, more generally, in any environment where its loss would yield a net fitness gain, the frequent mutations occurring in the population that tend to eliminate the functional coded element will turn adaptive.

How Frequently Are Loss-of-FCT and Gain-of-FCT Mutations Adaptive?

ALTHOUGH THE rates of appearance of loss-of-FCT and modification-of-function mutations that degrade protein activity are always expected to be much greater than the rate of appearance of gain-of-FCT mutations, a separate question concerns what fractions of the mutations in those categories are adaptive. That is, although loss-of-FCT mutations might appear rapidly, if they do not yield a selective benefit—if they are not adaptive—then they will not usually spread in a population. In the same vein, a gain-of-FCT mutation may eventually appear that builds some new genetic feature such as, say, a transcription factor binding site, yet if the feature is not adaptive within the organism's genetic context, it will not be selected.[16]

These are empirical questions that are difficult to answer conclusively. However, the data and experiments discussed in this review offer some insights. In the most open-ended laboratory evolution experiment, that of Lenski,[17] in which no specific selection pressure was intentionally brought to bear, all of the adaptive mutations so far identified have either been loss-of-FCT or modification-of-function ones, with strong reasons to think that most of the modification-of-function mutations diminished protein activity. Except in cases where specific genetic features were first removed, plus antibiotic gene capture by f1, all adaptive mutations in laboratory evolution experiments with viruses also seem to be loss-of-FCT or modification-of-function. Thus, in general laboratory evolutionary situations (that is, where a microorganism was under a general selective pressure rather than a specific one), adaptive loss-of-FCT or modification-of-function mutations were always available. That was not the case for gain-of-FCT mutations.

One objection might be that the above examples are artificial. They concern laboratory evolution, and it may be that diminished expression of some pre-existing, commonly held genes gives an organism an advantage over its conspecifics in such a constant environment, but not in the varied and changing environments of nature. It is true that the laboratory is an artificial environment, and the opportunities for some events that occur at irregular intervals in the wild, such as lateral gene transfer, are essentially nonexistent. This is clearly an area that needs to be addressed in more detail. Further laboratory evolution studies with more complex environments or cultures of mixed species would shed light on the extent to which such factors affect opportunities for adaptation by gain- or loss-of-FCT mutations. Nonetheless, results arguably similar to those that have been seen in laboratory evolution studies to date have also been seen in nature, such as the loss of many genes by *Yersinia pestis* (after, of course, acquisition of new genetic material in the form of several plasmids), and the loss-of-FCT mutations that have spread in human populations in response to selective pressure from malaria. A tentative conclusion suggested by these results, then, is that the complex genetic systems that are cells will often be able to adapt to selective pressure by effectively removing or diminishing one or more of their many functional coded elements.

A second possible objection is that many of the reviewed experiments were conducted on comparatively small populations of microbes for relatively short periods of time, so that, although loss-of-FCT and modification-of-function mutations might be expected to occur, there simply was not much opportunity to observe gain-of-FCT mutations. After all, one would certainly not expect new genes with complex new properties to arise on such short timescales. Although it is true that new complex gain-of-FCT mutations are not expected to occur on short timescales, the importance of experimental studies to our understanding of adaptation lies elsewhere. Leaving aside gain-of-FCT for the moment, the work reviewed here shows that organisms do indeed adapt quickly in the laboratory—by loss-of-FCT and modification-of-function mutations. If such adaptive mutations also arrive first in the wild, as they of course would be expected to do, then those will also be the kinds of mutations that are first available to selection in nature. This is a significant addition to our understanding of adaptation. That knowledge is also a necessary prerequisite for elucidating the nature of long-term adaptation, because consideration of how long-term adaptation proceeds must take into account how organisms adapt in the short term.

Returning to the topic of gain-of-FCT mutations, although complex gain-of-FCT mutations likely would occur only on long timescales unavailable to laboratory studies, simple gain-of-FCT mutations need not take nearly as long. As seen in Table 1, a gain-of-FCT mutation in sickle hemoglobin is triggered by a simple point mutation, which helps code for a new protein binding site. It has been estimated that new transcription-factor binding sites in higher eukaryotes can be formed relatively quickly by single point mutations in DNA sequences that are already near-matches.[18] In general, if a sequence of genomic DNA is initially only one nucleotide change removed from coding for an adaptive functional element, then a single simple point mutation could yield a gain-of-FCT. As seen in Table 5, several laboratory studies have achieved thousand- to million-fold saturations of their test organisms with point mutations, and most of the studies reviewed here have at least single-fold saturation. Thus, simple gain-of-FCT adaptive mutations that had sufficient selective value to outcompete more numerous loss-of-FCT or modification-of-function ones would be expected to have been observed in most experimental evolutionary studies if they had been available.

A third objection could be that the time and population scales of even the most ambitious laboratory evolution experiments are dwarfed when compared to those of nature. It is certainly true that over the long course of history many critical gain-of-FCT events occurred. However, that does not lessen our understanding, based on work by many laboratories over the course of decades, of how evolution works in the short term, or of how the incessant background of loss-of-FCT mutations may influence adaptation.

Conclusion

Adaptive evolution can cause a species to gain, lose, or modify a function. Therefore, it is of basic interest to determine whether any of these modes dominates the evolutionary process under particular circumstances. The results of decades of experimental laboratory evolution studies strongly suggest that at the molecular level loss-of-FCT and diminishing modification-of-function adaptive mutations predominate. In retrospect, this conclusion is readily understandable from our knowledge of the structure of genetic systems, and is concisely summarized by the first rule of adaptive evolution. Evolution has a myriad of facets; this one is worthy of some notice.

84. A Reply to Jerry Coyne

"The First Rule of Adaptive Evolution: A Reply to Jerry Coyne, Evolution News and Views, Discovery Institute, December 16, 2010.

At his blog, Why Evolution is True,[19] Jerry Coyne, professor of evolutionary biology at the University of Chicago, has been analyzing my recent paper, "Experimental Evolution, Loss-of-Function Mutations, and 'The First Rule of Adaptive Evolution,'" which appears in the latest issue of the *Quarterly Review of Biology*.[20] Although I usually don't respond to blog posts I will this time, both because Coyne is an eminent scientist and because he does say at least one nice thing about the paper.

First, the nice thing. About halfway through his comments Professor Coyne writes: "**My overall conclusion:** Behe has provided a useful survey of mutations that cause adaptation in short-term lab experiments on microbes (note that at least one of these—Rich Lenski's study—extends over several decades)."

Thanks. Much appreciated.

Next, he turns to damage control. Directly after the mild compliment, Coyne registers his main complaint about the paper: the conclusions supposedly can only be applied to laboratory evolution experiments, and say little about (Darwinian) evolution in nature. Coyne writes, "But his conclusions may be misleading when you extend them to bacterial or viral evolution in nature, and are certainly misleading if you extend them to eukaryotes (organisms with complex cells), for several reasons." Below I deal with Coyne's three reasons in turn.

Professor Coyne's first objection is: "In virtually none of the experiments summarized by Behe was there the possibility of adapting the way that many bacteria and viruses actually adapt in nature: by the uptake of DNA from other microbes. Lenski's studies of *E. coli*, for instance, and Bull's work on phage evolution, deliberately preclude the presence of other species that could serve as vectors of DNA, and thus of new FCTs."

Coyne is simply wrong here, at least about phages (bacterial viruses). Viruses grow in other organisms—the cells they infect. Thus, pretty much by definition they are in contact with other microbes for much of their life cycle, and it is thought that sometimes viruses acquire genes from their host cells. In fact, in one report by Bull's group[21] I reviewed, in which the gene for bacteriophage T7 ligase was intentionally removed at the start of an experiment, the investigators reported they initially expected the missing gene to be replaced: "At the outset, our expectation from work in other viral systems was that the loss of ligase activity would remain so deleterious to T7 that recovery to high fitness would require the genome to acquire new sequences through recombination or gene duplication and to replace ligase function by divergence of those sequences."

Unexpectedly, however, "This hope was not realized, and compensatory evolution occurred through point mutations and a deletion." Thus it seems that Professor Coyne's expectations about what is required for a gain-of-FCT event are not universally shared among scientists.

Coyne is of course correct that in experiments in which just one species of bacteria is present, the cells cannot acquire DNA from other species of bacteria. Yet those who conducted such experiments often had expectations for evolution much different from his. In the 1970s and 1980s many workers thought that gene duplication or recruitment plus divergence would allow bacteria to diversify the foodstuffs they could metabolize. Out of many such experiments, only one seemed to work by gain-of-FCT. Professor Coyne's dismissal of such experiments is pure hindsight.

Professor Coyne doesn't mention that in the review I argue that results from nature are consonant with conclusions drawn from lab experiments. I wrote:

> One objection might be that the above examples are artificial. They concern laboratory evolution.... Nonetheless, results arguably similar to those that have been seen in laboratory evolution studies to date have also been seen in nature, such as the loss of many genes by *Yersinia pestis* (after, of course, the acquisition of new genetic material in the form of several plasmids), and the loss-of-FCT mutations that have spread in human populations in response to selective pressure from malaria. A tentative conclusion suggested by these results is that the complex genetic systems that are cells will often be able to adapt to selective pressure by effectively removing or diminishing one or more of their many functional coded elements.

Coyne's second objection is the following: "In relatively short-term lab experiments there has simply not been enough time to observe the accumulation of complex FCTs, which take time to build or acquire from a rare horizontal transmission event."

I addressed that very point in my review:

> Furthermore, although complex gain-of-FCT mutations likely would occur only on long timescales unavailable to laboratory studies, sim-

ple gain-of-FCT mutations need not take nearly as long. As seen in Table 1, a gain-of-FCT mutation in sickle hemoglobin is triggered by a simple point mutation, which helps code for a new protein binding site. It has been estimated that new transcription-factor binding sites in higher eukaryotes can be formed relatively quickly by single point mutations in DNA sequences that are already near matches.... In general, if a sequence of genomic DNA is initially only one nucleotide removed from coding for an adaptive functional element, then a single simple point mutation could yield a gain-of-FCT. As seen in Table 5, several laboratory studies have achieved thousand- to million-fold saturations of their test organisms with point mutations, and most of the studies reviewed here have at least single-fold saturation. Thus, one would expect to have observed simple gain-of-FCT adaptive mutations that had sufficient selective value to outcompete more numerous loss-of-FCT or modification-of-function mutations in most experimental evolutionary studies, if they had indeed been available.

Yes, complex gain-of-FCT events would not be expected to occur, but simple GOF events would. Yet they didn't show up.

Professor Coyne then proceeds to put words in my mouth: "What he's saying is this: 'Yes, gain of FCTs could, and likely is, more important in nature than seen in these short-term experiments. But my conclusions are limited to these types of short-term lab studies.'"

No, that is not what I was saying at all. I was saying that, no matter what causes gain-of-FCT events to sporadically arise in nature (and I of course think the more complex ones likely resulted from deliberate intelligent design), short-term Darwinian evolution will be dominated by loss-of-FCT, which is itself an important, basic fact about the tempo of evolution.

Above I quoted Coyne talking about "complex FCTs, which take time to build or acquire from a rare horizontal transmission event." Yet cells aren't going to sit around twiddling their thumbs until that rare event shows up. *Any* mutation which confers an advantage at *any* time will be selected, and the large majority of those in the short term will be LOF. Ironically, Coyne seems to underestimate the power of natural selection, which "is daily and hourly scrutinising, throughout the world, every variation, even the slightest." A process which scrutinizes life "daily and hourly," as Darwin wrote, isn't going to wait around for some rare event.

Professor Coyne's third objection is: "Finally, Behe does not mention—and I think he should have—the extensive and very strong evidence for adaptation via gain-of-FCT mutations in eukaryotes."

As I show in Table 1 of the review, we have wonderful evidence of what Darwinian evolution has done to a multicellular eukaryotic species—*Homo sapiens*—in response to strong selective pressure from malaria over the past ten thousand years. A handful of mutations have been selected. The mutations are classified as: one GOF (the sickle mutation); two modification of functions; and five LOFs. That's pretty much the proportion of what one sees in bacteria and larger viruses, so there is no reason to think that short-term evolution in eukaryotes has a substantially different spectrum of adaptive mutations than for prokaryotes.

Coyne wants to focus on long-term evolution:

While [eukaryotes] may occasionally acquire genes or genetic elements by horizontal transfer, we *know* that they acquire new genes by the mechanism of *gene duplication and divergence*: new genes arise by duplication of old ones, and then the functions of these once-identical genes diverge as they acquire new mutations.... Think of all the genes that have arisen in eukaryotes in this way and gained novel function: classic examples include genes of the immune system, *Hox* gene families, olfactory genes, and the globin genes.

Unfortunately, Professor Coyne isn't making a critical distinction here. While we may know (or at least have very good evidence that is consonant with the idea) that new genes have arisen by duplication and divergence of old ones in eukaryotes, we do not know that happens by a Darwinian mechanism of random mutation and natural selection. And if some duplicate genes do arise and diversify by Darwinian processes, we do not know that explains all or even most of them. After all, while the long-term processes that Professor Coyne envisions are taking their

sweet time to come together, the fast and dirty short-term adaptive processes will dominate. That's what we know from the great efforts put into
experimental evolutionary studies by many investigators over decades.

And as I point out in the *QRB* paper, all of this can be neatly summarized by The First Rule of Adaptive Evolution: *Break or blunt any
functional coded element whose loss would yield a net fitness gain.*

85. MORE FROM JERRY COYNE

"More from Jerry Coyne," Evolution News and Views,
Discovery Institute, December 24, 2010.

A T HIS BLOG, UNIVERSITY OF CHICAGO PROFESSOR OF EVOLUTION-
ary biology Jerry Coyne has commented[22] on my reply (above) to
his analysis of my new review in the *Quarterly Review of Biology.*[23] This
time he has involved two other prominent scientists in the conversation.
I'll discuss the comments of one of them in this post and the other in a
second post. The first one is University of Texas professor of molecular
biology James J. Bull, who works on the laboratory evolution of bacterial viruses (phages). I reviewed a number of Bull's fascinating papers in
the recent *QRB* publication. Coyne solicited Prof. Bull's comments and
put them up on his blog. Bull says several nice things about my review,
but agrees with Prof. Coyne that he wouldn't expect "novelty" in the lab
evolutionary experiments he and others conducted, and he thinks they
are not a good model of how evolution occurs in nature. (I wonder if he
mentions this in his grant proposals....)

Prof. Bull states that bacteriophage T7 (which he used in his studies) avoids taking up DNA from its host, *E. coli,* so it really isn't an example of a system where novel DNA was available to the phage, despite
his initial hopes that it would be. (In the paper describing the work he
and his co-authors wrote, "At the outset, our expectation from work in
other viral systems was that the loss of ligase activity would... require the
[T7] genome to acquire new sequences through recombination or gene
duplication.") But, he writes in his new post, "what we failed to point out

in our paper, and is fatal to MB's criticism, is the fact that T7 degrades *E. coli* DNA, so even if the phage did incorporate an *E. coli* gene, it might well destroy itself in the next infection." This reasoning strikes me as overlooking an obvious problem, and overlooking an obvious solution to the problem.

First, the problem. If T7, and presumably other bacteriophages, find it advantageous to have a mechanism that excludes host DNA from being incorporated into the phage genome, doesn't this drastically cut down the opportunity for the very mixing of cross-species DNA that Coyne and Bull tout as the Darwinian solution to the problem of developing complex new functions? I suppose they could respond that, well, maybe the phages can't exclude other, non-host DNA, so that's where novel DNA would come from. But it seems host DNA would be by far the DNA the phages contacted the most. But if that is essentially excluded as a source, then the sorts of compensatory mutations that Professor Bull observed in his experiments are still by far the most likely ones to occur in nature. (And the grant application is saved!) It's a matter of rate. The adaptive mutations that come along first will be selected first, and clearly point mutations and deletions come along very rapidly in phage populations.

Next, the solution. If a phage has a mechanism that is preventing it from taking up DNA that could be advantageous to it (such as the gene for a DNA ligase in the case of the experiment of Rokyta et al.[24]) then all it has to do is break that mechanism and the opportunity for acquiring DNA is now opened to it. After all, breaking things is what random mutation does best, and, as I reviewed, many of the reported adaptive mutations in lab evolution experiments resulted from broken genes. Broken genes can also be neutral mutations. In the majority of the cultures of *E. coli* that Richard Lenski has grown for 50,000 generations, "mutator" strains took over. A mutator strain is one which has lost at least part of its ability to repair its DNA. If *E. coli* can toss out part of its repair ability with impunity, why couldn't T7 lose its inability to take up some helpful host DNA?

Professor Bull suggests that lab evolution experiments which use whole cells and viruses aren't needed to show the power of Darwinian processes because that is apparent in experiments using "directed" evolution. I strongly disagree with his assessment. In directed evolution, workers use an experimental setup so that a single, particular gene or protein must mutate to be adaptive. "Directed" evolution is a much, much more artificial system than ones that use whole cells and/or viruses, as he did. In response to some selective pressure, a cell has potentially very many more ways to adapt to deal with it than does a single protein—a cell has thousands of genes and thousands of regulatory elements that can potentially help the cell adapt by gain- or loss-of-function, or tweaking of pre-existing function. On the other hand, directed evolution artificially constrains the system to mutate the component that the experimenter chooses. It seems to me a bit inconsistent for someone to claim that single species of cells (and/or viruses) are insufficiently complex to produce gain-of-function mutations by Darwinian processes, but that artificially constraining mutation/selection to single genes or proteins shows it clearly. Seems to me this is exactly backwards.

In his post Professor Bull describes an experiment he did with co-workers which, they hoped, would mimic the process of gene duplication and divergence. They placed two copies of the same gene, each on its own kind of plasmid, into the same cell. The gene produced a protein that could disable one kind of antibiotic very well, and disable a second kind of antibiotic rather poorly. In the presence of both antibiotics, they expected one of the copies of the gene to stay about the same, degrading the first antibiotic. They expected the second copy of the gene to accumulate point mutations which would help it become more efficient at degrading the second kind of antibiotic (from other publications such mutants were already known to exist.) The system, however, had its own ideas. Bull says that contrary to expectations, one of the genes was deleted and the other gene accumulated point mutations so that it did a decent job degrading both antibiotics.

Professor Bull writes, "This study merely illustrates that the conditions favoring the maintenance of two copies undergoing evolutionary divergence are delicate." Skeptic that I am, instead of "delicate," I would say it illustrates that the conditions are "rare." That is, it demonstrates very nicely that having two copies of a gene under what seem to be ideal conditions for adaptive divergence is not enough. (A similar result using a different system was recently obtained by Gauger et al.[25]) Other factors enter into the result as well. Since we don't know exactly what those other factors are, or how rare they make successful duplication/ divergence events, we should not automatically assume that the occurrence of duplicated and diverged genes in nature happened by unguided, Darwinian processes.

86. EVEN MORE FROM JERRY COYNE

"Even More from Jerry Coyne," Evolution News and Views, Discovery Institute, January 12, 2011.

IN MY LAST ESSAY (ABOVE) I REPORTED THAT UNIVERSITY OF CHIcago evolutionary biologist Jerry Coyne, who had critiqued my recent *Quarterly Review of Biology* article[26] concerning laboratory evolution studies of the last four decades and what they show us about evolution, had asked several other prominent scientists for comments. I replied to those of experimental evolutionary biologist John Bull. In a subsequent post[27] Coyne discussed a recent paper by the group of fellow University of Chicago biologist Manyuan Long on gene duplication in fruit flies.[28] After a bit of delay due to the holidays, I will comment on that here.

Try as one might to keep Darwinists focused on the data, some can't help reverting to their favorite trope: questioning Darwinism simply must be based on religion. Unfortunately Professor Coyne succumbs to this. Introducing his blog post, he writes:

What role does the appearance of new genes, versus simple changes in old ones, play in evolution? There are two reasons why this question has recently become important.... The first involves a scientific controversy.... The second controversy is religious. Some advocates of intelligent design (ID)—most notably Michael Behe in a recent paper—have implied not only that evolved new genes or new genetic "elements" (e.g., regulatory sequences) aren't important in evolution, but that they play *almost no role at all*, especially compared to mutations that simply inactivate genes or make small changes, like single nucleotide substitutions, in existing genes. This is based on the religiously-motivated "theory" of ID, which maintains that new genetic information cannot arise by natural selection, but must installed [*sic*] in our genome by a magic poof from Jebus. [*sic*]

Anyone who reads the paper, however, knows my conclusions were based on the reviewed experiments of many labs over decades. Even Coyne knows this. In the very next sentence he writes, schizophrenically, "I've criticized Behe's conclusions, which are based on laboratory studies of bacteria and viruses that virtually eliminated the possibility of seeing new genes arise, but I don't want to reiterate my arguments here." Yet if my conclusions are based on "laboratory studies," then they ain't "religious," even if Coyne disagrees with them.

Professor Coyne is so upset that he imagines things that aren't in the paper. (They are "implied," you see.) So although I haven't actually written it, supposedly I have "implied not only that evolved new genes or new genetic 'elements'... aren't important in evolution, but that they play *almost no role at all*." [Coyne's emphasis]

"Play almost no role at all"? When I first read these "implied" words that Coyne wants to put in my mouth, I thought the argumentative move rang a bell. Sure enough, check out the Dilbert comic strip from November 1, 2001, where Dilbert complains that a coworker "changed what I said into a bizarre absolute." If one person says that an event is "very unlikely," and an interlocutor rephrases that into "so, you say it's logically impossible and would never happen even in an infinite multiverse," well then, the second fellow is setting up a straw man.

Contrast Coyne's imagined "implications" with what I actually wrote in the review. Considering possible objections to my conclusions I noted:

A third objection could be that the time and population scales of even the most ambitious laboratory evolution experiments are dwarfed when compared to those of nature. It is certainly true that, over the long course of history, many critical gain-of-FCT events occurred. However, that does not lessen our understanding, based upon work by many laboratories over the course of decades, of how evolution works in the short term, or of how the incessant background of loss-of-FCT mutations may influence adaptation.

Although I think that statement is clear enough in the context of the paper, let me say it differently in case some folks are confused. Loss-of-function mutations occur relatively rapidly, and LOF mutations can be adaptive. Gain-of-function mutations can be adaptive, too, but their rate of occurrence (including the rate of gene duplication-plus-divergence that Coyne is discussing) is much less. Thus, whenever a new selective pressure pops up, LOF adaptive mutations (if such there be in the particular circumstance) can appear most swiftly, and will likely dominate short-term adaptation. So when a GOF mutation eventually appears, it will likely be against the altered genetic background of the selective pressure ameliorated by the adaptive LOF mutation(s). In order to understand how evolution works in the long term, we must take that into consideration.

Professor Coyne notes that the new genes studied by Professor Long "arise quickly, at least on an evolutionary timescale." But adaptive LOF mutations arise quickly even on a laboratory timescale. For example, as I note in my *QRB* review, in one experiment adaptive mutations in *E. coli* cultures due to loss-of-function mutations in the *rpoS* gene "occurred, and indeed spread at rapid rates within a few generations of establishing glucose-limited chemostats." A few generations for *E. coli* can be on the order of hours. The gene duplications studied by Professor Long occur on the order of millions of years. Admittedly the situation in nature is more complex than in the laboratory. Nonetheless, whatever selective pressures the gene duplications encounter when they eventually show up will already have been substantially altered by adaptive LOF mutations. That's a very important point for evolutionary biologists to keep in mind.

I have never stated, nor do I think, that gene duplication and diversification cannot happen by Darwinian mechanisms, or that "they play *almost no role at all*" in the unfolding of life. (As a matter of fact, I discussed several examples of that in my 2007 book *The Edge of Evolution*.) That would be silly—why would anyone with knowledge of basic biochemical mechanisms deny that, say, the two gamma-globin coding regions on human chromosome 11 resulted from the duplication of a single gamma-globin gene and then the alteration of a single codon? What I don't think can happen is that duplication/divergence by Darwinian mechanisms can build new, complex interactive molecular machines or pathways. Assuming (since he is in fact critiquing them) Professor Coyne has been attentive to my arguments, one background assumption that he may have left unexpressed is that he thinks the newer duplicated genes discovered by Professor Long's excellent work represent such complex entities, or parts of them.

There is no reason to think so. A gene can duplicate and diversify without building a new machine or network, or even changing function much. The above example of the two gamma-globin genes shows that duplication does not necessarily result in change in function. The examples of delta- and epsilon-globin, which, like gamma-globin, presumably also resulted from the duplication of an ancestral beta-like globin gene, show that sequence can diversify further, but function remain very similar. Even myoglobin, which shares rather little sequence homology with the other globins, has not diverged much in biochemical function.

In his recent work Professor Long discovered that many of the new genes were essential for the viability of the organism—without the gene product, the fruit flies would die before maturity. Perhaps Professor Coyne thinks that that means the genes necessarily are parts of complex systems, or at least do something fundamentally new. Again, however, there is no reason to think so. The notion of "essential" genes is at best ambiguous. We know of examples of proteins that surely appear necessary, but whose genes are dispensable. The classic example is myoglobin. It is also easy to conceive of a simple route to an "essential" duplicate gene that does little new. Suppose, for example, that some gene was duplicated. Although the duplication caused the organism to express more of the protein than was optimum, subsequent mutations in the promoter or protein sequence of one or both of the copies decreased the total activity of the protein to pre-duplication levels. Now, however, if one of the copies is deleted, there is not enough residual protein activity for the organism to survive. The new copy is now "essential," although it does nothing that the original did not do.

To sum up, the important point of "Experimental Evolution, Loss-of-Function Mutations, and 'The First Rule of Adaptive Evolution'" is not that anything in particular in evolution is absolutely ruled out. Rather, the point is that short-term adaptation tends to be dominated by LOF mutations. And, tinkerer that it is, Darwinian evolution always works in the short term.

Here's an analogy that some people might find amusing and helpful. Think of GOF mutations (such as the gene duplication/divergence that Professor Coyne discusses) as the "snail mail" of evolution. And think of LOF mutations as the e-mail, texting, and phone calls of evolution. In a busy world, by the time a real letter shows up at someone's or some business's door, a lot of communication concerning the subject of the slow letter would already have happened by faster means, and the more important the topic, the more fast communication there likely would have been. That speedy communication can quite easily change the context of the letter, and either render it moot or at least less important. It is certainly possible that on occasion the slow letter will arrive with its impact unaffected by other messages, but it would be foolish to ignore the effect of the fast channels of communication.

87. MUCH ADO ABOUT YEAST

"More Darwinian Degradation: Much Ado about Yeast,"
Evolution News and Views, Discovery Institute, 2012.

RECENTLY A PAPER BY RATCLIFF ET AL.[29] ENTITLED "EXPERIMENTAL Evolution of Multicellularity" appeared and received a fair amount of press attention, including a story in the *New York Times*. The authors discuss their results in terms of the origin of multicellularity on earth.

The senior author of the paper is Michael Travisano of the University of Minnesota, who was a student of Richard Lenski's in the 1990s. The paper, published in *PNAS*, was edited by Lenski. The gist is as follows.

The authors repeated three steps multiple times: 1) they grew single-celled yeast in a flask; 2) they briefly centrifuged it; and 3) they took a small amount from the bottom of the flask to seed a new culture. This selected for cells that sedimented faster than most. After a number of rounds of selection the cells sedimented much faster than the beginning cells. Examination showed that the fast-sedimenting cells formed clusters due to incomplete separation of replicating mother-daughter cells.

The cell clusters also were 10% less fit (that's quite an amount) than the beginning cells in the absence of the sedimentation selection. After further selection it was seen that some cells in clusters would "commit suicide" (apoptosis), which apparently made the clusters more brittle and allowed chunks to break off and form new clusters. (The beginning cells already had the ability to undergo apoptosis.)

It seems to me that Richard Lenski, who knows how to get the most publicity out of exceedingly modest laboratory results, has taught his student well. In fact, the results can be regarded as the loss of two pre-existing abilities: 1) the loss of the ability to separate from the mother cell during cell division; and 2) the loss of control of apoptosis.

The authors did not analyze the genetic changes that occurred in the cells, but I strongly suspect that if and when they do, they'll discover that functioning genes or regulatory regions were broken or degraded.

This would be just one more example of evolution by loss of pre-existing systems, at which we already knew that Darwinian processes excel. The apparently insurmountable problem for Darwinism is to build new systems.

88. Malaria and Mutations

"Malaria and Mutations," Uncommon Descent (website), May 9, 2008.

AN INTERESTING PAPER APPEARED RECENTLY IN THE *New England Journal of Medicine*.[30] The workers there discovered some new mutations which confer some resistance to malaria on human blood cells in the lab. (Their usefulness in nature has not yet been nailed down.) The relevance to my analysis in *The Edge of Evolution* is that, like other mutations that help with malaria, these mutations, too, are ones which degrade the function of a normally very useful protein, called pyruvate kinase. As the workers note:

> Heterozygosity for partial or complete loss-of-function alleles... may have little negative effect on overall fitness (including transmission of mutant alleles), while providing a modest but significant protective effect against malaria. Although speculative, this situation would be similar to that proposed for hemoglobinopathies (sickle cell and both α-thalassemia and β-thalassemia) and G6PD deficiency...

This conclusion supports several strong themes of *The Edge of Evolution*, themes that reviewers have shied away from. First, even beneficial mutations are very often degradative mutations. Second, it's a lot faster to get a beneficial effect (if one is available to be had) by degrading a gene than by making specific changes in genes. The reason is that there are generally hundreds or thousands of ways to break a gene, but just a few to alter it beneficially without degrading it. And third, that random mutation plus natural selection is incoherent. That is, separate mutations are often scattered; they do not add up in a systematic way to give new, interacting molecular machinery.

Even in the professional literature, sickle cell disease is still called, along with other mutations related to malaria, "one of the best examples of natural selection acting on the human genome."[31] So these are our best examples! Yet breaking pyruvate kinase or G6PD or globin genes in thalassemia does not add up to any new system. Then where do the elegant nanosystems found in the cell come from? Not from random mutation and natural selection, that's for sure.

89. Puppy Dog Eyes for Darwin

"Puppy Dog Eyes for Darwin," Evolution News and Science Today, Discovery Institute, June 19, 2019.

A NEW STUDY[32] PUBLISHED IN THE *Proceedings of the National Academy of Sciences* reports on the canine musculature that under-lies "puppy dog eyes." The study has gotten tons of affectionate press, probably because puppies are so darned cute.

The story is also cast in an evolutionary framework. The research-ers found that a particular muscle abbreviated LAOM is almost always present and well-developed in dogs but is often missing or much less well-developed in wolves. The muscle allows dogs to raise their inner eyebrows in a way that humans find attractive. Since dogs are descended from wolves, the change is ipso facto "evolution." The authors go on to speculate that the trait was selected over time by humans adopting the dogs that seemed cutest.

Well, okay, why not. However, the study is simply a description of dog and wolf anatomy. No studies were done to find the genetic altera-tions that led to the doggy ability. Thus, it tells us pretty much nothing about the evolutionary mechanism. As I discuss in my new book, *Darwin Devolves*, many studies have shown that the genetic changes leading to the traits of various dog breeds—curly coat, shortened muzzles and legs, and more—are largely degradative. That is, the mutations mostly break or blunt pre-existing genes. For example, one mutation in dogs leads to increased muscle mass;[33] it's due to "a two-base-pair deletion in the third exon of MSTN leading to a premature stop codon at amino acid 313." In other words, the increased muscle mass is due to a broken gene.

Puppy dog eyes can melt the hearts of us humans but it's entirely possible—and quite likely in my view—that the mutation behind the muscle for "puppy dog eyes" is also degradative.

90. LENSKI HAS NO RESPONSE

"Woo-hoo! In Science Review of *Darwin Devolves*, Lenski
Has No Response to My Main Argument," Evolution News
and Views, Discovery Institute, February 7, 2019.

SCIENCE HAS JUST PUBLISHED A REVIEW[34] OF *DARWIN DEVOLVES*, more than two weeks before the book's official release date. (I suppose they wanted to be the first on the block to take a shot at it.) Let me first say this—Woo-hoo!! I'm simply ecstatic about the review. Not because it's favorable—it surely isn't. But because it is so embarrassingly, cringe-inducingly weak. It's the equivalent of a reviewer being rendered speechless, but soldiering on because he's been assigned to write 700 words—gotta say *something*. And it's co-authored by no less than Richard Lenski, member of the National Academy of Sciences and world-renowned investigator behind the 60,000-generation long-term evolution experiment (LTEE), to which I devoted most of Chapter 7!

In a few days I will offer a detailed rebuttal. But the overwhelmingly important point to notice right up front is that the reviewers (Lenski plus Josh Swamidass over at *Peaceful Science* and John Jay College biologist Nathan Lents) have absolutely no response to the very central argument of the book. The argument that I summarized as an epigraph on the first page of the book so no one could miss it. The one that I included in the title of a 2010 *Quarterly Review of Biology* article upon which the book is based. The one for which I chose the most in-your-face moniker that I could think of (consistent with the professional literature) to goad a response: The First Rule of Adaptive Evolution: *Break or blunt any gene whose loss would increase the number of offspring.* The rule summarizes the fact that the overwhelming tendency of random mutation is to degrade genes, and that very often is helpful. Thus natural selection itself acts as a powerful *de*-volutionary force, increasing helpful broken and degraded genes in the population.

And they had no response! That's because there is in fact *nothing* that can alleviate that fatal flaw in Darwinism. Much more to come soon.

91. A RESPONSE TO LENSKI

A brief prefatory comment for readers of this anthology: The following essay is my response to a review of *Darwin Devolves* by the very eminent biologist Richard Lenski and two co-authors, which appeared in the very prominent journal *Science* — two weeks in advance of the publication of the book itself. In such a setting, and with plenty of authors to divvy up the work, one would expect at the very least that the reviewers would make sure to get the facts right. Alas, a major theme of the review is that I had ignored past articles from fellow scientists that were critical of my writing, that I had forged ahead monomaniacally, heedless of all those perspicacious corrections. In fact, as you can of course see from the collection of reprinted essays in this book, I have engaged critics for over two decades, including on the very points that Lenski and friends bring up in their review, writing detailed, substantive rebuttals to what all too often were superficial, half-baked polemics.

"Train Wreck of a Review: A Response to Lenski et al. in *Science*,"
Evolution News and Views, Discovery Institute, February 14, 2019.

L AST WEEK *SCIENCE* UNEXPECTEDLY PUBLISHED A SCATHING PRE-publication review,[35] by Richard Lenski and two co-authors, of my book *Darwin Devolves*. I have already posted a short, gleeful reply noting their almost complete lack of a response to the book's main argument, but I had planned to say more. This lengthier post will address such points as they do make, grouped into four themes: supposed counter-examples they cite; stale arguments they bring up; Lenski's own evolution work; and a clear conclusion to draw.

For readers who don't have time to plow all the way through, here are the take-home lessons:

+ Gene-level counter-examples cited by the reviewers are shamelessly question-begging; the reviewers simply gesture at genes and assume they were produced and/or integrated into living systems by random processes, but neither the reviewers nor anyone else has even tried to show that is possible.

+ Organ-level counter-examples cited by the reviewers as produced by exaptive processes are similarly question-begging.

+ Criticisms of my earlier books cited by the reviewers were similarly question-begging and/or relied on vague, imaginative stories.

+ The reviewers are either unaware of or ignore my many detailed replies to earlier criticisms and to papers the reviewers themselves cite.

+ As noted in my previous post, the reviewers don't even attempt to grapple with the main argument of the book, that beneficial degradative mutations will rapidly, relentlessly, unavoidably outcompete beneficial constructive mutations at every time and population scale.

Supposed Counter-Examples

THE REVIEWERS mention two ways in which they think evolution might build novelties—exaptation and gene duplication. I'll focus first on exaptation.

I. Exaptation

The reviewers write: "Missing from Behe's discussion is any mention of exaptation, the process by which nature retools structures for new function.... The feathers of birds, gas bladders of fish, and ossicles of mammals have similar exaptive origins."

On the contrary, in Chapter 3, "Synthesizing Evolution," which summarizes and critiques both Darwin's original theory and neo-Darwinism, I directly discuss Ernst Mayr's invocation of exaptation (which he calls "change of function"). I write:

> Yet how do such elegant new biological features arise? Two broad ways that evolutionary novelties have been envisioned to occur, writes Mayr, are by "intensification of function" and "change of function." In a change of function, a structure that was used for one purpose is adapted to serve a different one; for example, lungs may have been converted to swim [gas] bladders in fish. This is an example of what has been called [by Mayr] the "principle of tinkering."[36]

I later castigate Mayr for vaporous hand-waving. (The "principle of tinkering," indeed! Not quite the same epistemic status as, say, Newton's Laws.) Yet, like Mayr, the reviewers don't even try to show—or cite anyone else who has tried to show—that any of the organs they mention could be produced even from pre-existing organs by a Darwinian mechanism, or by any unguided unintelligent mechanism. Despite my explicitly faulting Mayr for it, the reviewers themselves indulge in the same unreflective hand-waving.

And the unreflective hand-waving quickly descends into opaque spasms. In *Darwin Devolves* I write that "the hardest problem of biology—how to explain the origin of the particular, sophisticated, functional structures of life—"[37] is invisible to most evolutionary biologists. The reviewers themselves illustrate my point with gusto: "Exaptation also challenges Behe's notion of 'devolution' by showing that loss of one function can lead to gain of another. The evolutionary ancestors of whales lost their ability to walk on land as their front limbs evolved into flippers, for example, but flippers proved advantageous in the long run."

In other words, the reviewers seem to "reason," the ancestors of whales had legs, yet whales now have flippers, *therefore* the change must have been driven by random mutation and natural selection. Have they never heard the phrase *non sequitur*? As strange as it may seem to people outside the charmed circle, many Darwinian biologists find it difficult to distinguish the question of *what* occurred in biology from the question of *how* it occurred. The reviewers are oblivious to the fact that neither legs nor flippers nor a transition between them have ever been explained non-trivially. The key question of the book—whether changes in the history of life occurred by chance or by design—seems incomprehensible to them.

II. Gene Duplication

Now for gene duplication. Lenski and companions note that I am very skeptical that gene duplication plus random mutation will lead to significant evolutionary innovations—that is, to something other than slight modification of the parent gene's original function. However, they don't mention that I didn't always think that way. In the book I specifically note I have changed my mind about whether a Darwinian evolutionary pathway can lead via gene duplication from a simple myoglobin-like precursor to the sophisticated oxygen-delivery system that is hemoglobin.[38] In 1996, I wrote in *Darwin's Black Box* that I thought it might. In 1996, I imagined that gene duplication plus random mutation/natural selection might indeed explain increases in complexity of that degree.[39] Yet, due to advances that I discuss in *Darwin Devolves*, I now think it doesn't. Why did I change my mind? The reviewers don't think readers need to hear any of that.

Instead, to show how wrong I am about the role of gene duplication/ random mutation/natural selection, they point to "overwhelming evidence that this underlies trichromatic vision in primates (8), olfaction in mammals (9), and developmental innovations in all metazoans through the diversification of HOX genes (10)." The cited articles are from the years 1999, 2003, and 1998, respectively. And, as anyone with even a passing acquaintance with the topic would guess, the articles simply describe the occurrence of the genes. The authors of the articles don't even try to argue—let alone experimentally investigate—that the diversification and integration of the genes into slightly different functions could have occurred through blind Darwinian processes.

The reviewers may as well have just taken my own example from the book and said, "Everyone knows Darwinian processes can produce complex hemoglobin from simpler myoglobin because myoglobin genes are similar to hemoglobin genes" and been done with it. Apparently they prefer to choose their own examples with which to beg the question.

Hemoglobin has much more sophisticated oxygen-binding capabilities than myoglobin. In contrast, the protein groups the reviewers cite—opsins (for vision), olfactory receptors (for smell), and Hox proteins—all do pretty much the same thing as each other: opsins absorb light, olfactory receptors bind odorant molecules, Hox proteins bind DNA. I have no objection to thinking the proteins within those separate classes arose by gene duplication. As for how they might be successfully folded into an organism's biology, that is a separate question. My guess is that new odorant receptors and opsins might fit in by chance plus selection. New Hox proteins would be much, much more difficult.

Not every kind of protein has the same role so, like it or not, distinctions must be made. In a section of Chapter 8 entitled "Evolution by Gene Duplication Revisited" I write:

> Everyone—including me—thought we knew a lot more than we did. Still, no one should now make the opposite mistake and leap to the conclusion that no development of protein function at all can occur by a classical Darwinian mechanism. As I mentioned in Chapter 6, a cichlid rhodopsin has apparently switched multiple times between two forms sensitive to different wavelengths of light, and a recent study of Andean wrens discovered a point mutation that caused its hemoglobin to bind oxygen more strongly. Those and similar simple examples are straightforward. However, whenever multiple amino acid substitutions or other mutations were needed to confer a substantially different activity on a duplicated protein, it can no longer be blithely assumed that the transition was navigated by Darwinian evolutionary processes. Some may have been, but many others not.[40]

The reviewers seem to be like the New World monkey species *Aotus trivirgatus*, mentioned in a paper they cite,[41] that has lost a functional opsin gene, leaving the species with monochromatic vision. The monkeys can see only in black and white, and the reviewers can see only in random mutation and natural selection. Yet there's no a priori reason that chance couldn't occur alongside design and vice versa. Those who assert that everything arose ultimately by chance carry the heavy burden of showing that is even feasible. Neither the reviewers nor any of the papers they cite even try to do so.

Worse, the reviewers don't even pause to consider how the degradative processes I highlight in *Darwin Devolves* would affect any of their imagined scenarios. Degradation of genes is orders of magnitude faster than constructive changes. Would any selective pressure that could theoretically be ameliorated by gene duplication instead first be lessened by faster, more numerous degradative changes? If so, then there might be no pressure left to select the slower mutation when it eventually arrived. And when a gene did happen to duplicate, would faster degradative mutations remove any selective pressure to change it further? *At each and every step*, faster and more numerous degradative mutations would compete fiercely with any of the steps dreamed up by the reviewers. Yet they can't seem to even entertain such thoughts.

Guided Laboratory Evolution

Lenski and co-authors cite a paper on experimental evolution: "And in 2012, Andersson et al. showed that new functions can rapidly evolve in a suitable environment (11). Behe acknowledges none of these studies..." It turns out that I reviewed that paper[42] shortly after it came out and posted my comments on Discovery Institute's website. The comments are easily found with a few mouse-clicks. So much for due diligence.

My post on the paper is entitled "To Traverse a Maze, It Helps to Have a Mind"[43] (collected in the present volume as essay 48). It highlights the fact that the investigators actively guided the system over bumps and around corners to the desired results:

> The investigators' exceptionally intelligent manipulations are relegated to the online supplemental materials. Reading a brief part of the supplemental Materials and Methods section, entitled "Selection for bifunctional *HisA* mutants," is sufficient to see the absurdity of taking the results as a model for undirected Darwinian evolution....

+ They deleted an enzyme that previous work showed could likely be replaced.
+ They added the necessary nutrient histidine because previous work showed that mutations conferring an ability to make tryptophan destroyed the ability to make histidine.
+ The added histidine would have shut off production of the protein, so they removed the genetic control element to keep it in production.
+ Later, once they found mutations to produce tryptophan, they removed histidine from the medium to encourage the production of mutations restoring histidine synthesis.

... Needless to say, this ain't how unaided nature works—unless nature is guiding events toward a goal.

Yes, "Andersson et al. showed that new functions can rapidly evolve in a suitable environment"—but *under the guidance of an intelligent agent.* As I write in *Darwin Devolves,* in his own terrific 60,000-generation evolution project, Lenski discovered the devastation wreaked by *unguided,* random mutation plus selection.

Let me emphasize: in reviewing a book expressly advocating intelligent design, Lenski et al. can't seem to distinguish between experiments where investigators keep their hands off and those where investigators actively manipulate a system. Perhaps they can't see the difference.

STALE ARGUMENTS, UNREAD REPLIES

THE REVIEWERS recycle old hand-waving critiques of my earlier books that were either grossly inadequate or irrelevant (or both) even when they were first published, let alone today.

First Instance: "Behe also ignores the fact that some of his prior arguments have been dismantled (2). He includes a lengthy appendix that argues that the blood clotting cascade is irreducibly complex, for example, but fails to mention Kenneth Miller's simple, elegant scheme for its stepwise evolution (3)."

Lenski's citation #3 is to a philosophy of biology anthology from 2009,[44] which reproduced a chapter by Miller originally written for an earlier (2004) book of essays published by Cambridge University Press called *Debating Design*.[45] The title of Miller's chapter was "The Flagellum Unspun," which as its name implies focused on the bacterial outboard motor; it contained just a passing reference to blood clotting. I, too, contributed a chapter to the earlier book,[46] which was also reproduced in the philosophy of biology anthology.[47] My chapter was titled "Irreducible Complexity: Obstacle to Darwinian Evolution" (essay 25 of the present volume). In it I defended both the blood clotting cascade and the flagellum against Darwinian objections, including those of Miller. The reviewers either did not see or chose to ignore my chapter. Let me emphasize: the reviewers fault me for not discussing an article Miller published in the very book in which I answered his objections with my own article.

In his 1999 book *Finding Darwin's God*[48] Miller did pen what one might with great charity call a "scheme" for the evolution of blood clotting, but it was so sketchy (sandwiched between a description of modern clotting cascades in vertebrates and invertebrates, the actual "scheme" comprised a *single paragraph*) it makes Ernst Mayr's "principle of tinkering" look like a paradigm of scientific rigor. Miller later posted on his own website[49] some material cut from his book that added more hand-waving steps. However, a similar, more clearly presented generic scenario for building a cascade was posted earlier on the Internet by a Harvard grad student.[50] I pointed out the severe problems with their scenarios twenty years ago (essay 15 in the present volume).[51] The reviewers don't mention it.

Second Instance: "... or the fact that a progenitor fibrinogen gene has been discovered in echinoderms (4)."

The investigator behind the cited reference #4 is a man named Russell Doolittle—an eminent scientist, now-retired professor of biochemistry at the University of California-San Diego, and member of the National Academy of Sciences, who spent the bulk of his fifty-plus-year career working on the blood clotting cascade. I discussed his work in

Darwin's Black Box in 1996,[52] and I argued there that the cascade is in fact irreducibly complex. In the appendix of *Darwin Devolves*[53] I recount an incident from 1997 where he wrote an essay in the MIT-published *Boston Review* to refute me. In the essay he triumphantly described some then-recent experimental work done by other researchers....

But it turns out that Doolittle misread the paper he was discussing.[54] In fact, (and as noted earlier in the present volume) mice missing the two cascade proteins are very sick: their blood doesn't clot; they hemorrhage; females die during pregnancy. Promising evolutionary intermediates they are not. Please read *Darwin Devolves* for the details. Here I will make just two short points. First, Russell Doolittle is a top expert on blood clotting and evolution. Yet we can quickly see from his mistake—pointing to a paper about dying mice—that he does not know how blood clotting could have evolved by a Darwinian process. (If he did know, he could easily have cited a paper describing it.) And if Russell Doolittle doesn't know, no one knows, no one at all—most certainly including Kenneth Miller and the reviewers. Second, and equally important, Russell Doolittle knows all about gene duplication, yet that knowledge was of no help at all in trying to actually explain the evolution of the blood clotting cascade.

The reviewers note that in 1990 (well before his mistaken essay quoted above was written) Doolittle discovered a gene in an invertebrate group that has a family resemblance to vertebrate fibrinogen.[55] In *Darwin Devolves* I extensively discuss Russell Doolittle's further fine work on the clotting cascade.[56] But that work overwhelmingly consists simply of searching sequence databases for whichever clotting proteins might be found in whichever species. Such work may produce evidence pertinent to questions about common descent but, as I've written many times, *evidence of common descent is not evidence for Darwin's mechanism*. Neither Doolittle nor anyone else has even tried to show that any imagined transitions could have occurred by random processes plus selection. (Neither have other proposed mechanisms, such as those of the so-called Extended Evolutionary Synthesis, been shown to be adequate, as I also explain in *Darwin Devolves*.) Here's an excerpt from the appendix of the book where I discuss a few of the formidable problems facing undirected Darwinian evolution of blood clotting by gene duplication:

> In order to even begin to understand how Darwinian processes might build a clotting cascade, or even just significantly modify a pre-existing one, huge roadblocks need to be addressed, such as how to maintain fine control on the fly while randomly changing a system. I wish luck to anyone with that....
>
> What's more, as we've seen throughout this book, random mutation easily breaks or degrades genes. Since the blood clotting cascade is a finely balanced system—a seesaw of opposing protein functions that either promote or inhibit clotting—altering the balance by degrading one factor should be as effective in the short term as by strengthening another (like taking a bit of weight off one side of the seesaw instead of adding a bit to the other). And since degrading proteins is much faster and easier, that should almost always win out....
>
> As for Professor Doolittle, so too for the great majority of evolutionary biologists. All of these fundamental problems seem truly to be invisible to them. Evidence of common descent is routinely confused for evidence of Darwin's mechanism.[57]

Third Instance: "Behe doubles down on his claim that the evolution of chloroquine resistance in malaria by random mutations is exceedingly unlikely because at least two mutations are required, neither of which is beneficial without the other. His calculations have already been refuted (5), and it has long been known that neutral and even deleterious mutations can provide stepping stones to future adaptations."

I responded to the cited reference #5 in 2009 with a letter[58] that was published in the journal *Genetics* along with a reply by the authors, Durrett and Schmidt.[59] I followed up that exchange with further responses that can easily be found on the web (and as essay 46 in the present collection).[60] The reviewers tell readers nothing about that. Briefly, in my replies I showed that Durrett and Schmidt, two mathematicians, misunderstood some of the biology of the system (which they acknowledged), and when that mistake was rectified their calculations agreed pretty closely with my own. As for "even deleterious mutations" providing stepping stones, I happily agree. In fact, a major point of my book *The Edge of Evolution* was that the development of chloroquine resistance by the malarial parasite likely required two mutations, the first of which was deleterious. (I am mystified as to why the reviewers can't see that their first sentence in the quote above answers the last half of their second sentence.) Yet the need for a neutral or deleterious step in an evolutionary pathway will greatly slow Darwinian evolution, as it did in the case of chloroquine resistance.

Fourth Instance: "Indeed, a 2014 study, unmentioned by Behe, reported discovery of two genetic paths through which malaria has evolved chloroquine resistance through multiple steps (6)."

I blogged extensively on that paper soon after it was published.[61] In fact, I touted that terrific study of Summers et al.[62] far and wide, because it confirmed my surmise in *The Edge of Evolution* that two mutations were required before any chloroquine resistance occurred. (How hard can it be to type "Behe" and "Summers" into a search engine?) As they wrote in their abstract, "A minimum of two mutations sufficed for (low) CQ transport activity." The need for two specific mutations neatly explains the approximately billionfold increase in difficulty for the parasite to evolve resistance to chloroquine versus other antimalarial drugs, such as atovaquone, which require only one. (The fact that there may be several paths to the resistant state is a red herring because, as the paper showed, all paths pass through the deleterious step. Otherwise the de novo origin of chloroquine resistance would be much more frequent than it is, similar to that of atovaquone.) It also illustrates the feebleness of the Darwinian mechanism when confronted with the need for even the tiniest amount of coordination—just two simple point mutations. The difficulties go up exponentially with the number of mutations required.

So let me emphasize that the reviewers fault me for not mentioning a paper that I had in fact publicly and extensively written about, and which strongly supports my arguments.

ALLEGED "DERIDING" OF LENSKI'S TERRIFIC LONG-TERM EVOLUTION EXPERIMENT

RICHARD LENSKI has done wonderful work with his now-thirty-year-long project that follows the growth and evolution of *E. coli* cultures in his laboratory.[63] However, he does not own the interpretation of those publicly reported results. He himself interprets them within a standard Darwinian framework. Fine, he is certainly free to do so. However, in *Darwin Devolves* I do not assume as he does that it's Darwinism all the way down. I explain that, while the results do confirm the ability of random mutation and natural selection to produce beneficial small-scale changes (mostly by degradation of pre-existing genes), they also show why larger constructive changes are beyond its reach.

Random Is as Random Does

Lenski and co-authors write in the review:

There are indeed many examples of loss-of-function mutations that are advantageous, but Behe is selective in his examples. He dedicates the better part of chapter 7 to discussing a 65,000-generation *Escherichia coli* experiment, emphasizing the many mutations that arose that degraded function—an expected mode of adaptation to a simple laboratory environment, by the way—while dismissing improved functions and deriding one new one as a "sideshow" (1).

I don't "dismiss improved functions." Rather, as I explicitly state, my focus is on distinguishing beneficial degradative mutations from beneficial constructive mutations. That is, do most beneficial mutations help by constructing new "functional coded elements" (abbreviated as "FCTs"—for example, genes or control elements) or by degrading old ones? That's simply an exercise in counting the number of beneficial mutations that fall into each category. As I originally wrote in *The Quarterly Review of Biology*,[64] by far the most frequent beneficial mutations reported in the literature are ones that degrade FCTs.—Sorry! Don't blame me! That's just the way it is.

I didn't intend to hurt Lenski's feelings by calling the widely reported citrate mutation[65] that his lab isolated a "sideshow." Rather, I just meant to put that mutation in perspective. Here's an extended quote from *Darwin Devolves* for context:

But the stark lesson of this chapter by far overrides any squabbling about the significance of this or that particular mutation. To see why, consider the other mutations the citrate-eater has suffered along its evolutionary journey. Like all of the culture lines, the citrate-using bacteria have lost the ability to metabolize ribose, suffered killing "mobile element" mutations to other genes, and fixed degradative point mutations in even more. And, like five other replicate cell lines, the citrate-user has turned into a mutator, with a greatly degraded ability to repair its DNA. Whatever the bug's fate from here, it has irrevocably lost the services of perhaps a dozen genes.

But that's not all. In order to best accommodate the gene rearrangement that gave it the talent to eat citrate, several other mutations were found that fine-tuned its metabolism. Even before the critical mutation occurred, a different mutation in a gene for a protein that makes citrate in *E. coli* degraded the protein's ability to bind another metabolite abbreviated NADH, which normally helps regulate its activity. Another, later mutation to the same gene decreased its activity by about 90%. Why were those mutations helpful? As the authors write, "When citrate is the sole carbon source, [computer analysis] predicts optimal growth when there is no flux through [the enzyme]. In fact, any [of that enzyme] activity is detrimental." And if something is detrimental, random mutation will quickly get rid of it. Further computer analysis by the authors suggested that the citrate mutant would be even more efficient if two other metabolic pathways that were normally turned off were both switched on. They searched and discovered that two regulatory proteins that usually suppress those pathways had been degraded by point mutations, the traffic lights now stuck on green.

Interesting as it is, the ambiguous citrate mutation that started the hoopla is a sideshow. The overwhelmingly important and almost completely unnoticed lesson is that genes are being degraded left and right, both when they directly benefit the bacteria and when they do so indirectly in support of another mutation. The occasional, particularly noticeable modification-of-function or gain-of-FCT mutation can't turn back the tide of damaging and loss-of-FCT ones.[66]

The critical distinction to grasp is that beneficial degradative mutations are a completely different sort of beast from beneficial constructive ones. Helpful degradative mutations will arrive very rapidly and in much larger numbers than constructive ones. That's simply because it is much easier and faster to degrade a gene than to improve one constructively. What's more, beneficial degradative mutations are relentless. They will appear before, during, and after any constructive mutation, and compete fiercely with it for selection. And once a degradative mutation is established in a population, for all intents and purposes the functional form of the gene will never come back.

Lenski is rightly proud of his work, and naturally wants to accentuate the positive. I, however, am interested in the question of whether Darwinian processes could have constructed the sophisticated machinery of life, so I consider the results from a different angle.

"Degraded function—an expected mode of adaptation to a simple laboratory environment, by the way"

Rationalizing Results

The "expectation" that degradative mutations would dominate laboratory evolution experiments is an entirely post-hoc rationalization. Consider, for example, that it was only in 2013 that *PLoS Genetics* published a paper titled "Bacterial Adaptation through Loss of Function" in which researchers systematically demonstrated that breaking genes could almost always be beneficial in some environment or other. I write in *Darwin Devolves*:[67]

> In the only work I've seen that does focus on loss-of-function mutations as a general class, interesting in its own right, in 2013 researchers from Princeton and Columbia universities surveyed the literature and then conducted experiments of their own to see which bacterial genes could be broken and the bug would grow better.[68] They showed that "at least one beneficial [loss-of-FCT] mutation was identified in all but five of the 144 conditions considered." In other words, a bacterium could improve its lot by breaking a gene in over 96% of environmental circumstances examined....

A brief comment on the original work by a news writer shows that the simple distinction between beneficial and constructive mutations has clicked for at least one person: "*This study changes the widely held view* [my emphasis] that loss-of-function mutations are maladaptive."[69]

Like many people, evolutionary biologists revise their expectations in light of experience (or experimental results). The reviewers are doing that here, perhaps unwittingly.

In the Wild, Too

What's more, I devote a good chunk of Chapter 7 to demonstrating that loss-of-function mutations dominate not only in "a simple laboratory environment" but also in the wild. In Chapter 9 I write:

In the real world, any possibly beneficial, degradative mutations will arrive rapidly, in force, to alleviate any selective pressure on an organism—eons before the first multi-residue feature even appears on the scene.... The result is that every degradative change and every damaging single-step mutation would be tested multiple times as a solution, or as part of a solution, to whatever selective pressure a species was facing, and, if helpful, would spread to fixation well before a beneficial multi-residue feature even showed up. Where Darwinian processes dominate, the biological landscape would be expected to be littered with broken-but-helpful genes, damaged-yet-beneficial systems, degraded-organisms-on-crutches, ages before any fancy machinery was even available. That's exactly what we saw in Chapter 7 with laboratory *E. coli*, natural *Yersinia pestis*, wild polar bears, tame dog breeds and all other organisms so far examined.[70]

Whatever mutation helps alleviate some selective pressure by any means will immediately begin to spread. Degradative mutations can often help relieve selective pressure, and they arrive orders of magnitude faster than constructive mutations, both in the laboratory and in the wild.

DRAWING A CONCLUSION

CAN ANY important conclusion be drawn from this train wreck of a review? Consider that Richard Lenski is perhaps the most qualified scientist in the world to review the argument of *Darwin Devolves*. Lenski has spent decades overseeing the most extensive, most acclaimed laboratory evolution experiment conducted to date, for which he was elected a member of the National Academy of Sciences. His own work is a major focus of *Darwin Devolves*. He could easily and casually have pointed out any problems with the argument all by himself, merely by wielding Darwin's putatively powerful theory and his own expertise. Yet he and his co-authors spend the entire review deriding the author, barely mentioning Lenski's own work, recounting old criticisms by other people, and leaning heavily on aged theoretical conjectures instead of new experimental results.

The implication is clear. Just as Russell Doolittle unwittingly showed, simply by his mistaken citation, that no explanation for the origin of the blood clotting cascade existed, so Lenski and his fellow reviewers show that there is no answer to the problems for unguided evolution described in *Darwin Devolves*. Although it serves a useful role in understanding changes at the margins of biology, as an explanation for the overarching structure of life Darwin's theory is defunct.

92. LESSONS FROM POLAR BEAR STUDIES

"Lessons from Polar Bear Studies," Evolution News and
Science Today, Discovery Institute, March 6, 2019.

THIS IS THE FIRST IN A SERIES OF POSTS RESPONDING TO THE EX-
tended critique of *Darwin Devolves* by Richard Lenski at his blog,
Telliamed Revisited.[71] Professor Lenski is perhaps the most qualified
scientist in the world to analyze the arguments of the book. He is the
Hannah Distinguished Professor of Microbial Ecology at Michigan
State University, a MacArthur ("Genius Award") Fellow, and a member of the National Academy of Sciences with hundreds of publications,
who also has a strong interest in the history and philosophy of science.
His own laboratory evolution work is a central focus of the book. I am
very grateful to Professor Lenski for taking time to assess *Darwin De-*

volves. His comments will allow interested readers to quickly gauge the relative strength of arguments against the book's thesis.

Although it was not the topic of his first post, I will begin with Lenski's discussion[72] of the example with which I open my book—the polar bear genome—because it illustrates some principles that will be useful going forward. For readers who don't have time to read to the end, here are a couple of take-home lessons:

+ Experimental evidence strongly supports my conclusion (disputed without good reason by Lenski and others) that highly selected mutations in the polar bear genome work by breaking or blunting pre-existing functions.

+ A "function" of a protein is a lower-level molecular feature or activity, such as being a gear or a tether; it should not be confounded with higher-level phenotypic effects, such as "lowers cholesterol" or "makes the organism happy." Ignoring the distinction leads to much confusion.

Where We Agree

AT THE beginning of *Darwin Devolves* I discuss work by researchers[73] who compared the genome of the brown bear (*Ursus arctos*) with that of the polar bear (*Ursus maritimus*). Those species separated from a common ancestor hundreds of thousands of years ago. By analyzing the DNA sequence data, the researchers were able to determine the genes whose selection most strongly adjusted the polar bear lineage to a frigid environment. One of those genes, called *APOB*, is involved in fat metabolism. As I wrote: "The scientists who studied the polar bear's genome detected multiple mutations in *APOB*. Since few experiments can be done with grumpy polar bears, they analyzed the changes by computer. They determined that the mutations were very likely to be damaging—that is, likely to degrade or destroy the function of the protein that the gene codes for."

In fact, about half of the mutations in the seventeen most highly selected polar bear genes were predicted to be damaging. What's more, since many genes had multiple mutations, I noted that about two-thirds to four-fifths of selected genes had suffered at least one damaging mutation. I used this example to set the stage for the main theme of the book, that Darwin's mechanism works chiefly by degrading pre-existing genetic information, which sometimes helps a species survive.

Echoing blogged arguments by his lesser-known co-authors of the appalling review of my book in *Science*,[74] Professor Lenski points out (as I repeatedly do in the book) that the computer analysis is a *prediction* that a particular mutation will or won't be damaging; it is not an experimental *demonstration*. In other words, the prediction could be wrong. Further, the program categorizes mutations into just three categories: *probably damaging, possibly damaging*, and *benign*. (Benign means simply that, as far as the program can tell, no damage has been done to the protein by that change; it does not mean the change is constructive.) Thus, as he stresses, the program is not set up to detect if in fact some new function had been gained by the protein. He goes on to emphasize that the polar bear is superbly adapted to its high-fat diet—much better in that regard than the brown bear. All of which, I happily agree, is true.

WHERE WE DISAGREE

BUT THEN, without benefit of supporting data, Lenski waxes strongly optimistic. He quotes an author of the study and then stresses his own view in bold face:

> In a news piece about this research, one of the paper's authors, Rasmus Nielsen, said: "The *APOB* variant in polar bears must be to do with the transport and storage of cholesterol… Perhaps it makes the process more efficient." In other words, **these mutations may not have damaged the protein at all, but quite possibly improved one of its activities, namely the clearance of cholesterol from the blood of a species that subsists on an extremely high-fat diet.**

Lenski is almost certainly wrong about the bolded text. Here's why. In 1995 researchers[75] knocked out (destroyed) one of the two copies of the *APOB* gene in a mouse model—the same gene as has been selected in polar bears. Although *APOB* is itself involved in the larger process of the transport of cholesterol, mice missing one copy of the APOB gene actually had *lower* plasma cholesterol levels than mice with two copies. (Mice missing both copies died before birth.) What's more, the researchers noted, "When fed a diet rich in fat and cholesterol, heterozygous mice were protected from diet-induced hypercholesterolemia."

The researchers admitted they did not know how it all came together—how that effect on the complex cholesterol-transport system resulted from breaking the gene. Nonetheless, there is no ambiguity about the mouse results. Simply by lowering the amount/activity of APOB, mice were protected from the effects of a high-fat diet. Deletion of one copy of the gene may have made the process of cholesterol removal more efficient, as Rasmus Nielsen speculated above about the polar bear, but it did so by decreasing the activity of mouse *APOB*.

Just to be extra clear about the relevance of the mouse results to the interpretation of the polar bear genome, let me state my reasoning explicitly. *Given the experimental results with mice, it is most parsimonious to think APOB is broken or blunted in polar bears.* For mice, having only half as much *APOB* activity protects them from a high-fat diet. For polar bears, having mutated *APOB* genes protects them from a high-fat diet. If those polar bear mutations decreased the activity of *APOB* by half or more, then we might expect a similar protective effect as was seen in the mouse. Given that computer analysis also estimated the *APOB* mutations in the polar bear as likely to be damaging, it is most reasonable to think the activity of the protein has been blunted by the mutations.

Thus, there is no good reason to speculate about possible new activities of the coded protein in the polar bear. Rather, the simplest hypothesis is that the mutations in the polar bear lineage that were judged by computer analysis as likely to be damaging did indeed blunt the activity of the *APOB* protein in that species—that is, made it less effective. That molecular loss gave rise to a happy, higher-level phenotypic result—an increased tolerance of polar bears for their high-fat diet.

THE WAY TO BET

THE CAVEATS mentioned above by Professor Lenski—about how computer-assignment of a mutation as "damaging" is not a guarantee, and that the protein may have secretly gained some positive new function—are correct. He is also quite right to say that without detailed biochemical and other experiments we cannot know for sure how the change affected the protein and the larger system at the molecular level. Nonetheless, computer methods of analyzing mutations are widely used because they are generally accurate. And they do not suddenly lose their accuracy when I cite their results. So, in the absence of specific information otherwise, that's the way for a disinterested scientist to bet. There is no positive reason—other than an attempt to fend off criticism of the Darwinian mechanism—to doubt the conclusion.

The *APOB* gene is exceptional in having such detailed research done on it. Most other genes haven't been so closely investigated. Nonetheless, in the absence of positive evidence to doubt a prediction for a specific case, the results of the computer analysis should be tentatively accepted for other genes to which it has been applied as well. Skepticism on the matter seems to stem less from the data than it does from reflexive defensiveness. (One of Lenski's co-reviewers actually talked himself into thinking that "it is entirely possible that *none* of the 17 most positively selected genes in polar bears are 'damaged.'"[76] Now *there's* a great opportunity for someone to make a few dollars with a friendly wager.)

LOWER-LEVEL FUNCTIONS VERSUS HIGHER-LEVEL PURPOSES

I'D LIKE to highlight one final critical point. Let me set it up with a homey analogy. When I was fourteen, I worked weekends at McDonald's, and sometimes I'd be assigned to operate the milkshake machine. The machine was broken down each night for cleaning. One of my tasks early in the morning before opening was to reassemble its parts. There were maybe a dozen parts to put together—sprockets, clamps, gaskets, and such. Shakes were very popular back then (mid 1960s) and made many customers happy for a while. Nonetheless, the function of the parts of a shake machine is not "to make people happy." The function of a sprocket or a clamp isn't even "to make a milkshake." Rather, they have lower-level mechanical duties that are subservient to the overarching higher purposes of the systems.

The same is true of *APOB*. Its function is *not* "to help polar bears survive," nor even "to clear cholesterol." Rather, it has one or more lower-level functions that are subservient to those higher purposes. Thus, the fact that cholesterol might be cleared more efficiently in polar bears does not at all mean that *APOB* hasn't been degraded, any more than breaking the off-switch of a shake machine so that it can't be inadvertently switched off during the lunch hour means some new improved function was added.

In both *Darwin Devolves* and my *Quarterly Review of Biology* paper[77] on which it is based, I repeatedly stressed the need to look *beneath* higher-level, phenotypic changes to associated underlying molecular-level mutations. Did they help by constructing or by degrading what I termed Functional Coded elemenTs (FCTs)? Helpful higher-level changes can often be misleading, because they might actually be based on degradative molecular changes. There is every reason to think that's what occurred in the evolution of the examples I cite in *Darwin Devolves*, definitely including the magnificent *Ursus maritimus*. The more effective clearance of its cholesterol allows the polar bear to thrive on a diet of seal blubber, but it is the result of a mutation that breaks or blunts *APOB*.

93. AN INSUPERABLE PROBLEM

"For Dreams of Darwinian Evolution, First Rule of Adaptive
Evolution Is an *Insuperable* Problem," Evolution News and
Science Today, Discovery Institute, March 14, 2019.

THIS IS THE SECOND IN A SERIES OF POSTS RESPONDING TO THE extended critique of *Darwin Devolves* by Richard Lenski at his blog, Telliamed Revisited....

In my first post I responded (out of order) to Professor Lenski's second post,[78] which discussed the polar bear genome. I showed that, regarding my argument that the polar bear's more efficient lipid metabolism (compared to the brown bear's) arose by degradative mutations, his skepticism was unfounded. Briefly, the *APOB* gene of polar bears is mutated with changes that were predicted by computer methods to be damaging. A 1995 study[79] showed that a mouse model that had one copy of the *APOB* gene knocked out actually had lower plasma cholesterol levels and increased resistance to hypercholesterolemia from a high-fat diet. If the polar bear mutations acted to lower the activity of its own *APOB*, a result similar to that for the mouse might be expected. Thus, there is no good reason to speculate about new functions, as Professor Lenski and others did.

Here I will go back and respond to his first post. For those who don't have time to read the whole thing, the take-home lessons are these:

+ Professor Lenski's contrasting of the frequency versus importance of evolutionary changes is misconceived and his illustrations are inapt.

+ Mindless evolution works only in the short term. That is an insuperable problem for long-term Darwinian progress.

Frequency Versus Importance

The theme of Richard Lenski's first post, "Does Behe's 'First Rule' Really Show that Evolutionary Biology Has a Big Problem?,"[80] can be summarized by the following excerpt:

> When it comes to the *power* of natural selection, what is *most frequent* versus *most important* can be very different things....
>
> Behe is right that mutations that break or blunt a gene can be adaptive. And he's right that, when such mutations are adaptive, they are easy to come by. But Behe is wrong when he implies these facts present a problem for evolutionary biology, because his thesis confuses frequencies over the short run with lasting impacts over the long haul of evolution. [emphasis in original]

As an illustration of the distinction he perceives between frequency and importance, Lenski notes that, while we may feel somewhat sick when infected by a common cold virus, if we get infected by the rarer HIV or Ebola viruses we may well die. The first type of infection is more frequent but the second type is more important. A second example of frequency versus importance that he uses is the stock market. A person who invested an equal amount of money in the stock of one hundred separate companies decades ago might get rich even if eighty of the companies eventually went bankrupt—if the other twenty companies did very well. Professor Lenski then writes of *Tiktaalik*—the famous fossil of a fish with apparent wrist joints—which may be an intermediate on the evolutionary path to all terrestrial vertebrates. Its occurrence was infrequent but certainly important in the history of life.

The problem, of course, is that none of his illustrations is relevant to the matter at hand. At the risk of being insufferably didactic, let me spell it out. The fact that there are various types of viruses that infect humans with varying frequencies and cause illnesses of varying severity says absolutely nothing about the evolution by Darwinian processes of a virus, let alone, say, of an eye or cellular machinery. Neither does the strategy that a person might use for placing bets on the stock market say anything pertinent. And the fact that widely different terrestrial vertebrates descended from an aquatic vertebrate does not even attempt to address the riddle of what drove that descent. It flagrantly begs the question of whether mindless processes are adequate for the task or whether intelligence was necessary, as I argue at book-length in *Darwin Devolves*. Worst of all, it doesn't even mention—let alone discuss in detail—the crucial, foundational, molecular level of evolution, which, ironically or not, is Lenski's own area of great expertise. In short, Professor Lenski—the most qualified scientist in the world to address it—doesn't even engage the question.

Death by a Thousand Cuts

NOT ONLY are Lenski's illustrations inapt, but the larger point he seemingly wants to make is simply wrong. Frequency itself can magnify the importance of an event, both inside and outside of an evolutionary context. There's no necessary distinction between the two. When it is multiplied by the number of opportunities to happen, the cumulative importance of an event can be greatly increased. For example, if a person's chance of dying from each instance of the flu is 5% and he is expected to contract the flu ten times in his lifetime, then he should be more worried about that than about contracting Ebola, provided he lives in a non-infested region. What's more, multiple separate frequent conditions can interact to yield a much increased, synergistic effect. If both your car's brakes and tires are worn, the windshield wipers don't work, and the gas tank has a leak, then you have a much greater risk of a serious accident than with any of the conditions alone.

In the context of evolution, the impact of frequency is much more pronounced—due to natural selection itself. That's because selection will quickly spread any beneficial mutation, even if the mutation degrades an organism's genetic patrimony. (After all, as Richard Dawkins and others have emphasized, natural selection is blind.) Thus, it will grab on to any mutation that helps at the moment, without regard to the long-term fate of the species, and comparatively quickly increase the numbers of that mutation over the generations until it is fixed in the population (that is, until all members of a species have it). When that happens, the original unmutated version of the gene is gone. Thus, any further potentially beneficial changes to come along must work with a degraded foundation. (A real-life example is noted by physicians who work on malaria and the many baleful effects of the helpful degradative mutations that were spread in human populations. As Richard Carter and Kamini N. Mendis note, "This burden is composed not only of the direct effects of malaria but also of the great legacy of debilitating, and sometimes lethal, inherited diseases that have been selected under its impact in the past."[81])

No Such Thing as "Importance"

Lenski is comparing reality to wishful thinking. He is contrasting a real process, mutation, with an imaginary attribute, importance. Although frequency is built right into the fundamental equations of molecular evolution as the mutation rate, there is no such thing in those equations—or in all of the Darwinian conception of evolution—as "importance."

To see that's the case, consider the basic equation for the expected waiting time to the first appearance of a mutation that will spread in a small population: $t = 1 / (2Nvs)$. The symbol v in the equation is the mutation rate, that is, the frequency at which mutations arise. Other symbols in the equation are: t, the waiting time; N, the population size; and s, the selection coefficient.

There is no term for "importance." Even the selection coefficient, s, doesn't measure "importance" in the sense that Professor Lenski uses it. That variable only measures relative survival of offspring, not whether a change is degradative or constructive, good for future development, neutral to it, or a roadblock. As I strongly emphasize in Chapter 8 of *Darwin Devolves*, beyond the immediate, very next mutational step, Darwin's mechanism is utterly blind to a species' survival, let alone to its putative future improvement. In Darwinian evolutionary scenarios the only factor leading to new sophisticated traits is the imagination of the storyteller. It surely isn't anything to be found in the mathematics of molecular evolution.

The Huge Degradative Advantage

As I initially discussed in a book chapter[82] and as I emphasize in *Darwin Devolves*, beneficial degradative mutations have a very strong, natural, built-in advantage over beneficial constructive ones, exactly because of their frequency of occurrence.

Let me explain briefly here. Consider two genes, either of which when mutated would be beneficial for an organism to meet some particular selective challenge. The first gene (call it A) would be helpful if it mutated (call the mutated protein A*) at a particular residue of the protein it coded for to give a new constructive feature (perhaps a helpful new binding site). The second gene (call it B) would be helpful if it mutated (to B*) so that its activity were substantially degraded or eliminated entirely. Yet there are orders of magnitude—a hundred to a thousand—more ways to degrade B than to improve A. That means that if neither mutation were originally present in the population of a species, B* would be expected to appear in only a hundredth to a thousandth of the time needed for A* to show up.

For example, if in this situation the time expected for a constructive mutation to arrive were a hundred thousand years, a degradative mutation would arrive in only one hundred to one thousand years. The result is that B* would have 99,000 to 99,900 years to spread through the population to fixation before A* even showed up. If both A* and B* relieve the same selective pressure, then when A* eventually did show up there would be no more pressure to relieve, since B* had done so long before. Thus degradative mutation B* has a built-in advantage simply because its mutation rate is much higher.

If a population is large enough to be expected to already contain some copies of A* and B*, then, simply because of the many more ways to break B to produce B*, there would be expected to be a hundred to a thousand times the number of the broken gene in the population compared to A*. That means it would have a hundred to a thousand times the chance of fixing before A*. Looking at it from a different angle, the selection coefficient for B* could be a hundred to a thousand times less than that for A* and still have an equal chance to fix in the population first—to fix a degraded gene.

EXPONENTIALLY WORSE

THE SITUATION grows exponentially worse if even just two independent changes are needed for what I call in the book a mini-irreducibly complex (mIC) feature, such as a disulfide bond. Here's what I wrote in Chapter 9, "The Revenge of the Principle of Comparative Difficulty":

> When responding to David Snoke and myself Michael Lynch wrote that, using the assumptions of his optimistic model, "adaptive multi-residue functions can evolve on time scales of a million years (or much less)." Okay, much less—let's say a hundred thousand years. But, as Richard Lenski's experimental work (described in Chapter 7) shows so clearly, beneficial damaging mutations evolve on a time scale of *weeks*. That's at least *a million times* faster than the *simplest* mIC features evolving by the *fastest* route imagined. To put that in perspective, damaging mutations are like packages delivered cross-country by FedEx; mutations to construct mIC features are like packages delivered by turtles.
>
> In the real world, any possibly beneficial degradative mutations will arrive rapidly, in force, to alleviate any selective pressure on an organism—eons before the first multi-residue feature even appears on the scene.... The result is that *every* degradative change and *every* damaging single-step mutation would be tested multiple times as a solution, or as part of a solution, to whatever selective pressure a species was facing, and, if helpful, would spread to fixation well before a beneficial multi-residue feature even showed up. Where Darwinian processes dominate, the biological landscape would be expected to be littered with broken-but-helpful genes, damaged-yet-beneficial systems, degraded-organisms-on-crutches, ages before any fancy machinery was even available. That's exactly what we saw in Chapter 7 with laboratory *E. coli*, natural *Yersinia pestis*, wild polar bears, tame dog breeds and all other organisms so far examined. [emphasis in original]

Darwin's mechanism of random mutation and natural selection ensures that any and all quick-and-dirty degradative fixes (I know, let's call it "tinkering") will have a much greater number of chances to take over a population ahead of simple constructive solutions, and an astronomically greater number of chances than even the most modestly cooperative solution. That will overwhelm any significant mindless constructive evolution.

A Brief Point

ONE FINAL point. In his second post (on the polar bear), Lenski writes concerning targets of degradative evolution: "The function has to be one that is not—or rather, *no longer*—useful to the organism. For example, eyes are no longer useful to an organism whose ancestors lived above ground, but which itself now lives in perpetual darkness in a cave" (emphasis in original). That's not the case. For example, the CCR5 gene codes for a protein involved in the immune system. HIV uses the protein as an attachment site to allow it to enter cells. Some populations of people are immune to HIV because the gene for the CCR5 protein has been broken. The working protein was useful, but apparently not as useful as immunity to HIV (or an historically related virus). So rather than Lenski's take, the correct general principle is summarized by The First Rule of Adaptive Evolution: *Break or blunt any functional coded element whose loss would yield a net fitness gain.* There only has to be a *net* benefit to a species to throw away a function. That bodes quite poorly for constructive Darwinian evolution.

94. The LTEE is Doing Great!

"Thanks, Professor Lenski, the LTEE Is Doing Great!," Evolution News and Science Today, Discovery Institute, March 29, 2019.

THIS IS THE THIRD IN A SERIES OF POSTS RESPONDING TO THE EXtended critique of *Darwin Devolves* by Richard Lenski at his blog, Telliamed Revisited....

Unintended Consequences

I HAVE already addressed several of the issues that Lenski has raised at his blog in his third post on *Darwin Devolves*, "Is the LTEE Breaking Bad?,"[83] in my responses[84] to the review by my Lehigh colleagues,[85] because they cited his work frequently. Nonetheless, repetition is a fine teaching tool. So here I will again speak to those issues and also address a few others.

In "Is the LTEE Breaking Bad?," Lenski agrees that the beneficial mutations seen in his Long-Term Evolution Experiment are overwhelmingly degradative or loss-of-function ones. Even so, that does not concern him because "the LTEE represents an ideal system in which to observe degradative evolution." Beneficial degradative changes are only to be expected there, it seems.

> The LTEE was designed (intelligently, in my opinion!) to be extremely simple in order to address some basic questions about the dynamics and repeatability of evolution, while minimizing complications. It was not intended to mimic the complexities of nature, nor was it meant to be a test-bed for the evolution of new functions. The environment in which the bacteria grow is extremely simple....
>
> Indeed, the LTEE environment is so extremely simple that one might reasonably expect the bacteria would evolve by breaking many existing functions. That is because **the cells could, without consequence, lose their abilities to exploit resources not present in the flasks, lose their defenses against absent predators and competitors, and lose their capacities to withstand no-longer-relevant extreme temperatures, bile salts, antibiotics, and more.**[86] [emphasis in original]

In other words, there are many tools in the robust *E. coli* genomic toolbox that wouldn't be needed in the Michigan State lab. It could lose them without immediate consequence. In fact, there may even be some benefit to losing them, either by simply saving the energy of making them, or by diverting resources to other pathways that are more heavily used in the lab environment.

Yet one inevitable consequence of losing genes is losing flexibility—while the parent bacterial strain could evolve to live in the lab environment, the mutant bacteria can't go home again. If their environment shifted back to the more complex, more hostile one from which they had been taken, they would no longer be competitive. Thus, as a strong rule, we would expect that where any species (not just *E. coli*), either in the lab or in the wild, adapt by losing genes, they simultaneously limit their ability to adapt to future environmental changes. Just as the lab *E. coli* is becoming more environmentally restricted as it loses more genes, the examples from the wild of the polar bear and *Yersinia pestis* that I cited in *Darwin Devolves* are, too.

SUPER-SPEED

As QUOTED above, Professor Lenski set up the LTEE to probe "basic questions about the dynamics... of evolution." One basic dynamic that the results show with great clarity is that beneficial degradative mutations occur orders of magnitude faster than beneficial non-degradative ones. Beneficial degradative mutations such as destruction of the ribose operon occurred in about a hundred generations.[87] On the other hand, the pokey non-degradative citrate mutation appeared only after 30,000 generations.[88] In terms of sheer growth in the lab environment, it helped the bacteria very much more than any of the degradative ones—the citrate mutants completely took over the colony overnight. Nonetheless, the citrate mutation appeared only well after maybe a dozen mutations that degraded genes had already swept to fixation, permanently restricting the bacterial strain.

Why is that? The citrate mutation was much slower to appear in Lenski's laboratory conditions, of course, because it required more exacting, less probable conditions that necessarily occur much less frequently. As a rule, all non-degradative mutations will be less frequent than degradative ones. So one very important dynamic result that the LTEE makes crystal-clear is that degradative mutations arrive very rapidly, constructive ones much more slowly. In other words, any and all helpful degradative changes will have the time they need to take over a population well before the arrival of particular non-degradative changes, such as the citrate mutation.

THE SIGNIFICANCE OF CITRATE

PROFESSOR LENSKI thinks the citrate mutation is a good example of the constructive power of evolution. Yet he considers it in isolation and overlooks the overwhelmingly degradative context surrounding that event.

To his credit, Behe does write about the lineage that evolved the ability to consume the citrate. However, he dismisses it as a "sideshow" (p. 365 [*sic*, actually p. 190]), because he refuses to call this new capability a **gain** of function. Instead, Behe writes (p. 362 [sic, actually p. 189]) that under his self-fulfilling scheme "the mutation would be counted as modification-of-function—because no new functional coded element was gained or lost, just copied." **In other words, Behe won't count any newly evolved function as a gain of function unless some entirely new gene or control region "poofs" into existence.** [emphasis in original]

I'm more than happy to agree that the citrate mutation was a very interesting development in the lab evolution project. However, as I've written before,[89] I called it a "sideshow" because it was preceded by many degradative mutations and succeeded by several more that enhanced its efficacy. Compared to degradative mutations, it was numerically swamped. In my view, by far the most important point is that helpful degradative mutations appear so fast compared to constructive ones that they will swarm after any change in the environment, quickly adapting an organism. That's exactly what we should expect from the First Rule of Adaptive Evolution. If a lucky constructive mutation—occasionally, eventually—comes along, too, well, that's nice, but it doesn't undo the damage done by poison-pill mutations.

THE LTEE IS POOF-LESS

As FOR Professor Lenski's complaint that "Behe won't count any newly evolved function as a gain of function unless some entirely new gene or control region 'poofs' into existence" (boldface removed), he is incorrect. In both my *Quarterly Review of Biology* paper and *Darwin Devolves* I counted the sickle cell mutation (a single amino-acid change) of hemoglobin as gain-of-FCT (FCT stands for "Functional Coded elemenT"), because it resulted in a new protein-protein binding site. As I explained at length in both places, I classified mutations according to their molecular effects—not their phenotypic effects—and according to whether they resulted in the gain or loss of such functional coded elements as:

promoters; enhancers; insulators; Shine-Dalgarno sequences; tRNA genes; miRNA genes; protein coding sequences; organellar targeting- or localization signals; intron/extron splice sites; codons specifying the binding site of a protein for another molecule (such as its substrate, another protein, or a small allosteric regulator); codons specifying a processing site of a protein (such as a cleavage, myristoylation, or phosphorylation site); polyadenylation signals; and transcription and translation termination signals.[90]

In order to properly evaluate Darwin's mechanism, one has to observe what it does at the molecular level in terms of coded genetic elements. Many careful studies document that it overwhelmingly degrades them. No new such elements were seen in the LTEE.

OVER 90 PERCENT DEGRADATIVE QUALIFIES AS "CHIEFLY"

PROFESSOR LENSKI tries to shield at least some of the beneficial mutations of the LTEE from being labeled as degradative.

In *Darwin Devolves*, Behe asserts (p. 344 [*sic*, actually p. 179]) that "it's very likely that all of the identified beneficial mutations worked by degrading or outright breaking the respective ancestor genes." He includes a footnote that acknowledges our work that suggests the fine-tuning of some protein functions, but there he writes (p. 609 [*sic*, actually p. 316]): "More recent investigation by Lenski's lab suggests that mutations in a small minority (10 of 57) of selected *E. coli* genes may not completely break them but rather, as they put it, 'fine-tune' them (probably by degrading their functions)." Why does Behe assert that fine-tuning of genes occurred "probably by degrading their functions"?

Perhaps it's because this assertion supports his claim, but more charitably I suspect the underlying reason is similar to the problematic inferences that got Behe into trouble in the case of the polar bear's genes. That is, if one assumes the ancestral state of a gene is perfect, then there's no room for improvement in its function, and the only possible functional changes are degradative.... It would be surprising if some proteins couldn't be fine-tuned such that their activities were improved under the particular pH, temperature, osmolarity, and other conditions of the LTEE.

Several comments. First, the polar bear. It turns out that Lenski and his unreliable sources were grasping at straws in trying to avoid the straightforward conclusion that the degradation of the polar bear *APOB* protein actually helped it to tolerate a high-fat diet. The same effect is seen in mice when one copy of the gene is knocked out.[91]

Second, I agree that one should expect some fine-tuning in the LTEE. Yet, if "only" forty-seven of fifty-seven beneficial mutations worked by degrading or destroying genes in the LTEE project, I would consider my argument to be quite fully vindicated. After all, I never claimed that all beneficial mutations must be degradative. The First Rule of Adaptive Evolution is probabilistic, not prescriptive.

Third, as for why I considered the remaining selected proteins to probably be degrading, the reasons can be found in Lenski's paper itself.[92] Three of the ten (nadR, pykF, and yijC) that met his team's criterion for being potentially fine-tuning (having the same point mutation selected at least twice in different replicate strains) could also be selected if they suffered loss-of-function mutations, indicating that degrading the protein's activity was beneficial. Another protein (spoT) suffered multiple different mutations, likely indicating it was being damaged at various points (which of course is much easier to do than to improve a protein). Several of the other proteins (atoC, hflB, infC, rpsD) acquired single amino acid substitutions that, I do agree, may be "fine-tuning" the proteins in the sense of marginally adjusting them to the conditions of the LTEE. However, they also nicely illustrate the difficulty of doing so, since they took tens of thousands of generations to appear, far longer than beneficial degradative mutations in other genes.

Thus, at the end of scores of thousands of generations, there is some hope that perhaps four of fifty-seven beneficial mutations—fewer than 10%—did not degrade their respective genes. As I wrote in *Darwin Devolves*, "Darwin's mechanism works chiefly by squandering genetic information for short-term gain."

THE ANT AND THE GRASSHOPPER

ONE FINAL, ominous point. The paper Lenski is referring to above is titled "Core Genes Evolve Rapidly in the Long-Term Evolution Experiment with *Escherichia coli*." The abstract reads in part:

> We asked whether the same genes that evolve rapidly in the long-term evolution experiment (LTEE) with *Escherichia coli* have also diversified extensively in nature. To make this comparison, we identified ~2000 core genes shared among 60 *E. coli* strains. During the LTEE, core genes accumulated significantly more nonsynonymous mutations than flexible (i.e., noncore) genes. Furthermore, core genes under positive selection in the LTEE are more conserved in nature than the average core gene.[93]

In other words, random mutation will just as mindlessly throw away even core genes (presumably more useful under more circumstances) than noncore genes, if that affords momentary advantage. There is no planning for or anticipating the future in Darwinism. As in the fable of the industrious and far-sighted ant and the short-sighted grasshopper, Darwin's mechanism is clearly the grasshopper.

IN GRATEFUL RECOGNITION

ANYONE WHO is interested in the topic of evolution should feel very grateful to Richard Lenski and his lab. Because of their work we can discuss the implications of real, relevant facts, rather than of unanchored speculations.

95. CAN'T ANYBODY HERE MAKE DISTINCTIONS?

"Can't Anybody Here Make Distinctions?," Evolution News and Science Today, Discovery Institute, April 12, 2019.

THIS IS THE FOURTH IN A SERIES OF POSTS RESPONDING TO THE extended critique of *Darwin Devolves* by Richard Lenski at his blog, Telliamed Revisited....

"Solid and Interesting"

In his fourth post, "Evolution Goes Viral! (And How Real Science Works),"[94] Professor Lenski revisits a series of experiments on the bacteriophage lambda begun by his lab around 2012. Briefly, bacteriophages are viruses that invade and eat bacterial cells. Lambda specializes in eating E. coli cells. In order to invade the cell, lambda has to bind to a specific bacterial membrane protein, dubbed LamB, to gain a foothold. The Michigan lab grew a strain of E. coli that had lost much (but not all) of its ability to make LamB, together in a culture with bacteriophage lambda. The lambda had a much more difficult time invading those bacterial cells than normal ones, since its docking site was much rarer.

Over time, however, lambda acquired mutations in the protein (called "J") that is responsible for binding LamB of E. coli. The mutations allowed it to bind to a second E. coli membrane protein, OmpF. Mutant phages could then invade cells that were unavailable to unmutated phages, so they prospered. When Lenski's then-student Justin Meyer investigated,[95] he saw that at least four specific amino acid changes had occurred, and all of them were necessary for the new ability to bind OmpF. Lenski's current post emphasizes that requirement for multiple mutations, so the new interaction seems to him to be irreducibly complex and beyond the "edge of evolution." What's more, nothing was broken, so that contradicts the main argument of *Darwin Devolves*, he thinks.

I disagree. As I commented on the paper soon after it was published, "As always, the work of the Lenski lab is solid and interesting, but is spun like a top to make it appear to support Darwinian evolution more than it does."[96] Other instances of viruses adapting to new hosts with surface proteins that are structurally similar (as are LamB and OmpF) to the old binding site had been known for a while. What's more, as Lenski's lab later showed,[97] the mutational pathway simply increased the initial strength of binding to the original protein until it also bound the structurally similar one. As I pointed out in *The Edge of Evolution* and emphasized in my reply[98] to Sean Carroll's unfavorable review[99] of the book in *Science*, the fact that some properties can develop gradually does not mean that all can (such as the book's major example of chloroquine resistance). Thus the new lambda function is neither irreducibly complex nor beyond the edge of evolution.

As far as nothing being broken, that, too, is incorrect. As I wrote at the time, in fact two *E. coli* genes were broken in order to escape infection, first by the original lambda and then by the mutated one.

But Not "Gain-of-FCT"

Professor Lenski seems to be annoyed with me over terminology. "Behe called lambda's new ability to infect via the OmpF receptor a modification of function, instead of a **gain of function**, based on his peculiar definition, whereby a gain of function is claimed to occur only if an entirely **new** gene 'poofs' into existence" (emphasis in original).

(I addressed Lenski's incorrect "poofs" assertion in my last post.) I admit to using inexact wording in the 2012 comment. However, as a rule I try to avoid the phrases "gain of *function*" and "loss of *function*" altogether because they are imprecise. As I wrote in *The Quarterly Review of Biology* in 2010,[100] I focus on what I term gain or loss of *FCT*. FCT is a (sort of) acronym for *Functional Coded elemenT*. As I wrote in my previous post in this series, an FCT can be any coded feature of a gene or protein, such as: promoters; enhancers; insulators; Shine-Dalgarno sequences; protein coding sequences; protein binding sites, organellar targeting or localization signals; intron/extron splice sites, etc. (Some of those could potentially be formed by the switch of a single nucleotide—no "poofing" required.) It is coded elements such as those that underlie life, and a critical evaluation of Darwin's mechanism to produce them has been sorely neglected in experimental evolution studies.

As I wrote in 2012,[101] in my classification scheme I would categorize the ability of mutated lambda to bind to OmpF as a "modification of function" precisely because it is using the very same site on the J protein that it uses to bind to LamB. If it used a different area of the protein— one that had not previously been involved in protein binding (such as is the case with the sickle hemoglobin mutation, which I label a gain-of-FCT)—or a protein other than J, then I would call it a "gain-of-FCT." I am not using "sleight of hand," as Professor Lenski complains. Rather, I am trying to make critical and overlooked distinctions. For some reason, that seems not to be welcomed.

What the Data Show

LENSKI WRITES in bold type, "To make his strange argument, Behe works very, very hard to convince readers that standard evolutionary processes are (i) really, really good at degrading functions, and (ii) really, really bad at producing anything new." Well, yes, I do. And I do so because that's what the data show—including the data of Lenski's own (terrific) LTEE.

In *Darwin Devolves* I quote James Shapiro's observation that "as many biologists have argued since the 19th Century, random changes would overwhelmingly tend to degrade intricately organized systems rather than adapt them to new functions."[102] If the results of modern science show those biologists to have been correct, why the reluctance to let people know?

96. Response to Lehigh Colleagues, Part 1

The Department of Biological Sciences at Lehigh University has a handful of evolutionary biologists on the faculty. As a rule, they are no more receptive to the argument for intelligent design than most other academic biologists in the country. As you might imagine, having to share a departmental affiliation with a heretic makes some of them uneasy. When *Darwin Devolves* was approaching its publication date, two of my colleagues requested a review copy from the publisher and wrote a review for the professional biology journal *Evolution*. It was quite unflattering. This elicited many Darwinian hoots on the internet of "Look, even Behe's own colleagues think he's way off base" (or words to that effect…). It's important to realize, however, that, although my colleagues are very smart and accomplished, they have no more answers to the questions I raise than do other biologists. If the noted evolutionary biologists Richard Lenski and Jerry Coyne sidestep the arguments of *Darwin Devolves*, then it's not surprising that no one else has answers either—no one in the world. Nonetheless, I'm grateful for the unflattering review because it gives another chance to see that even very smart biologists don't know how Darwinian processes could build the elegance and complexity of life.

"A Response to My Lehigh Colleagues, Part 1," Evolution News and Science Today, Discovery Institute, March 22, 2019.

R ECENTLY, IN THE JOURNAL *EVOLUTION*, TWO OF MY COLLEAGUES IN the Lehigh University Department of Biological Sciences published a seven-page critical review of *Darwin Devolves*.[103] As I'll show below, it pretty much completely misses the mark. Nonetheless, it is a good illustration of how sincere-yet-perplexed professional evolutionary biologists view the data, as well as how they see opposition to their views, and so it is a possible opening to mutual understanding. This is the first of a three-part reply.

I'd like to begin by enthusiastically affirming that the co-authors of the review, Greg Lang and Amber Rice, are terrific young scientists. Greg's research is on the experimental laboratory evolution of yeast, and he's an associate editor at the *Journal of Molecular Evolution*. Amber studies the evolutionary effects of the hybridization of two species of chickadee, and she's an associate editor for *Evolution*. Not surprisingly, the review is well written and the authors have done a lot of homework, not only reading the book itself but also digging into other material I have written and relevant literature. What's more, Greg and Amber are both salt-of-the-earth folks, cheerful, friendly, and great colleagues. The additional Lehigh people they cite in the acknowledgements section share all those qualities. There is no reason for anyone to take any of the remarks in their review as anything other than their best honest professional opinions of the matter. So let's get to the substance of the review.

"Two Critical Errors of Logic"

AFTER INTRODUCTORY remarks, Lang and Rice begin by deferring to the review of my book in *Science*[104] and a follow-up web post[105] to show that the book contains "a few factual errors and many errors of omission." (I along with others have dealt at length with those already.[106]) Instead, in their own review they focus on what they see as two logical errors of the book: 1) that I wrongly equate "the prevalence of loss-of-function mutations to the inevitable degradation of biological systems and the impossibility of evolution to produce novelty"; and 2) that I wrongly confuse proteins with machines, and use that misguided metaphor to mislead readers. I'll take those two points and their many sub-parts mostly in turn.

They begin with logical error #1 by deriding the First Rule of Adaptive Evolution as a "quality sound bite" that is "simplistic and untruthful to the data." Recall that the First Rule states, "Break or blunt any functional gene whose loss would increase the number of a species' offspring." Also recall that I explained, in both the book and the journal article[107] where it was first published, that it is called a "rule" in the sense of being a rule of thumb, not an unbreakable law, and it is called the "first" rule because that is what we should generally expect to happen first to help a species adapt, simply because there are many more ways to break a gene than to build a new constructive feature.

As you might imagine, I have read the *Evolution* review closely. Yet nowhere do the authors even try to show why the First Rule isn't a correct statement. They point to mutations that are not degradative, but fail to show quantitatively that those other types will arise faster than degradative ones. In fact, the other types are expected to be orders of magnitude slower.

The reviewers agree that the First Rule is fine for explaining many results from the experimental evolution of microbes such as bacteria and yeast, but they balk at extending it beyond the lab. In fact, they actively argue that lab results really can't tell us much about the real world: "No deletion is beneficial in all environments and beneficial loss-of-function mutations that arise in experimental evolution are unlikely to succeed if, say, cells are required to mate, the static environment is disturbed, or glucose is temporarily depleted." All of those situations, of course, will be common outside a laboratory.

One big fly in their argument, however, is that they overlook the results from non-laboratory evolution that I give in the book. Every species that has been examined in sufficient detail so far shows the same pattern as seen in lab results. For example, I open the book with a discussion of polar bear evolution. About two-thirds to three-quarters of the most highly selected genes that separated the polar bear from the brown bear are estimated by computer methods to have experienced mutations that were functionally damaging. (Some other reviewers questioned this. I showed why they are mistaken here.[108]) Similar results were seen for the woolly mammoth.[109] Neither of those species evolved in the laboratory. Except for the sickle mutation (which itself is a desperate remedy), all mutations selected in the wild in humans for resistance to malaria are degradative.[110] Dog breed evolution, which has been touted as a great stand-in for selection in the wild,[111] is mostly degradative, and lots of breeds have health problems.[112]

What's more, we might well ask, if it doesn't mimic the world realistically, why do federal funding agencies award grants to those who study laboratory evolution? Lang and Rice aver that it does indeed give lots of helpful information: "Collectively, experimental evolution has yielded new insights into the tempo of genotypic and phenotypic adaptation, the role of historical contingency in the evolution of new traits, second-order selection on mutator alleles, the power of sex to combine favorable (and purge deleterious) mutations, the dynamics of adaptation, and the seemingly unlimited potential of adaptive evolution."

True, those phrases are press-release fodder. As I have shown in numerous posts, the results are much more modest than the headlines make them sound.[113] But I don't go anywhere near as far as Lang and Rice themselves go in deflating those claims, since they also insist that the lab results are all based on an unnatural situation—on the supposed unnatural prevalence of degradative mutations in artificial environments. So, why should we trust that the results reflect what would happen in nature? How can the reviewers with any consistency accept some of the lab results but not others?

It astounds me to see how quickly lab evolution researchers disavow the importance of their own life's work when some outsider draws an unwelcome inference. But perhaps we can still save the day. Maybe all of the researchers' results point to some important lessons about unguided evolution. In fact, there's no reason to think that many lab evolution experiments are different in relevant ways from how nature behaves.

Lab Reflects Nature

The first objection Lang and Rice raise against extrapolating results from lab evolution studies to evolution in the wild is that the environments are critically different: "Loss of function mutations are expected to contribute disproportionately to adaptation in experimental evolution, where selective pressures are high and conditions are constant, or nearly so."

That sounds a little off. After all, selective pressures in raw nature can be pretty stringent, too, if, for example, 85% of a vertebrate species died in a single year due to altered weather conditions.[114] And, as we saw above, species in a complex changing natural environment, such as the polar bear and humans, show evolutionary behavior similar to that seen in laboratories. What's more, the conditions in lab evolution are generally far from simple or "constant." That's because by far the greatest complexity in any organism's environment is not due to the temperature or solution conditions in which it finds itself. Rather it is due to the presence of other organisms, including others of its own species. If any individual suddenly gains a selective advantage—even in otherwise quite constant environmental conditions—then its progeny have a splendid chance to outcompete the progeny of all other organisms.

Let's look at several examples from the best-known lab evolution experiment—that of Richard Lenski, who has been growing *E. coli* for over
thirty years to observe how it adapts.[115] I'll begin with the best-known
mutation of that long study—the development of mutant bacteria that
could eat citrate in the presence of oxygen, which the ancestor strain
could not do.[116] I'll skip over the molecular details here (I've commented
elsewhere[117]) and concentrate on the bottom line. Even though the environment had been constant, *overnight* the citrate mutant strain outgrew
its brethren and took over the flask.

The initial citrate mutation was not degradative; rather, it involved a
rearrangement of the bacterium's DNA.[118] Nonetheless, soon after that
initial non-degradative change, several additional mutations occurred in
other genes in support of the citrate utilization pathway. All appear to
have broken their respective genes.[119] Thus, as the First Rule of Adaptive
Evolution would lead one to expect, a helpful, non-degradative mutation
that took tens of thousands of generations to appear was quickly fine-
tuned by mutations that broke several genes.

Let me stress that the genes that were broken following the initial citrate mutation had been helpful to the bacterium up to that point. They
were apparently doing useful tasks. However, once the citrate mutation
came along, the environment changed and they became a net burden, so
out they went. Thus even useful genes, when circumstances change, will
easily be tossed overboard by random mutation and natural selection to
maximize the net benefit of even a non-degradative change.

We can derive another important lesson from the story of the citrate
mutation. At the beginning of the *E. coli* evolution project, the starting
bacteria were genetically pretty uniform (except for marker genes and
such), because they came from a pure strain. (That is indeed one source
of real constancy that the reviewers may have had in mind.) The bacteria
then diverged from each other mostly by degradative mutations, because
those were the quickest beneficial changes to hand in the new environment in which they found themselves. Yet the aftermath of the initial
citrate mutation shows the same behavior. That is, the mutant rapidly

took over the flask, yielding a new pure strain, and the new strain further adapted to its new environment by beneficial degradative evolution. We should expect the same behavior after selection on any gene in any species. Any non-neutral change in *any* organism's genome represents a de facto new environment and, as the First Rule states, will tend strongly to be fine-tuned by the most rapidly occurring beneficial mutations. Of course, degradative mutations occur most rapidly.

(The only expected exception to this situation would be if no genes are available that can helpfully be degraded. An example may be the development of chloroquine resistance by the malaria parasite *Plasmodium falciparum*, which occurred mainly by multiple point mutations in the PfCRT protein.[120])

MUTATING A MUTATOR

A SECOND example of fine-tuning by degradation in the Lenski experiment can be seen in the rise and slight fall of mutator strains—that is, bacteria that have lost much of their ability to repair their DNA. From the beginning of the E. *coli* lab evolution project, Lenski separately grew a dozen different test flasks of bacteria in order to be able to ask questions about the replicability of evolution. Six out of the twelve replicate strains eventually became mutators (because a gene involved in DNA repair broke),[121] with mutation rates more than a hundred times those of non-mutators.

There is some question about whether those loss-of-function mutations helped the bacteria directly or by making other beneficial mutations appear faster.[122] (That's what the reviewers are referring to above as "second-order selection on mutator alleles.") Whatever the resolution of that second-order question, one mutator led to a first-order effect. The Lenski lab noticed that, after a while, the mutation rate of one of the mutator strains had decreased by half. Upon investigation they determined that the mutation rate had been reduced by breaking a second gene that is involved in DNA repair.[123] Thus, a problem caused by breaking one gene was partially offset by breaking a different gene. That's what random mutation and natural selection do.

Let me emphasize that, like the genes broken to fine-tune the citrate mutation, the second gene involved in repair had been useful. It was performing a beneficial function. It was not superfluous. Nonetheless, since the environment changed with the appearance of the mutator mutation, the net benefit of getting rid of the gene apparently outweighed the benefit of keeping it. So out it went. The bacterium is now better adapted to its current environment, but certainly less flexible than it had been.

A laboratory is not nature, but we do lab experiments to understand how nature behaves. Lab evolution experiments show that whenever the environment changes, microorganisms will adjust with whatever helpful mutations come along first. Both simple math and relevant experiments indicate that by far those will mostly be degradative mutations.

97. Response to Lehigh Colleagues, Part 2

"A Response to My Lehigh Colleagues, Part 2," Evolution News and Science Today, Discovery Institute, March 25, 2019.

Recently two of my Lehigh University Department of Biological Sciences colleagues published a seven-page critical review[124] of *Darwin Devolves* in the journal *Evolution*. As I'll show below, it pretty much completely misses the mark. Nonetheless, it is a good illustration of how sincere-yet-perplexed professional evolutionary biologists view the data, as well as how they see opposition to their views, and so it is a possible opening to mutual understanding. This is the second of a three-part reply. It continues directly from Part 1 (immediately above).

A Limited Accounting of Degradation

Greg Lang and Amber Rice cite a number of articles to show that loss-of-function mutations are just a small minority of those found in studies of organisms:

However, the truth is that loss-of-function mutations account for only a small fraction of natural genetic variation. In humans only ~3.5% of exonic and splice site variants (57,137 out of 1,639,223) are putatively loss of function, and a survey of 42 yeast strains found that only 242 of the nearly 6000 genes contain putative loss-of-function variants. Compared to the vast majority of natural genetic variants, loss-of-function variants have a much lower allele frequency distribution.

Yet those three studies they cite all search only for mutations that are pretty much guaranteed to totally kill a gene or protein. For example, one paper says:

> We adopted a definition for LoF variants expected to correlate with complete loss of function of the affected transcripts: stop codon-introducing (nonsense) or splice site-disrupting single-nucleotide variants (SNVs), insertion/deletion (indel) variants predicted to disrupt a transcript's reading frame, or larger deletions.[125]

That's akin to counting only burnt-out shells of wrecked cars as examples of accidents that degrade an auto, while ignoring fender benders, flat tires, and so on. There are many more mutations that would not be picked up by the researchers' methods that nonetheless would be expected to seriously degrade or even destroy the function of a protein. Since the rates leading to the kinds of mutations in the cited papers are likely to be at least ten-fold lower than general point mutations in the gene[126] (which, again, the study passed over) there may be many more genes—perhaps five- to ten-fold more (about a quarter to a half of mutated genes)—that have been degraded or even functionally destroyed. Further research is needed to say for sure. (I know which way I'll bet.) The remaining fraction of mutated genes in the population is likely to consist mostly of selectively neutral changes, neither helping nor hurting the organism, and not contributing anything in themselves to the fitness of the species.

Replenishing the Gene Store

The reviewers then point to work showing that, while some genes are indeed degraded over the life of a species, new genes arise by duplication or horizontal gene transfer to replenish the supply. Thus there is a continuous supply of raw material for new evolution. But there are at least three serious problems they overlook. First, assuming a generation time of one year, the rate of duplication of any particular gene is estimated to be about one per ten million years per organism (although there is much uncertainty[127]) and, although it is frequent in prokaryotes,[128] in eukaryotes the rate of horizontal gene transfer is much less.[129] The rate of any particular gene suffering a degradative mutation is expected to be about a hundred times faster than duplication. Thus, every gene that could help by being degraded would have an average of one hundred chances to do so for every one chance another gene would have that could help by duplicating. Second, as its name indicates, gene duplication yields just an extra copy of a gene, with the same properties as the parent gene. Thus, the extra copy would have to twiddle its thumbs for another expected ten million years or so, all the while trying to dodge inactivating mutations,[130] before acquiring a second mutation that might differentiate it a little bit in a positive way from the first—exactly like an unduplicated gene. In other words, gene duplication is no magic bullet to eliminate the problems of unguided evolution.

In their review Lang and Rice write that perhaps the very fact that there were two copies of a particular gene would itself be helpful, because of the extra activity it would add to the cell. I agree that is possible. However, it is special pleading, because most duplicated genes would not be expected to behave that way.[131] For every extra restriction put on the gene that is supposed to duplicate (such as partial duplication, duplication that joins it to another gene, and so on), a careful study of the topic must adjust the mutation rate downward, because fewer genes/events are expected to meet those extra restrictions.

The third and most serious problem Lang and Rice overlook is that they assume without argument that a duplicated gene would be able to integrate into an organism's biology strictly by Darwinian (or at least unintelligent) processes. Yet not all genes or functions are the same, so critical distinctions must be made. As I have written in detail in Chapter

8 of *Darwin Devolves* and in response to other reviewers,[132] some genes with simpler duties may have been able to do so but others not. For example, currently duplicated genes for proteins called opsins are involved in color vision of humans. Yet all those proteins do pretty much the same thing, so duplication of an opsin gene would not be expected to disrupt an organism's current biology too much. On the other hand, duplicated developmental genes, such as those for Hox proteins, would be expected to have a much more difficult time of it; they would be much more likely to cause birth defects than to help.

Since the question we are discussing is not about simple common descent, but rather about whether such fantastic development as we see in life could be produced without intelligent guidance, then a proponent of Darwinian evolution has to show that chance could fold in genes for even the most difficult pathways, if the question is not to be begged. No one has ever even tried to show that.

A Fourth Nasty Problem

One of the papers the reviewers reference in this section is Shen et al.[133] If you look up that paper you find the following two (of four) "Highlights" listed on the first page: "Reconstruction of 45 metabolic traits infers complex budding yeast common ancestor" and "Reductive evolution of traits and genes is a major mode of evolutionary diversification." It must take Darwinian tunnel vision to cite a paper that emphasizes how a complex ancestor gave rise to simpler yeast species by losing abilities over time as support for arguing that Darwinian evolution can build complexity.

The Shen et al. results point strongly to a fourth nasty problem with the notion of gene duplication as replacement for older, degraded genes. On average, degradation would be expected to remove variety in the kinds of genes, whereas even successful duplication and integration of a new gene just increases a pre-existing gene type. Over time that will diminish gene diversity.

The Shen et al. paper isn't alone in noticing the phenomenon of genome reduction. As sequencing data becomes more plentiful and accurate, more papers are being published that show the importance of loss-of-function from more complex states in evolution.[134] As one group writes of mammalian development, "Our results suggest that gene loss is an evolutionary mechanism for adaptation that may be more widespread than previously anticipated. Hence, investigating gene losses has great potential to reveal the genomic basis underlying macroevolutionary changes."[135] Another group comments, "These findings are consistent with the 'less-is-more' hypothesis, which argues that the loss of functional elements underlies critical aspects of human evolution."[136]

STANDING VARIATION

LANG AND Rice ding me for disrespecting standing variation. Standing variation consists of the mutant genes that are already present in a population and can be called upon by natural selection to help a species adapt to changed environmental circumstances, obviating the need for a new mutation. For example, the most highly selected mutant gene associated with thick- versus thin-beaked Galápagos finches did not first arise when Peter and Rosemary Grant were studying the finches in the 1970s. It actually arose about a million years ago[137] and has been present in the group ever since. Ancient standing variation also seems to be behind the very rapid evolution of cichlid fish in Lake Victoria.[138] I discuss both of those examples in *Darwin Devolves*.

The reviewers write that the fact that helpful mutations have been around in the population for millenia "does not lessen the instrumental role of standing genetic variation in adaptation to new environments." I heartily agree, and never wrote otherwise. There are two major problems, however, for the reviewers' position. The first is that evolution by natural selection of standing variation does not address the primordial question that my book focused on—how complex structures arise, particularly at the molecular level. The second major problem is that standing variation nicely illustrates how preexisting slapdash mutations actively inhibit more complex ones. That is, rapid beneficial degradative mutations can become standing variation.

For example, that mutant protein that is most strongly associated with thin- versus thick-beak genes in Darwin's finches, ALX1, has only two changed amino acid residues out of 326 compared to the wild type protein. Both of those are predicted by computer analysis to be damaging to the protein's function.[139] Yet apparently no better solution to the task of changing finch beak shape has come along in a million years, even though an enormous number of mutations would be expected to occur in the bird population during that time.

Why not? Well, consider that an army platoon that takes an unoccupied hill has a much easier task than an opposing force that later wants to displace them. Similarly, a likely big factor in finch evolution is that the quick and dirty mutations have already been established. So in order to supplant them, a new mutation would have to be better right away than the fixed ones. That is, its selection coefficient compared to mutation-free ALX1 would have to be greater than the damaging ones. There is no known correlation, however, between the strength of the selection coefficient and whether a mutation is constructive or degradative. Thus, we have no reason to think standing variation would be supplanted.

Recognizing that hurdle could go a long way toward understanding the reason for stasis in evolution or, put another way, the reason for the equilibrium in punctuated equilibrium. And the generality of punctuated equilibrium[140] reminds us that the same situation—quick and dirty mutations either stalling or completely preventing constructive ones—is expected to be very frequent *on Darwinian principles*.

ECOLOGICAL CHANGES

LANG AND Rice emphasize the importance of the ecological diversification and behavioral changes of Darwin's finches, as opposed to just changes in body shape.

Darwin's finches are an icon of evolution for good reason, having radiated into numerous ecological niches and developed diverse resource specializations (including at least one case—feeding on mature leaves—that is, to the best of our knowledge, unknown in other bird *orders*, much less families). By adopting a restrictive definition of fundamental biological change, Behe dismisses all corresponding behavioral, digestive, and physiological adaptations. [emphasis in original]

Species limits and relationships of the Galápagos finches remain uncertain.[141] Yet the massive study by Lamichhaney et al.,[142] in which the complete genomes of 120 Galápagos finches were sequenced (over *100 billion* nucleotides), including representatives of every separate species and population, found that the most highly selected finch gene was *ALX1*, which, again, is associated with thick versus thin beaks. If those alterations of the finches' behavioral and feeding habits required genetic changes, they eluded discovery. Perhaps the ecological changes are mostly the result of nongenetic modifications.

The authors of the review point to the example of the evolution of stickleback fish in freshwater lakes that have reduced defensive armored plates compared to saltwater varieties: "The causative variants are likely *cis*-regulatory changes that decreased expression of [the gene] *Eda* in developing armor, but not in other tissues. *Darwin Devolves* accepts as evidence only de novo protein evolution, a restriction Behe uses to support his 'First Rule' and claim that 'Darwinian evolution is self-limiting.'"

They have misunderstood the First Rule. There's nothing in "break or blunt any functional gene" that confines degradative mutations just to protein coding regions. If it would benefit a species to reduce the activity of a gene by messing up its control elements instead of its coded protein sequence, that works too. The first mutation that comes along to helpfully suppress a gene's activity is the one with the best chance of being established in a population.

The very next sentence the reviewers write is this: "Narrow by definition and unsupported by the data, Behe's First Rule does not stand up to scrutiny." On the contrary, the scrutiny itself doesn't stand up.

98. Response to Lehigh Colleagues, Part 3

"A Response to My Lehigh Colleagues, Part 3," Evolution News
and Science Today, Discovery Institute, March 26, 2019.

RECENTLY TWO OF MY LEHIGH UNIVERSITY DEPARTMENT OF BIO-logical Sciences colleagues published a seven-page critical review[143] of *Darwin Devolves* in the journal *Evolution*. As I'll show below, it pretty much completely misses the mark. Nonetheless, it is a good illustration of how sincere-yet-perplexed professional evolutionary biologists view the data, as well as how they see opposition to their views, and so it is a possible opening to mutual understanding. This is the third of a three-part reply. It continues directly from Part 2 (immediately above).

OF COURSE PROTEINS ARE MACHINES

A BASIC difference between the views of Greg Lang and Amber Rice and my own concerns the nature of the molecular foundation of life. They object that I consider many biochemical systems to be actual machines. They quote a line from *Darwin Devolves* stating that protein systems are "literal machines—molecular trucks, pumps, scanners, and more." They write disapprovingly that the book claims that "rod cells are fiber optic cables… The planthopper's hind legs are a 'large, in your face, interacting gear system.'" They do concede that I didn't make up those claims about the machine-like nature of the systems out of whole cloth: "Most of the analogies in *Darwin Devolves* are not Behe's creation—he has done well to scour press coverage and the scientific literature for relatable metaphors; and he is generous with their use." Nonetheless, they say, "reality remains: proteins are not machines, a flagellum is not an outboard motor."

On this point they are simply wrong. "Molecular machine" is no metaphor; it is an accurate description. Unless Lang and Rice are arguing obliquely for some sort of vitalism—where the matter of life is somehow different from nonliving matter—then of course proteins and

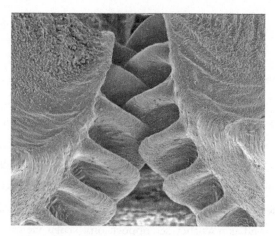

Figure 7.1. The gears of the planthopping insect *Issus coleoptratus.*

systems such as the bacterial flagellum are machinery. What else could they be? Although they aren't made of metal or plastic like our everyday tools, protein systems consist of atoms of carbon, oxygen, nitrogen, and so on—the same kinds of atoms as are found in inorganic matter, nothing special.

A dictionary definition of "machine" is "an assembly of interconnected components arranged to transmit or modify force in order to perform useful work."[144] Take a look for yourself at the gears of the planthopper[145] in Figure 7.1, at the fiber-optic cells of the retina,[146] and at the bacterial flagellum in Figure 1.4. Do you think they fit that dictionary definition? Just like arms, legs, and jaws at the macro-level of life, all of which are organized to perform tasks and work by mechanical forces, so too are the molecular systems at the foundation of life.

We need not concern ourselves here with the question of whether life is more than just machinery,[147] a view for which I have much sympathy. Nevertheless, we can speak literally of the machinery of life in the same way that we can speak literally of the chemical elements of life (carbon, hydrogen, oxygen, etc.), all without implying that life is just machinery, or just chemical elements.

Biologists routinely use the phrase "molecular machine" (just do a search of PubMed or Google Scholar), and have done so for a long time. For example, consider from 1997, "The ATP Synthase—a Splendid Molecular Machine,"[148] and from 1999, "The 26S Proteasome: A Molecular Machine Designed for Controlled Proteolysis."[149] Organic chemists have also long used the term, albeit for much more modest assemblages than are found in life, though ones that won the 2016 Nobel Prize in Chemistry. As the Royal Swedish Academy of Sciences announced then: "The Nobel Prize in Chemistry 2016 is awarded to Jean-Pierre Sauvage, Sir J. Fraser Stoddart and Bernard L. Feringa for their design and production of molecular machines. They have developed molecules with controllable movements, which can perform a task when energy is added."[150]

Lang and Rice do not *argue* that proteins are not machines. Rather, they simply *declare* "proteins are proteins, and not machines," list a few things some proteins do, and assume that makes it obvious they can't be machines: "Proteins are promiscuous. They moonlight, by chance interacting with other cellular components to effect phenotype outside their traditionally ascribed roles." Well, now. So can a nut or bolt be "promiscuous" by, say, holding together various kinds of machines? Can a mousetrap "moonlight" as a tie clip?[151] What exactly is it about those features they list that contradicts the dictionary definition of a machine? Or contradicts the evidence of your own eyes when viewing the images of protein machinery cited above?—Nothing at all.

Hand-Waving at Irreducible Complexity

Lang and Rice use their misunderstanding of molecular machinery as a basis for attacking irreducible complexity: "By acknowledging the reality that proteins are proteins, and not machines, we immediately recognize the shortcomings of irreducible complexity." How so? They quote my definition of IC as "a single system composed of several well-matched, interacting parts that contribute to the basic function, wherein the removal of any one of the parts causes the system to effectively cease functioning." They then object: "The concept of irreducible complexity is flawed for two reasons. First, it considers a system only in its current state and assumes that complex interdependency has always existed. Second, irreducible complexity does not consider that proteins perform multiple functions and, therefore, evolutionary paths that seem unlikely when considering only one function may be realized through a series of stepwise improvements on another function."

They are wrong on both counts. There is nothing in the definition of IC that requires what they say it does there. Irreducible complexity does focus on the current state of a system, but it does not assume that "complex interdependency has always existed." Rather, it strongly implies (although it does not absolutely rule) that the complex interdependency did not arise by Darwinian processes and instead required intelligent input to produce. IC also does not require that proteins do not perform multiple functions. In fact, in 1996 in *Darwin's Black Box*, I pointed out several that do, and showed why that does not help at all in explaining IC.

On top of those mix-ups, the reviewers then forget the definition of irreducible complexity as "a single system" that they stated just in the previous paragraph, and begin to write of it in terms of "essential genes":

> Simply because a system in its current form is irreducibly complex is not evidence that it did not evolve by random mutation and natural selection. Essentiality of a gene or protein is relative to its current state. For two closely related strains of yeast, between 1% and 5% of genes that are essential in one strain are dispensable in the other. Conditional essentiality is not simply due to the presence of second copy (or a close paralog) of the gene in one strain but not the other; rather, conditional essentiality is a complex trait involving two or more modifying loci.

All of that may be true and interesting, but it is beside the point. A gene that is essential for the growth of an organism is not at all the same thing as a component of a single system that is required for the system's function. For example, the motor is a necessary component of the irreducibly complex bacterial flagellum—in its absence, the molecular machine cannot work. But it is not necessary for the growth of *E. coli* in Richard Lenski's laboratory evolution project. On the other hand, hemoglobin may be essential for human life, but it is not part of a "single" irreducibly complex system. Cells and organisms are composed of *many* molecular machines and biochemical pathways; they do not constitute a "single system" in the sense of the definition of irreducible complexity. I have written explaining such confusions long ago.[152] It's little wonder that the reviewers don't appreciate the problem IC presents to undirected evolution, since they misunderstand the concept itself.

Lang and Rice are welcome to think up all the evolutionary pathways they want, and join the long line of critics who have tried and (as I briefly show in *Darwin Devolves* and at much greater length elsewhere)[153] failed over the years to figure out how irreducibly complex functional systems could be produced by random, undirected processes. Yet they don't even try. The penultimate section of the long review is entitled "Two Examples to Illustrate the Evolution of Complexity." (Hmm. Do you, dear reader, notice that an important word is missing from that title?) The reviewers dredge up a couple of old experiments that, at the time they were first published, were furiously spun as grave problems for irreducible complexity. (One sure way to get an otherwise unremarkable paper noticed over the past several decades has been to claim it refutes those bothersome ID folks.) However, the results were actually quite modest and the relevance to ID claims nonexistent, as I wrote at the time (essays 54 and 69 of the present volume). Suffice it to say here that the first example cited by the reviewers (an investigation of a complex molecular machine called a vacuolar ATPase) at the very best concerned sideways evolution; no new functions, let alone new complex machinery, were involved.[154] With the second example, a virus in a lab evolution experiment swapped out a binding site for a certain protein in the membrane of *E. coli* for a binding site for a second, homologous protein. In the process several *E. coli* genes were broken to help the bacterium survive.[155]

Color me unimpressed. Yet it's reasonable to think that, in preparation for writing the review, Lang and Rice would search for the very best studies that have so far been produced that they think challenge irreducible complexity and intelligent design. Perhaps they found them.

TOWARD MUTUAL UNDERSTANDING

IN THE Conclusion of their review Lang and Rice raise a plaintive cry, first celebrating in general the skepticism of scientists and then bemoaning the skepticism of the public toward grand Darwinian claims:

> Scientists—by nature or by training—are skeptics. Even the most time-honored theories are reevaluated as new data come to light.... [O]ver 150 years after *On the Origin of Species*—less than 20% of Americans accept that humans evolved by natural and unguided processes. It is hard to think of any other discipline where mainstream acceptance of its core paradigm is more at odds with the scientific consensus.

> Why evolution by natural selection is difficult for so many to accept is beyond the scope of this review; however, it is not for a lack of evidence: the data (only some of which we present here) are more than sufficient to convince any open-minded skeptic that unguided evolution is capable of generating complex systems.

Perhaps I can help. After all, I used to believe that a Darwinian process did indeed build the wonders of life; I had no particular animus against it. Yet I believed it on the say-so of my instructors and the authority of science, not on hard evidence. When I read a book criticizing Darwin's theory from an agnostic viewpoint[156] it startled me, and I then began a literature search for real evidence that random mutation and natural selection could really do what was claimed for them. I came up completely empty. In the over thirty years since then, I've only become more convinced of the inadequacy of Darwinism, and more persuaded of the need for intelligent design at ever-deeper levels of biology, as detailed in my books.

Clearly Greg and Amber honestly disagree. How to explain that? To help answer, let's first consider a different scientific discipline—physics. The history of physics offers powerful lessons that widespread agreement on even the most basic ideas in a field is no guarantee that there is sufficient evidence to support the theory, or indeed that there is any evidence for it at all. Just ask James Clerk Maxwell, who wrote the article "Æther" in 1878 for the ninth edition of the Encyclopædia Britannica: "Whatever difficulties we may have in forming a consistent idea of the constitution of the æther, *there can be no doubt* that the interplanetary and interstellar spaces are not empty, but are occupied by a material substance or body, which is *certainly* the largest, and probably the most uniform body *of which we have any knowledge.*"[157] [emphasis added]

Maxwell, one of the greatest physicists of all time, calculated the density—to three significant figures—of the æther, a substance that doesn't exist. If that doesn't make the case for the peril of over-reliance on theory—and the need for profound scientific humility—nothing will.

But surely no branch of contemporary science could go so far astray, could it?—Maybe. In the past few years a theoretical particle physicist named Sabine Hossenfelder has made a splash by criticizing the reliance of other theoreticians on a gauzy concept of "beauty" to guide their calculations. She thinks that pretty much the whole field has been barking up the wrong tree for thirty years. Last summer she released a book on the topic, *Lost in Math*, which was favorably reviewed in *Nature*.[158] She also maintains a blog, BackReaction, and holds forth regularly and entertainingly. Recently she put up a typically insightful, acerbic post, "Particle Physicists Excited over Discovery of Nothing in Particular."[159] The first reader to comment at the site wrote sympathetically, "I believe it's hard for anyone on the inside of a tribe to see the limitations of their own thinking. One has to step outside of the protection ring of orthodoxy."

Respect the Views of the Public

Precisely. It's hardly news that a group can share strong views on topics of mutual interest to its members, which many on the outside find less than compelling. Theoretical particle physicists, lawyers, members of the military, union members, business people, clergy, and on and on. It would be hard to find a group that didn't have such shared views. Of course, that includes evolutionary biologists and scientists in general. I would like to delicately suggest that a large chunk of the disconnect (although certainly not the only factor) between the public and biologists over evolution is that, as a rule, biologists share a commitment to Darwin's theory that the general public does not. That shared commitment leads biologists (and scientists in general) to require substantially less evidence to persuade them of the theory's verity and scope than someone outside the tribe.

Contra Lang and Rice, it's preposterous to say that the data "are more than sufficient to convince any open-minded skeptic that unguided evolution is capable of generating complex systems." Unless one defines a skeptic of Darwin's theory (the most prominent proposed "unguided" explanation) as closed-minded, a quick visit to the library will disabuse one of that notion.[160] Even in their own review, at best the authors argue that they see no obstacle to Darwinian processes producing functional complex systems; they surely don't demonstrate that it can. And of all the relevant literature in books and journals, the two papers they pointed to as examples of the power of Darwin's mechanism are quite modest indeed. When my first book, *Darwin's Black Box*, was published in 1996 it elicited comments by bona fide evolutionary biologists such as: "There are presently no detailed Darwinian accounts of the evolution of any biochemical system, only a variety of wishful speculations,"[161] and "There is no doubt that the pathways described by Behe are dauntingly complex, and their evolution will be hard to unravel.... We may forever be unable to envisage the first proto-pathways."[162] It's hard to reconcile such statements with an assertion that the data are "more than sufficient."

As quoted earlier, Lang and Rice write: "Scientists—by nature or by training—are skeptics. Even the most time-honored theories are re-evaluated as new data come to light." That claim wouldn't survive even a short trip through the history of science, which is of course replete with people (that's another name for scientists) dogmatically defending their cherished ideas from the slings and arrows of skeptical unbelievers. Few scientists are as emotionless as Mr. Spock, and to maintain otherwise is little more than group-flattery. More to the point, scientists do not have a corner on the market for skepticism. In all walks of life that trait has its uses. A banker evaluating a loan, a voter listening to a politician's speech, a teacher wondering whether the dog really did eat this student's homework, a judge considering whether a defendant is indeed remorseful, an historian evaluating the direction of her academic field—pretty much everyone is skeptical when they smell a rat. And "pretty much everyone" is another way to say "the public."

For what it's worth, my advice on this matter for Lang, Rice, and others with similar views is to respect the opinions of the public, even if one disagrees with them and thinks them ill-founded, because, when it comes to the grand claims for Darwin's theory, many folks think they smell a rat and are prudently exercising their skepticism. Indeed, instead of blaming the public, they should consider the possibility that perhaps the evidence for the vast scope of Darwin's theory really isn't as strong as biologists over the years have been telling each other.

99. Bullet Points for Jerry Coyne

"Bullet Points for Jerry Coyne," Evolution News and Science
Today, Discovery Institute, March 12, 2019.

UNIVERSITY OF CHICAGO EVOLUTIONARY BIOLOGIST JERRY COYNE reviewed *Darwin Devolves* for this past Sunday's *Washington Post*.[163] As you might expect, it's written in the venerable style of Richard Dawkins's review of *The Edge of Evolution* for the *New York Times* back in 2007: long on sneering, smearing, and assertion; short to nonexistent on telling readers what the book's actual arguments are. Alas, Coyne's piece has too little intellectual content to sustain any real engagement. So I'll simply proceed from its beginning to its end.

Coyne: "'Intelligent design' arose after opponents of evolution repeatedly failed on First Amendment grounds to get Bible-based creationism taught in the public schools.... intelligent design, which scientists have dubbed 'creationism in a cheap tuxedo.'"

Me: Good idea—let's link the author to a scorned group right at the start and smear his motives.

Coyne: "Behe does not rely on the Bible as a science textbook. Rather, he admits that evolution occurs by natural selection sifting new mutations and that all species are related via common ancestors."

Me: But we'll call him a "creationist" anyway, to milk that epithet for all it's worth.

Coyne: "Scientists… pointed out numerous scenarios in which a system fitting Behe's definition of 'irreducible complexity' could evolve in a step-by-step manner (one is the hormone pathway studied by my Chicago colleague Joe Thornton)."

Me: I spent half of Chapter 8 on Thornton's work, discussing why it shows deep and unexpected problems for Darwinian evolution. Coyne not only doesn't summarize my argument, he doesn't even tell readers I make one.

Coyne: "They then adduced clear evidence from many complex biochemical systems that these scenarios had actually occurred."

Me: I showed in the appendix that no evidence beyond handwaving has been published since *Darwin's Black Box*. Again, not even a mention by Coyne that I dispute his claim.

Coyne: "These systems… embody an absurd, Rube Goldberg-like complexity that makes no sense as the handiwork of an engineer but makes perfect sense as a product of a long and unguided historical process."

Me: Wow, the great theologian Jerry Coyne has determined that God wouldn't have done it that way—no need for actual evidence that Darwin's mechanism can do the job. We all anxiously await the unveiling of Coyne's superior designs for a clotting cascade and a flagellum.

Coyne: "Behe's rationale for designed mutations is circular. He claims that biochemical pathways are designed rather than evolved because they're based on the 'purposeful arrangement of parts.' But which arrangements are those designed with a purpose? They're simply the pathways that Behe sees as too complex to have evolved."

Me: So Coyne can't think of a purpose for an eye? Or for the leg gears of the planthopper? Or for the supercharged flagellum of the magneto-tactic bacterium MO-1? That's funny—the authors of the science papers on those systems that I cite in the book seem to have had no trouble identifying their purposes.

Coyne: "Perhaps Behe's most ludicrous claim is this: Evolution within the lowest levels of biological classification—genera and species—might be purely Darwinian, but the origin of higher-level groups—families, orders and so on—requires designed mutations. Yet as every biologist knows, groupings above the level of species are purely subjective."

Me: Can Coyne tell the difference between a plant and an animal? Between a bird and a fish? A cat and a dog? Sure, as I discuss in the book, a classification system is a human invention and so it inevitably has uncertainties, ambiguities, and mistakes. But implying that biological classification reflects nothing real is disingenuous at best.

Coyne: "Behe selectively gives a handful of examples in which mutations have produced broken genes that are nevertheless useful, but he simply ignores the large number of adaptive mutations that do not inactivate genes. These include duplications, in which a gene is accidentally copied twice, with the copies diverging in useful ways (this is how primates acquired our three-color vision, as well as different forms of hemoglobin)."

Me: I wrote a section in Chapter 8 titled "Evolution by Gene Duplication Revisited" in which I explain why duplication and diversification by Darwinian processes may account for some things but not for others. I specifically explain why I changed my mind about sophisticated hemoglobin, concluding that duplication and diversification could not have built it, since it would require much more modification starting from a simple myoglobin-like gene than would mere duplication of opsin (color-vision) genes. Coyne doesn't even let readers know I discuss it.

Coyne: "Behe also argues that evolution is self-limiting because natural selection 'adjust[s] a biological system to its current function' and thus 'works to block the system from taking up a significantly different function.' But… Think of how feathers, which probably evolved to conserve body heat in dinosaurs, opened up the possibility of flight—leading to all the diverse birds on Earth."

Me: It never ceases to amaze me that Darwinists like Coyne are unable to separate the question of what happened from the question of how it happened. Okay, flightless dinosaurs had feathers and birds can now fly. So what exactly is the evidence that it happened by a Darwinian process? What is the evidence that a Darwinian process could even, say, differentiate owls and crows from a common ancestor? I argue at length in the book that unintelligent processes aren't remotely up to those tasks. Without any substantive counter-argument, Coyne simply responds like a kid on a playground: "Yes they can too do that!"

Coyne: "Like his creationist kin, Behe devotes his time not to giving evidence for intelligent design but to attacking evolutionary biology."

Me: Gee, Coyne must have missed Chapter 10 in *Darwin Devolves* ("A Terrible Thing to Waste"), as well as Chapters 8 and 9 in *Darwin's Black Box* ("Intelligent Design" and "Questions About Design"), and Chapter 11 in *The Edge of Evolution* ("All the World's a Stage"). I explain at length in those chapters and elsewhere that the work of a mind—design—is evinced precisely by the purposeful arrangement of parts, such as is found in abundance in life. For pretty much the entirety of recorded history until Darwin, almost everyone thought life was designed exactly for that reason—the arrangement of parts for a purpose—as I discuss in the preface to the book. Contrary to Coyne, it is Darwin's audacious assertion—that complex interactive functional structures could be produced by random variation and natural selection—that has gone unsupported by pertinent evidence. Coyne's unwillingness or inability to grasp the argument for design does not mean the argument hasn't been made.

Coyne: "Since humans are placed in the same family as other great apes (Hominidae), Behe's theory predicts that we arose without a designer's intervention. But here he backpedals, asserting that there are 'excellent reasons to suspect those differences [between humans and other apes] are well beyond Darwinian processes.' Sadly, he doesn't give these reasons, but I'd guess they stem from the Christian belief that Homo sapiens is a special creation of God."

Me: Actually, they stem from our personal awareness that we can reason, speak, think abstractly, and so on—in other words, that we have self-reflective, reasoning minds—which arguably is the most profound attribute in the world. By the way, I also wrote in the book that there are good reasons to doubt that giraffes could arise from a shorter-necked relative like the okapi, even though they are in the same biological family. For some reason Coyne doesn't ascribe my skepticism there to Christian belief.

Coyne: "In 1998, the Discovery Institute drafted the 'Wedge Document,' a secret plan (leaked in 1999) to spread Christianity in America by teaching intelligent design and fighting materialism.... Well, now it's 20 years on, and despite the efforts of Behe and other neo-creationists, intelligent design has been discredited as science and outed as disguised religion."

Me: Yes, the horrible threat of a group trying to persuade people of its ideas by writing books and articles has so far been countered by brave folks like Jerry Coyne, who use the kind of overwhelming evidence and impeccable logic showcased in his book review.

Coyne is quite the prominent evolutionary biologist, and has been antagonistic to intelligent design arguments for decades. If Darwin's theory were actually the powerful idea it's claimed to be, Coyne should have been able to counter design easily, simply by summarizing its arguments and showing how Darwin and his intellectual heirs deal with them. Yet he can't even bring himself to mention what those arguments are. Instead he tries to whip up hysteria against a book that argues for what most people already believe. That speaks volumes about the actual strength of Darwin's theory.

Part Eight: Can We Talk?

"Can we talk?" That was the late comedienne Joan Rivers' signature phrase. She meant, can we be frank with each other? Can we get to the bottom of this? Can we move past pretenses and appearances and concerns about what the children will think and have an honest conversation? Of course, in her hands the expression was an invitation to laugh. Yet it is a serious question in many areas of life—and unfortunately, the answer seems to be trending toward no.

Sure, these days there are many more avenues available to get a message out, so historically marginalized groups can at least have a presence in the public square. You'd think this greater ease of communication would result in increased intellectual diversity. But malignant groups (terrorists, scam artists, and too much more) can poison discussion channels. Worse, in this age of instant communication—24-hour news channels, video-recording cell phones, the Twitter-verse, and whatever else has come along by the time this book is published—hecklers and gatekeepers seem to have more power to put a chill on discussion than purveyors of serious ideas have to get a hearing.

Newspapers, opinion journals, and organizations whose business models require them to maintain an unsullied reputation in their community can be exquisitely sensitive to controversy and quickly learn to avoid touchy topics. It can take a lot of time and effort to put together an argument that is intended to be persuasive; it takes only seconds to post abuse online—just check the comment boxes of any online news or opinion site. Social media also provides individuals new tools to retreat into specialized opinion ghettos, where one only hears from like-minded individuals. These are places where lazy arguments rarely get tested by thoughtful opponents. So, while we can still hope for increased intellectual diversity, the age of social media and instant communication might in fact usher in an era of decreased intellectual diversity.

Intelligent design is one topic badly in need of adult discussion. But the history of conversation regarding ID is a mixed bag. On the one hand I and other ID proponents have had books brought out by respected publishing firms and op-ed submissions accepted by major newspapers. On the other hand, a number of our speaking engagements have been cancelled under pressure, campaigns have been waged to shut off our avenues of publication, and efforts have been made—sometimes successfully—to dismiss design proponents from their jobs. Finding opportunities to speak effectively as an ID advocate is a chancy proposition, and one fraught with danger.

The need for open discussion is perhaps most acute in public schools. Most of us gain much of our understanding about how the world works in school. Of course life teaches us many lessons as we age, and some people go on to read widely on their own as adults. Yet not everything we absorb in our callow years gets reexamined. So, on average, school lessons will have a tremendous influence on how the aggregate public views the world.

What, then, should schools teach about how life arose and developed over time? The topic is fraught with the most profound implications, touching not only scientific matters, but philosophical, theological, and social ones, too. Surely when intelligent and well-informed people disagree on a topic of such importance, open inquiry is the order of the day. Unfortunately, that hasn't been the case in most of our schools and universities.

Discovery Institute's Center for Science and Culture, a leading institutional hub for the theory of intelligent design, doesn't even recommend that intelligent design be introduced into public school science biology classes. They don't want to politicize ID in this way. Instead, they make a much more modest recommendation. They urge that evolutionary theory be taught in a balanced manner, with students exposed not just to the evidence for Darwinism but also to some of the evidence from the peer-reviewed scientific literature against it. The approach encourages critical thinking for aspiring scientists, a capacity crucial to scientific progress. And it has been tested in the courts and has consistently met with approval there. But for defenders of the status quo, even this is too much. The dogmatic Darwinists demand that only an airbrushed case for their theory be presented to students—no whisper of dissent in the textbooks or other assigned reading. No uncooperative facts allowed.

Nonetheless, I'm serenely confident that the Darwinian paradigm will collapse and the idea of the intelligent design of life will eventually carry the day, even in the snootiest intellectual circles, even in state-controlled schools. This will happen not because of any particular thing that I or any other ID proponent has done or will do. Rather, it's simply because that's the way the biological data are headed. Virtually every week stunningly sophisticated new features of life are being uncovered by science, features that in any other context we would immediately recognize as designed. This evidence cannot forever be denied.

This section contains a selection of pieces I've written for diverse audiences over the past two decades on the topic of whether intelligent design can and should be discussed in diverse venues.

Let's talk.

100. Teach Evolution and Ask Hard Questions

"Teach Evolution and Ask Hard Questions,"
New York Times, August 13, 1999.

The debate leading the Kansas Board of Education to abolish the requirement for teaching evolution has about the same connection to reality as the play *Inherit the Wind* had to the actual Scopes trial. In both cases complex historical, scientific, and philosophical issues gave way to the simplifying demands of the morality play. If the schoolchildren of Kansas and other states are to receive a good science education, however, then we'll have to forgo the fun of demonizing each other, take a deep breath, and start making a few distinctions.

Regrettably, the action of the Kansas board makes that much more difficult. Not only are teachers there now discouraged from discussing evidence in support of Darwin's theory, results questioning it won't be heard either.

For example, let's look at three claims of evidence for Darwinian evolution often cited by high school textbooks. First, as the use of antibiotics has become common, mutant strains of resistant bacteria have become more common, threatening public health. Second, dark-colored variants of a certain moth species evaded predation by birds because their color matched the sooty tree trunks of industrial England. Third, the embryos of fish, amphibians, birds, and mammals look virtually identical in an early stage of development, becoming different only at later stages.

A relevant distinction, however, is that only the first example is true. The second example is unsupported by current evidence, while the third is downright false. Although light- and dark-colored moths did vary in expected ways in some regions of England, elsewhere they didn't. Further, textbook photographs showing moths resting in the day on tree trunks, where birds supposedly ate them, run afoul of the fact that the moths are active at night and don't normally rest on tree trunks. After learning about the problems with this favorite Darwinian example, an evolutionary scientist wrote in the journal *Nature* that he felt as he did as a boy when he learned there was no Santa Claus.

The story of the embryos is an object lesson in seeing what you want to see. Sketches of vertebrate embryos were first made in the late nineteenth century by Ernst Haeckel, an admirer of Darwin. In the years that followed, apparently nobody verified the accuracy of Haeckel's drawings. Prominent scientists declared in textbooks that the theory of evolution predicted, explained, and was supported by the striking similarity of vertebrate embryos. And that is what generations of American students have learned.

Recently, however, an international team of scientists decided to check the drawings' reliability. They found that Haeckel had, well, taken liberties: the embryos are significantly different from each other. The head of the research team observed that "it looks like it's turning out to be one of the most famous fakes in biology."[1] What's more, the embryonic stages shown in the drawings are actually not the earliest ones. The earliest stages show much greater variation.

If I were teaching a high school biology course, I certainly would want my students to understand Darwin's theory of evolution by natural selection, which explains antibiotic resistance and a lot of other things. I would want them to know the many similarities among organisms that are interpreted in terms of common descent, as well as to understand the laboratory experiments that show organisms changing in response to selective pressure.

But I would also want them to learn to make distinctions and ask tough questions. Questions we might discuss include these:

If it's so difficult to demonstrate that small changes in modern moths are the result of natural selection, how confident can we be that Darwinian selection drove large changes in the distant past? If supposedly identical embryos were touted as strong evidence for evolution, does the recent demonstration of variation in embryos now count as evidence against evolution? If some scientists relied for a century on an old, mistaken piece of data because they thought it supported the accepted theory, is it possible they might even now give short shrift to legitimate contrary data or interpretations? Discussing questions like these would help students see that sometimes a theory actively shapes the way we think, and also that there are still exciting, unanswered questions in biology that may require fresh ideas.

It's a shame that Kansas students won't get to take part in such a discussion. We should make sure that the students of other states do.

Emotions run very deep on the subject of evolution, and while the morality play generally casts religious people as the ones who want to limit discussion, some scientists on the "rational" side could fit that role, too. But if we want our children to become educated citizens, we have to broaden discussion, not limit it.

Teach Darwin's elegant theory. But also discuss where it has real problems accounting for the data, where data are severely limited, where scientists might be engaged in wishful thinking, and where alternative—even "heretical"—explanations are possible.

101. The National Academy of... Religion?

"The National Academy of ... Religion?," *Ethics and Medics* 25, no. 2 (February 2000): 1–3.

T HE REJECTION LAST AUGUST BY THE KANSAS STATE BOARD OF Education of the National Academy of Science's standards on teaching evolution[2] stirred a national controversy. Although the conflict was portrayed in the press through the usual lens of biblical fundamentalism versus scientific progress, and though some elements of that could indeed be seen, many people opposing the standards viewed the conflict not as a defense of a specific faith, but as an effort to slow the spread of an aggressive scientism. The National Academy, it was argued, was going far beyond the data to impress on schoolchildren a materialistic view of the world. In this essay I evaluate that claim by examining some recent Academy publications: *Science and Creationism* and *Teaching About Evolution and the Nature of Science.*[3] The booklets provide a window into the Academy's reasoning, and both are intended to influence the way science teachers teach. (Indeed, a copy of *Science and Creationism* is included free when ordering the science standards from the National Academy Press.)

Confusing Definitions

A CAREFUL reading of the publications identifies a basic source of confusion: unusual and expansive definitions of both evolution and creationism. In the Academy's usage, evolution is not simply a name for a biological theory, but for any change. "Evolution in the broadest sense explains that what we see today is different from what existed in the past."[4] The definition, however, is constructed to also encompass Darwin's particular mechanism: "Biological evolution concerns changes in living things during the history of life on Earth.... Over time, biological processes such as natural selection give rise to new species."[5] Thus the unarguable—change over time—is conflated with the highly problematic—natural selection as evolution's mechanism.

Creationism also means different things at different points in the Academy's writings. *Teaching About Evolution and the Nature of Science* defines "creation science" in the way it is commonly understood, as "the conviction that God created the universe—including humans and other living things—all at once in the relatively recent past."[6] In *Science and Creationism*, however, a new class of beliefs is subsumed into the definition:

> Other advocates of creation science are willing to accept that Earth, the planets, and the stars may have existed for millions of years. But they argue that the various types of organisms, and especially humans, could only have come about with supernatural intervention, because they show "intelligent design." In this booklet, both these "Young Earth" and "Old Earth" views are referred to as "creationism" or "special creation."[7]

The definition of creationism thus travels a long way from biblical literalism to encompass any "intervention" (the Academy does not define the term) by God in the history of life. It has also traveled a long way from what science can properly assert. Indeed, unlike its vigorous factual defense of an ancient age for the earth, the only objection that the Academy offers to the activity of God in an ancient universe is a philosophical one: "intervention is not subjectable to meaningful tests."[8]

At points the Academy seems to become frustrated and ascribes hidden motives to those who disagree with it. "The arguments of creationists are not driven by evidence that can be observed in the natural world.... The explanation is seen as unalterable, and evidence is sought only to support a particular conclusion by whatever means possible."[9]

Certainly any view can be held dogmatically, but to brand everyone who disagrees with you as intellectually dishonest does not help to defuse controversy.

SCHOLARLY STANDARDS

ANOTHER DIFFICULTY with the Academy's publications is, surprisingly, a disregard of scholarly standards. For example, it asserts that "intelligent design proponents argue that... the many steps that blood goes through when it clots, are so irreducibly complex that they can function only if all the components are operative at once."[10] Readers are then reassured that "blood clotting has been explained"[11] within a Darwinian framework. Yet no citations are provided for someone who might want to check those or other claims. Indeed, the Academy does not even mention where these intelligent design arguments may be found, or who makes them.

Perhaps I may be forgiven for thinking that in this section the Academy had my writings in mind, since my recent book *Darwin's Black Box*[12] coined the phrase "irreducible complexity" and argued for intelligent design, using the blood clotting cascade as an example. Yet mixed with my reasoning in the Academy's account are elements I do not recognize at all. Immediately before alluding to my arguments, the Academy asserts that proponents of intelligent design argue structural complexity is "proof" of "the direct hand of God" in "specially creating" "organisms" "as they are today."[13] I believe none of those things.

The purpose of the scholarly practice of citing one's sources is to prevent the making of unsupported claims and the caricaturing of one's intellectual opponents. It is puzzling that the prestigious National Academy of Sciences disregards such minimal standards.

The Academy's reticence about mentioning the publications of Darwinian skeptics extends to its compilation of reading resources. Although it lists scores of books, including popular works by such aggressive atheists as biologist Richard Dawkins and philosopher Daniel Dennett, no book arguing against Darwinian evolution is found. Not even books by noted biologists such as Stuart Kauffman or Brian Goodwin, who question the adequacy of natural selection from a strictly materialistic perspective, are cited to break the Darwinian monopoly. Teachers following the Academy's lists would never read opponents of Darwinism making their own case in their own words; students instructed by such teachers would have little idea why there was any controversy at all.

LIMITING DISCUSSION

THE ACADEMY'S alignment of science with an effectively materialistic philosophical position is seen most clearly in its discussion of the origin of life. Modern origin-of-life studies can be dated to the early 1950s when Stanley Miller, then a graduate student at the University of Chicago, mixed together in a flask gases thought to have been present on the primitive earth, sparked them with an electric spark to simulate lightning, and saw that some amino acids were produced. Over the years many scientists have followed Miller's path. Nonetheless, despite considerable effort science is currently at a loss to explain how life might have been started by random chemical processes.

In its discussion, however, the National Academy shifts the focus away from the negative experimental results to the attitude of origin-of-life researchers: "For those who are studying the origin of life, the question is no longer whether life could have originated by chemical processes involving nonbiological components. The question instead has become which of many pathways might have been followed to produce the first cells."[14]

Evidently the Academy wants to convey to teachers and their students the attitude that thinking about the origin of life should stay strictly within the confines of "chemical processes involving nonbiological components." That attitude, however, is not demanded by the physical evidence and is not even the way some scientists are approaching the problem. In his recent book probing life's origin, physicist Paul Davies concluded, "My personal belief, for what it is worth, is that a fully satisfactory theory of the origin of life demands some radically new ideas."[15]

The Academy, however, encourages students to think along convention-al lines.

The Academy's Religious Philosophy

By far the most troubling aspect of the Academy's writings is its implic-it endorsement of a specific religious philosophy—"theistic evolution." By that term the Academy appears to mean the assumption that, while a God may or may not have established the laws of nature, any further in-teraction is ruled out, or at least must be scientifically indistinguishable from the action of random physical laws. The Academy writes: "Many religious persons, including many scientists, hold that God created the universe and the various processes driving physical and biological evo-lution.... This belief, which sometimes is termed "theistic evolution," is not in disagreement with scientific explanations of evolution. Indeed, it reflects the remarkable and inspiring character of the physical universe revealed by cosmology, paleontology, molecular biology, and many other scientific disciplines."[16]

The Academy's writing here could easily be viewed as a quasi-gov-ernmental agency with considerable public influence acting to promote a specific religious tenet through the public schools—that God would not act in nature. While most Church-State conflicts occur at the pe-riphery of a "religious establishment," it seems that this comes very close to the center. Whether or not one thinks that theistic evolution is a fair description of the relationship of God to the world, most American citi-zens would balk at having that position sponsored by the public schools.

Conclusion: Include Other Views

Documents published by the National Academy of Sciences support those who argue that the Academy has overstepped its role. Instead of simply offering the public an unbiased assessment of scientific evidence, it appears to be attempting to steer science education into channels that offer the least resistance to Darwinian theory. In so doing it has blurred definitions, overstated science's progress in some areas, ignored or misrepresented challenges to Darwinism, and become disturbingly entangled in religious matters.

A major step to improve this situation would be easy to implement: persons with different viewpoints on evolution should be included on the committee which drafts Academy publications on teaching evolution. Government panels typically try to represent the diverse views of the citizenry, since a committee made up exclusively of persons sharing similar views can share shortcomings as well. The Academy needs help to avoid the blind spots of scientific materialism.

102. A LETTER TO MICROBE

When the journal *Microbe* published a letter concerning the evolution of the bacterial flagellum, I sent the following letter to help them understand intelligent design.

"Unpublished letter to *Microbe*," Uncommon Descent (website), October 12, 2007 and September 22, 2013.

IN "EVOLUTION OF THE BACTERIAL FLAGELLUM," WONG ET AL.[17] SEEK to counter arguments of intelligent design proponents such as myself that the flagellum did not evolve by random mutation and natural selection. Unfortunately, their otherwise-fine review misunderstands design reasoning and so fails to engage that issue. The critical passage from Wong et al. is the first paragraph:

> Proponents of the intelligent design (ID) explanation for how organisms developed claim that the bacterial flagellum (BF) is irreducibly complex. They argue that this structure is so complicated that it could not have emerged through random selection but had to be designed by an intelligent entity. One part of this claim is that each flagellar component is used solely for the purpose of making a flagellum that, in turn, is used only for motility. Further, each flagellar protein is assumed to have appeared independently of the other component proteins.

Although the first two sentences are overly simplistic, they are more or less correct; the last two sentences, however, are quite wrong. (The authors cite no references for these latter claims.) It is no part of the design argument that each component of an irreducibly complex structure must be used solely for that purpose, nor that each part must arise independently. In my 1996 book *Darwin's Black Box*, which brought the concept of irreducible complexity to wide public attention, I pointed out the fact that, for example, proteins of the blood clotting cascade share sequence homology with each other and with other serine proteases, and the fact that ciliary proteins such as tubulin are involved in other tasks in the cell. Yet I explained that neither sequence homology nor multiple functions showed how integrated systems containing many parts could be put together by small random steps. Unfortunately, Wong et al. spend their efforts addressing their own erroneous assertions. They fail to address the only pertinent question, the question of whether random, unintelligent processes—even when filtered by natural selection—could plausibly build a structure such as the flagellum.

To address the adequacy of random processes plus selection would require rigorous experiments or calculations showing that the intricate, functional structures are not too improbable given the evolutionary resources available. Recent work bears negatively on this difficult question. In long-term laboratory evolution experiments over tens of thousands of generations,[18] cultures of E. coli were repeatedly seen to lose the ability to make ribose and maltose, and to repair their DNA. Some mutations shut down expression of their flagellar genes, apparently to conserve energy. No selected mutations were observed which could plausibly be argued to be the incipient stages of some new, complex functional system.

Similar kinds of results are seen in other well-studied evolutionary systems. For example, in response to strong pressure from the malarial parasite, the human genome has suffered a handful of positively selected yet degradative mutations,[19] including ones that render nonfunctional the genes for glucose-6-phosphate dehydrogenase, the alpha and beta chains of hemoglobin, band 3 protein, and others. Again, no selected mutations were observed which could plausibly be argued to be the incipient stages of some new, complex functional system.

To a skeptic such as myself, this does not look like the sort of process which could build complex molecular machinery. Those who would argue persuasively against intelligent design must address this basic issue.

The *Microbe* Editor's Email Response

THANK YOU for your recent letter on Dr. Saier's article. We are declining to publish the letter as it does not address the main points of the article. The article and *Microbe*'s decision to publish it were not intended to address the broad question of the validity of evolution. ASM has published its Statement on the Scientific Basis for Evolution which summarizes the views of the Society on the topic:

> Knowledge of the microbial world is essential to understanding the evolution of life on Earth. The characteristics of microorganisms—small size, rapid reproduction, mobility, and facility in exchanging genetic information—allow them to adapt rapidly to environmental influences. In microbiology, the validity of evolutionary principles is supported by [1] readily demonstrated mutation, recombination and selection, which are the fundamental mechanisms of evolution; [2] comparisons based on genomic data that support a common ancestry of life; and [3] observable rates of genetic change and the extent of genomic diversity which indicate that divergence has occurred over a very long scale of geologic time, and testify to the great antiquity of life on Earth. Thus, microorganisms illustrate evolution in action, and microbiologists have been able to make use of the microbes' evolutionary capacity in the development of life-improving and life-saving innovations in medicine, agriculture, and for the environment. By contrast, proposed alternatives to evolution, such as intelligent design and other forms of creationism, are not scientific, in part because they fail to provide a framework for useful, testable predictions. The use of the supposed "irreducible complexity" of the bacterial flagellum as an argument to endow nonscientific concepts with what appears to be legitimacy, is spurious and not based on fact. Evolution is not mere conjecture, but a conclusive discovery supported by a coherent body of integrated evidence. Overwhelmingly, the scientific community, regardless of religious belief, accepts evolution as central to an understanding of life and the life sciences. A fundamental aspect of the practice of science is

to separate one's personal beliefs from the pursuit of understanding of the natural world. It is important that society and future generations recognize the legitimacy of testable, verified, fact-based learning about the origins and diversity of life.

ME AGAIN

THANKS VERY much for your email. However, your statement that my letter "does not address the main points of the article" is quite difficult to understand. The clearly stated purpose of the Wong et al. article is to refute intelligent design reasoning. My letter shows that the authors misunderstand design reasoning, so that the supposed refutation addresses a straw man. How can that not be "address[ing] the main points"?

If you will read my letter with attention, you will notice that I did not question "the validity of evolution." In fact, although many people are confused on this point, the concept of intelligent design has no proper quarrel with the validity of evolution understood as change over time or common ancestry. Intelligent design is quite compatible with descent with modification. It simply argues that some facets of life resulted from intelligent planning or direction, rather than relying exclusively on random events, as Darwinian theory proposes. My discussion in the letter of the results of random mutations in experiments on E. coli, and in human/malaria evolutionary warfare, underscores this point.

Most publications consider it responsible journalism to publish letters by well-known advocates of views attacked in articles. The purpose, of course, is to avoid misleading readers of the journal by unknowingly misstating or caricaturing a position. In order that your readers will not form a mistaken view of what the intelligent design argument actually states, I ask you to reconsider the decision not to publish my letter.

THE RESPONSE FROM MICROBE

[CRICKETS.]

103. "The Great Debate,"
The Biochemist

"Evolution: The Great Debate," letter to the editor,
The Biochemist (April 2009): 55.

UNIVERSITY OF WISCONSIN HISTORIAN RON NUMBERS HAS observed: "I suspect most people don't understand what intelligent design is. I've talked to enough people who didn't understand it to make me really suspicious, and there are even people who lecture about it and don't know what they're talking about. I don't define it; what I do is let the advocates of intelligent design speak for themselves."[20]

It is unfortunate that Kevin Padian and Nick Matzke didn't follow that good advice in their recent article in *Biochemical Journal* (reprinted in *The Biochemist*), "Darwin, Dover, 'Intelligent Design' and Textbooks." Rather than letting ID advocates speak for themselves, they concocted their own notion, that "the whole point of ID is to establish that miraculous supernatural intervention was required in the history of life.... ID proponents attempt scientifically to demonstrate their proposal, i.e. repeated miraculous intervention in the history of life."[21]

Padian and Matzke don't know what they're talking about. The basic claim of ID is simply that empirical evidence can lead one to a conclusion of design in nature. It says nothing about "repeated miraculous intervention." I made this quite clear in my most recent book, *The Edge of Evolution*, in a section entitled "No Interference."

"The assumption that design unavoidably requires 'interference' rests mostly on a lack of imagination," I wrote. "There's no reason that the extended fine-tuning view I am presenting here necessarily requires active meddling with nature any more than the fine tuning of theistic evolution does."[22]

(Theistic evolution seems to be what Padian and Matzke meant by their phrase "your father's 'ID'".[23] Theistic evolution is also officially tolerated as a scientifically compatible viewpoint by no less than the National Academy of Sciences.[24]) It is true that many ID proponents do not think, as I do, that complete common ancestry is correct, and do think, as I don't, that miraculous interventions have taken place. Yet they understand, as Padian and Matzke apparently do not, that the basic argument for design does not reach that far—design focuses on the endpoint, not the process. Arguments about process require further evidence beyond that for the bare conclusion of design. That should not be a difficult point to grasp.

I sympathize with Padian and Matzke—if the judge at the Dover trial had ruled for my side, I probably would try to puff up his stature, too. But he didn't, so I looked more closely at his ruling. They call him an "independent thinker... who hewed to no party line." Well, he must be the only independent thinker who copied his independent thoughts directly from someone else's party-line document. A month before the judge issued his opinion, the plaintiffs' attorneys submitted to him a 161-page brief, detailing exactly how they wished him to rule. Here's a brief excerpt from the lawyer's document: "Professor Behe testified that the strength of an analogy depends on the degree of similarity entailed in the two propositions. If this is the test, intelligent design completely fails."[25]

And now this from the "judge's" decision: "Professor Behe testified that the strength of the analogy depends upon the degree of similarity entailed in the two propositions; however, if this is the test, ID completely fails."[26]

How much independent thought does it take to decide that those two sentences should be joined with the conjunction "however"? (By the way, I did not testify that ID was based on an analogy; I never even used the word "analogy." That's just one of a number of the lawyers' mischaracterizations of my testimony copied into the judge's opinion.)

Although it's understandable that Padian and Matzke wish that the judge had comprehended and then agreed with their academic arguments (I wished the same for my testimony), in fact there is no evidence that the judge understood any of the expert academic testimony in the court. Whenever his opinion touched on any expert testimony—either scientific, philosophical, or theological, either for the defendants or for the plaintiffs—he simply copied the text from the plaintiffs' attorneys' brief. (When discussing school board meetings or newspaper editorials, the judge used his own words.) As Padian and Matzke rightly observed, my testimony on the biochemical complexity of the flagellum "appeared to make the judge's eyes glaze over."[27] There is no reason to think that this former head of the Pennsylvania Liquor Control Board grasped the distinction between flagellin and prothrombin, gene duplication and point mutation, Thomas Aquinas and David Hume, or random mutation and common descent.

I realize that the concept of intelligent design strikes many scientists as strange. But whenever the topic is a fundamental one, such as how the universe or life arose, some possibilities are bound to strike us as strange. For example, the bioinformatician Eugene Koonin recently proposed that the origin of life is not the result of some lawlike process, but is simply a tremendously unlikely accident, that nonetheless is destined to occur repeatedly in an infinite universe.[28] Whether one believes life had a lawlike beginning or not, it is clear that Koonin's strange-sounding proposal is possible, and its discussion should not be prohibited. The same goes for the strange-sounding proposal of intelligent design.

Michael J. Behe
Lehigh University

104. LETTER TO SCIENCE, 2009

"Unpublished Letter to *Science*," Uncommon Descent (website), May 21, 2009.

THE MAY 1ST ISSUE OF *SCIENCE* CONTAINED A "NEWS FOCUS" AR-ticle entitled "On the Origin of the Immune System."[29] While describing some then-current work in the area, the author, John Travis, made liberal use of myself as an unreasonably skeptical foil. I wrote the following letter to the editor of *Science* pointing out inaccuracies in the story but, wouldn't ya know, they didn't think the letter would be of sufficient interest to their readers to print it.

To the editor:

In his article "On the Origin of the Immune System," John Travis makes the same mistake as did the judge in the 2005 Dover trial—badly confusing the notions of intelligent design, common descent, and evolution. Citing the courtroom theatrics of the lawyers who piled a stack of textbooks and articles in front of me, Travis quotes me as remarking, "They're wonderful articles.... They simply just don't address the question I pose." Unfortunately, Travis seems uninterested in what that question might be. Instead he cheers, "Score one for evolution."

Although some news reporters, lawyers, and parents are confused on the topic, "intelligent design" is not the opposite of "evolution." As some biologists before Darwin theorized, organisms might have descended with modification and be related by common descent, but the process might have been guided by some form of intelligence or teleological driving force in nature. Darwin's chief contribution was not the simple idea of common descent, but the hypothesis that evolution is driven completely by ateleological mechanisms, prominently including random variation and natural selection. Intelligent design has no proper argument with the bare idea of common descent; rather, it disputes the sufficiency of ateleological mechanisms to explain all facets of biology. Those who fail to grasp such distinctions are like people who can't distinguish between the ideas of Darwin and, say, Lamarck.

In the courtroom scenario Travis recounts, I was testifying that science has not shown that a Darwinian mechanism could account for the immune system. Travis's article itself confirms that is still true. He cites some biologists who think the adaptive immune system arose in a "big bang"; he quotes other scientists who assert, "There was never a big bang of immunology." He discusses vertebrate immunologists who think they know what the selective advantage of the system is; he quotes invertebrate immunologists who feel otherwise. So are we to think that its history is uncertain and even its selective advantage is unknown, yet the mechanism by which the adaptive immune system arose is settled?

In my court testimony I cited the then-new article by Klein and Nikolaidis, "The Descent of the Antibody-Based Immune System by Gradual Evolution,"[30] which first disputed the big bang hypothesis. In it the authors candidly remark, "Here, we sketch out some of the changes that the emergence of the AIS entailed and speculate how they may have come about." Valuable as it might be to science, however, speculation is not data, let alone an experimental result. Students are poorly served when they are not taught to distinguish among them.

105. RESPONSE TO PRINCETON

"Response to Princeton University President Shirley Tilghman,"
letter to the editor, *Princeton Alumni Weekly*, March 8, 2006.

IN THE JANUARY ISSUE, PRESIDENT TILGHMAN[31] URGES THAT SKEP-tics of Darwinism such as myself engage in debate "respectfully." Unfortunately, she doesn't follow her own advice. Instead she employs inflammatory rhetoric, warning darkly of an "assault" by "Christian fundamentalists." Now, I'm a Roman Catholic who learned Darwin's theory in parochial school and still thinks it explains much of biology, although certainly not all. So do I count as a "fundamentalist"? Does the writing of books exploring intelligent design, like my *Darwin's Black Box*, constitute an "assault"? Is it "respectful" to characterize the action of citizens petitioning their legislators as an "assault"?

I don't think so. In fact, I think President Tilghman's address itself illuminates why so many people are so suspicious of Darwin's theory. When a theory has to be defended with emotionally charged calls-to-arms, it makes people smell a rat. If a scientist has good evidence in hand for a theory, she should simply state it. If she doesn't, she should plainly admit it.

106. Bloggingheads TV and Me

"Bloggingheads TV and Me," Evolution News and
Views, Discovery Institute, August 28, 2009.

I'VE JUST BEEN THROUGH THE WEIRDEST BOOK-RELATED EXPERIENCE
I've had since a Canadian university professor with a cocked rat trap
chased me around after a talk I gave a dozen years ago, threatening to
spring it on me. Last week I got the following email bearing the title "Invitation to Appear on Bloggingheads TV" from a senior editor at that
site:

Hi, Michael—

I'd like to invite you to appear on Bloggingheads.tv, a web site that hosts
video dialogues between journalists, bloggers, and scholars. We have a
partnership with the *New York Times* by which they feature excerpts
from some of our shows on their site.

Past guests include prominent thinkers such as Paul Krugman,
Paul Ehrlich, Frans de Waal, David Frum, Richard Wrangham, Francis Fukuyama, Robert Kagan, and Michael Kinsley. Here is one of our
recent shows, a dialogue between Paul Nelson, of the Discovery Institute, and Ron Numbers, of Wisconsin-Madison: https://blogging-
heads.tv/videos/2175 .

I'm hoping that you might be interested in participating, as well.
First-time participants often report how refreshingly unconstrained
they find the format—how it lets them present their views with a depth
and subtlety not possible on TV or radio. We'd love to have you join us.
If you're available, please let me know, and we can see about arranging a
taping. Thank you for your time.

He seemed like such a nice fellow, so after a couple days I emailed
him back to say, sure, I'd be glad to. The editor responded, okay, sometime next week, your discussion partner will be John McWhorter of the
Manhattan Institute. I had never heard of McWhorter, so I googled his
name and saw that he's a linguist who often writes on race matters. I
didn't know what to expect because I know some conservatives (which
he seemed to be from his bio) don't like ID one bit.

Everything was arranged for the taping Tuesday afternoon. When the interview started, I was surprised and delighted to learn that McWhorter was actually a fan of mine. He said (I'm paraphrasing here) he loved *The Edge of Evolution* and wanted the book to become better known. He said that this was one of the few times that he initiated an interview at Bloggingheads. He said he was familiar with criticisms of the book and found them unpersuasive. He said that Darwinism just didn't seem to him to be able to cut the mustard in explaining life, and he had yet to read a good, detailed explanation for a large evolutionary change. He also said that he had never believed in God, but that *The Edge of Evolution* got him thinking.

In return I summarized my arguments from *The Edge of Evolution*, talked about protein structure, addressed his objections that intelligent design is "boring" and a scientific dead-end, and so on. At the end of the taping I thought, gee, those folks at Bloggingheads TV are a real nice bunch.

The next day I emailed the Bloggingheads editor to ask when the show would go on. He answered right back that at that very moment it had been activated, and thanks for participating. I clicked the link, and there was the show. I thought I looked older on screen than I am (my beard isn't really that white), but emailed some friends anyway to let them know the interview was up. That evening I got an email from one of them saying that he couldn't find the interview—it had been yanked.

Let me emphasize this, dear readers. Here we are living in the land of the free and the home of the brave. And yet a web site puts up an interview with an (ahem...) somewhat controversial figure, pulls it back down within hours, erases it, sends it down the memory hole. Why might that be? There would seem to be two possibilities: 1) maybe we aren't quite as free as we think, or 2) maybe not quite as brave.

I bet on possibility #2. Because of the magic of the internet, it turns out that shortly after the show's posting, the comments section of the site was overrun by "bitterly virulent" (in the words of one principal in this saga) cyber bullies, some murmuring darkly about a grim future for Bloggingheads.

After I found out the video was removed I emailed John McWhorter and the editor to ask for an explanation, and John emailed back that he himself requested the video to be pulled because people thought he was too easy on me, which was supposedly contrary to that old Bloggingheads spirit. I find that quite implausible (other shows on the site feature discussions between people who agree on many things). Rather, I suspect the folks at the website weren't expecting the vitriolic reaction, began to worry about their good names and future employment prospects, pictured themselves banished to a virtual leper colony, panicked, and folded.

Well, mobs, including internet mobs, are scary things, and it's understandable to panic when they unexpectedly show up at your door. But if you're going to set up a website to air discussions about contentious issues of the day, you should have a whole lot more guts than displayed by Bloggingheads TV.

There's a now-famous archival photograph of the Soviet Communist dictator Joseph Stalin with a shorter man standing just to the right of him, and then a second photograph identical to the first, but with this shorter man airbrushed out of the shot—removed by Stalin's team of expert photography doctors after the fellow had fallen victim to one of Stalin's many purges.[32] I posted the two versions of the archival photograph in the original version of this essay at Evolution News and joked, "Below is a time-lapse picture of my Bloggingheads interview. I'm the guy on the right." Tongue in cheek, of course, but in service of a serious point: the way Bloggingheads behaved was a bit Orwellian.

107. Bloggingheads Explains

The out-of-town senior editor of Bloggingheads, Robert Wright, eventually restored the interview upon his return, but he and science writer George Johnson of the *New York Times* tried to mollify the pitchfork-carrying crowd with a more-in-sorrow apologia about the incident.

"Bloggingheads Explains," Evolution News and Views, Discovery Institute, September 18, 2009.

IN A NEW SEGMENT, BLOGGINGHEADS CHIEF ROBERT WRIGHT AND Bloggingheads correspondent George Johnson go on for seventy-five minutes about the trauma of a pair of heretics (me and Paul Nelson, on separate segments) appearing on their site.[33] I would urge everyone who doesn't have pressing matters to attend to, such as washing your hair, to tune in for the full time. It's really fascinating in its way to see two grown men in such a hand-wringing lather. It's also fascinating to see that neither of them in seventy-five minutes offers a reason for the correctness of their own views, or the wrongness of ours. The closest they come is when George Johnson invokes the hoary "methodological naturalism."

One little segment was particularly rich. Johnson is faint with indignation that, in a post commenting on Bloggingheads originally pulling my interview with John McWhorter, I put up the well-known picture of Joseph Stalin standing with a group of people and a second photo in which one poor bloke's image had been removed. Why, lamented Johnson, I was comparing poor Bloggingheads to a murderous regime! These ID folks revel in their supposed persecution!

In that post my only remark concerning the picture (which appeared at the end) was "Below is a time-lapse picture of my Bloggingheads interview. I'm the guy on the right." (That is, the fellow who had disappeared.) Now, it may be that George Johnson has never heard the phrase "tongue-in-cheek"; if so he should look it up. Otherwise he may be confused that the guy on the right actually doesn't look like me, and that the interview wasn't really filmed next to a Russian river.

Interestingly, at one point when he was explaining why he had the interview put back up, Robert Wright said that if you take down an interview that had been put up, why, it looks "like you're trying to re-write history." Well, now, what picture might someone post if he was trying to illustrate Wright's astute point?

108. WHETHER INTELLIGENT DESIGN IS SCIENCE

"Whether Intelligent Design Is Science: A Response to the Opinion of the Court in Kitzmiller vs Dover Area School District," Discovery Institute, 2006.

O N DECEMBER 20, 2005, JUDGE JOHN JONES ISSUED HIS OPINION IN the matter of Kitzmiller,[34] in which I was the lead witness for the defense. There are many statements of the Court with which I disagree scattered throughout the opinion. However, here I will remark only on section E-4, "Whether ID is Science."[35]

The Court finds that intelligent design (ID) is not science. In its legal analysis, the Court takes what I would call a restricted sociological view of science: "science" is what the consensus of the community of practicing scientists declares it to be. The word "science" belongs to that community and to no one else. Thus, in the Court's reasoning, since prominent science organizations have declared intelligent design to not be science, it is not science. Although at first blush that may seem reasonable, the restricted sociological view of science risks conflating the presumptions and prejudices of the current group of practitioners with the way physical reality must be understood.

On the other hand, like myself most of the public takes a broader view: "science" is an unrestricted search for the truth about nature based on reasoning from physical evidence. By those lights, intelligent design is indeed science. Thus there is a disconnect between the two views of what "science" is. Although the two views rarely conflict at all, the dissonance grows acute when the topic turns to the most fundamental matters, such as the origins of the universe, life, and mind.

Below I proceed sequentially through section E-4. Statements from the opinion are in italics, followed by my comments.

COMMENTARY

Court: (1) "ID violates the centuries-old ground rules of science by invoking and permitting supernatural causation."

Me: It does no such thing. The Court's opinion ignores, both here and elsewhere, the distinction between an implication of a theory and the theory itself. As I testified, when it was first proposed, the Big Bang theory struck many scientists as pointing to a supernatural cause. Yet it clearly is a scientific theory, because it is based entirely on physical data and logical inferences. The same is true of intelligent design.

Court: (2) "The argument of irreducible complexity, central to ID, employs the same flawed and illogical contrived dualism that doomed creation science in the 1980s."

Me: "The dualism is "contrived" and "illogical" only if one confuses ID with creationism, as the Court does. There are indeed more possible explanations for life than Darwinian evolution and young earth creationism, so evidence against one doesn't count as evidence for the other. However, if one simply contrasts intelligent causes with unintelligent causes, as ID does, then those two categories do constitute a mutually exclusive and exhaustive set of possible explanations. Thus evidence against the ability of unintelligent causes to explain a phenomenon does strengthen the case for an intelligent cause."

Court: (3) "ID's negative attacks on evolution have been refuted by the scientific community."

Me: To the extent that the Court has in mind my own biochemical arguments against Darwinism, and to the extent that "refute" is here meant as "shown to be wrong" rather than just "controverted," then I strongly disagree, as I have written in a number of places. If "refute" is just intended to mean "controverted," then that is obvious, trivial, and an injudicious use of language. A "controversial" idea, such as ID, by definition is "controverted."

Court: (4) "ID is predicated on supernatural causation, as we previously explained and as various expert testimony revealed.... (21:96-100 (Behe); P-718 at 696, 700 ('implausible that the designer is a natural entity')."

Me: Again, repeatedly, the Court's opinion ignores the distinction between an implication of a theory and the theory itself. If I think it is implausible that the cause of the Big Bang was natural, as I do, that does not make the Big Bang Theory a religious one, because the theory is based on physical, observable data and logical inferences. The same is true for ID.

Court: (5) "ID proponents primarily argue for design through negative arguments against evolution, as illustrated by Professor Behe's argument that 'irreducibly complex' systems cannot be produced through Darwinian, or any natural, mechanisms. (5:38-41 (Pennock); 1:39, 2:15, 2:35-37, 3:96 (Miller); 16:72-73 (Padian); 10:148 (Forrest))."

Me: In my remark here I will focus on the word "cannot." I never said or wrote that Darwinian evolution "cannot" be correct, in the sense of somehow being logically impossible, as the court implies (referencing exclusively to the plaintiffs' expert witnesses). In its use of the word "cannot" the Court echoes the unfair strategy of Darwinists to force skeptics to try to prove a negative, to prove that Darwinism is impossible. However, unlike in mathematics or philosophy, in science one cannot conclusively prove a negative. One can't conclusively prove that Darwinism is false any more than one can conclusively prove that the "ether" doesn't exist. With this unfair strategy, rather than demonstrating empirical plausibility, Darwinists claim that the mere logical possibility that random mutation and natural selection may in some unknown manner account for a system counts in their favor.

In the history of science no successful theory has ever demonstrated that all rival theories are impossible, and neither should intelligent design be held to such an unreasonable, inappropriate standard. Rather, a theory succeeds by explaining the data better than competing theories do.

Two things I would now add: The word "cannot" could be changed to "are extremely unlikely to have been produced," and that would be an accurate description of the aspect of intelligent design that argues *against* the Darwinian mechanism. The ID position also offers positive evidence for intelligent design, which the court's definition elides.

Court: (6) "Professor Behe admitted in 'Reply to My Critics' that there was a defect in his view of irreducible complexity because, while it purports to be a challenge to natural selection, it does not actually address 'the task facing natural selection.' (P-718 at 695). Professor Behe specifically explained that '[t]he current definition puts the focus on removing a part from an already functioning system,' but '[t]he difficult task facing Darwinian evolution, however, would not be to remove parts from sophisticated pre-existing systems; it would be to bring together components to make a new system in the first place.'"

Me: I "admitted" this "defect" in the definition of irreducible complexity in the context of discussing (in passing, in a long article) a zany hypothetical example that Robert Pennock concocted in his book *Tower of Babel*. Pennock, a philosopher, wrote that a complex watch could be made by starting with a more complex chronometer (a very precise timepiece used by sailors) and carefully breaking it!—So therefore a watch isn't irreducibly complex! As I testified, I have not bothered to address Pennock's point because I regard the example as obviously and totally contrived—it has nothing to do with biologically relevant questions of evolution. That the words of my article are quoted by the Court without any reference to the context of Pennock's silly example appears invidious and is certainly confused.

Court: (7) "Although Professor Behe is adamant in his definition of irreducible complexity when he says a precursor 'missing a part is by definition nonfunctional,' what he obviously means is that it will not function in the same way the system functions when all the parts are present."

Me: Yes, it's obvious that's what I meant because that's exactly what I wrote in *Darwin's Black Box*: "An irreducibly complex system cannot be produced directly (that is, by continuously improving the initial function, which continues to work by the same mechanism)..."[36] If it doesn't work the same way when a part is missing, then it can't be produced directly, which is just what I wrote. Nonetheless, I do agree that, for example, a computer missing a critical part can still "function" as, say, a door stop. That hardly constitutes a concession on my part.

Court: (8) "Professor Behe excludes, by definition, the possibility that a precursor to the bacterial flagellum functioned not as a rotary motor, but in some other way, for example as a secretory system. (19:88-95 (Behe))."

Me: I certainly do not exclude that bald possibility merely by definition. In fact in *Darwin's Black Box* I specifically considered those kinds of cases. However, I classified those as indirect routes. Indirect routes, I argued, were quite implausible: "Even if a system is irreducibly complex (and thus cannot have been produced directly), however, one cannot definitely rule out the possibility of an indirect, circuitous route. As the complexity of an interacting system increases, though, the likelihood of such an indirect route drops precipitously."[37]

University of Rochester evolutionary biologist H. Allen Orr agrees that indirect evolution is unlikely:

> We might think that some of the parts of an irreducibly complex system evolved step by step for some other purpose and were then recruited wholesale to a new function. But this is also unlikely. You may as well hope that half your car's transmission will suddenly help out in the airbag department. Such things might happen very, very rarely, but they surely do not offer a general solution to irreducible complexity.[38]

There is no strict *logical* barrier to a Darwinian precursor to a bacterial flagellum having functioned as a secretory system and then, by dint of random mutation and natural selection, turning into a rotary device. There is also no absolute *logical* barrier to it having functioned as, say, a structural component of the cell, a light-harvesting machine, a nuclear reactor, a space ship, or, as Kenneth Miller has suggested, a paper weight. But none of these has anything to do with its function as a rotary motor, and so none of them explain that actual ability of the flagellum.

A bare assertion that one kind of complex system (say, a car's transmission) can turn into another kind of complex system (say, a car's airbag) by random mutation and natural selection is not evidence of anything, and does nothing to alleviate the difficulty of irreducible complexity for

Darwinism. Children who are taught to uncritically accept such vaporous assertions are being seriously misled.

Court: (9) "Notably, the NAS has rejected Professor Behe's claim for irreducible complexity by using the following cogent reasoning:

> [S]tructures and processes that are claimed to be "irreducibly" complex typically are not on closer inspection.... The evolution of complex molecular systems can occur in several ways. Natural selection can bring together parts of a system for one function at one time and then, at a later time, recombine those parts with other systems of components to produce a system that has a different function. Genes can be duplicated, altered, and then amplified through natural selection. The complex biochemical cascade resulting in blood clotting has been explained in this fashion."

Me: Well, that's a fine prose summary of the theory, but there is precious little experimental evidence that random mutation and natural selection can do what the NAS claim they can do. As I testified, in the nineteenth century prominent physicists overwhelmingly believed in the ether, not because of positive evidence for it, but because their theories of light required it. The "ether," however, does not exist. Nor do experiments exist that demonstrate the power of natural selection to make irreducibly complex biochemical systems, either directly or indirectly—proclamations of the National Academy notwithstanding. Again, children who are taught to mistake assertions for experimental demonstrations are being seriously misled.

Court: (10) "Professor Behe has applied the concept of irreducible complexity to only a few select systems: (1) the bacterial flagellum; (2) the blood clotting cascade; and (3) the immune system. Contrary to Professor Behe's assertions with respect to these few biochemical systems among the myriad existing in nature, however, Dr. Miller presented evidence, based upon peer-reviewed studies, that they are not in fact irreducibly complex."

Me: In this section, despite my protestations, the Court simply accepts Miller's adulterated definition of irreducible complexity in which a system is not irreducible if you can use one or more of its parts for another purpose, and disregards careful distinctions I made in *Darwin's Black Box*. The distinctions can be read in my Court testimony. In short, the Court uncritically accepts strawman arguments.[39]

Court: (11) "In fact, on cross-examination, Professor Behe was questioned concerning his 1996 claim that science would never find an evolutionary explanation for the immune system. He was presented with fifty eight peer-reviewed publications, nine books, and several immunology textbook chapters about the evolution of the immune system; however, he simply insisted that this was still not sufficient evidence of evolution, and that it was not 'good enough.' (23:19 (Behe))."

Me: Several points:

1. Although the opinion's phrasing makes it seem to come from my mouth, the remark about the studies being "not good enough" was the cross-examining attorney's, not mine.

2. I was given no chance to read the publications, and at the time considered the dumping of a stack of papers and books on the witness stand to be just a stunt, simply bad courtroom theater. Yet the Court treats it seriously.

3. The Court here speaks of "evidence for evolution." Throughout the trial I carefully distinguished between the various meanings of the word "evolution," and I made it abundantly clear that I was challenging Darwin's proposed mechanism of random mutation coupled to natural selection. Unfortunately, the Court here, as in many other places in its opinion, ignores the distinction between evolution and Darwinism.

I said in my testimony that the studies may have been fine as far as they went, but that they certainly did not present detailed, rigorous explanations for the evolution of the immune system by random mutation and natural selection—if they had, that knowledge would be reflected in more recent studies that I had had a chance to read (see below).

4. This is the most blatant example of the Court's simply accepting the plaintiffs' say-so on the state of the science and disregarding the

opinions of the defendants' experts. I strongly suspect the Court did not itself read the "fifty eight peer-reviewed publications, nine books, and several immunology textbook chapters about the evolution of the immune system" and determine from its own expertise that they demonstrated Darwinian claims. How can the Court declare that a stack of publications shows anything at all if the defense expert disputes it and the Court has not itself read and understood them?

In my own direct testimony I went through the papers referenced by Professor Miller in his testimony and showed they didn't even contain the phrase "random mutation"; that is, they assumed Darwinian evolution by random mutation and natural selection was true—they did not even try to demonstrate it. I further showed in particular that several very recent immunology papers cited by Miller were highly speculative—in other words, that there is no current rigorous Darwinian explanation for the immune system. The Court does not mention this testimony.

Court: (12) "We find that such evidence demonstrates that the ID argument is dependent upon setting a scientifically unreasonable burden of proof for the theory of evolution."

Me: Again, as I made abundantly clear at trial, it isn't "evolution" but Darwinism—random mutation and natural selection—that ID challenges. Darwinism makes the large, crucial claim that random processes and natural selection can account for the functional complexity of life. Thus the "burden of proof" for Darwinism necessarily is to support its special claim—not simply to show that common descent looks to be true. How can a demand for Darwinism to convincingly support its express claim be "unreasonable"?

The nineteenth century ether theory of the propagation of light could not be tested simply by showing that light was a wave; it had to test directly for the ether. Darwinism is not tested by studies showing simply that organisms are related; it has to show evidence for the sufficiency of random mutation and natural selection to make complex, functional systems.

Court: (13) "As a further example, the test for ID proposed by both Professors Behe and Minnich is to grow the bacterial flagellum in the laboratory; however, no-one inside or outside of the IDM, including those who propose the test, has conducted it. (P-718; 18:125-27 (Behe); 22:102-06 (Behe))."

Me: If I conducted such an experiment and no flagellum were evolved, what Darwinist would believe me? What Darwinist would take that as evidence for my claims that Darwinism is wrong and ID is right? As I testified to the Court, Kenneth Miller claimed there was experimental evidence showing that complex biochemical systems could evolve by random mutation and natural selection, and he pointed to the work of Barry Hall on the lac operon. I explained in great detail to the Court why Miller was exaggerating, was incorrect, and was making claims that Barry Hall himself never did. *However, no Darwinist I am aware of subsequently took Hall's experiments as evidence against Darwinism.* Neither did the Court mention it in its opinion.

The flagellum experiment the Court described above is one that, if successful, would strongly affirm Darwinian claims, and so should have been attempted long ago by one or more of the many, many adherents of Darwinism in the scientific community. That none of them has tried such an experiment, and that similar experiments that were tried on other molecular systems have failed, should count heavily against their theory.

Court: (14) "We will now consider the purportedly 'positive argument' for design encompassed in the phrase used numerous times by Professors Behe and Minnich throughout their expert testimony, which is the 'purposeful arrangement of parts.'... As previously indicated, this argument is merely a restatement of the Reverend William Paley's argument applied at the cell level. Minnich, Behe, and Paley reach the same conclusion, that complex organisms must have been designed using the same reasoning, except that Professors Behe and Minnich refuse to identify the designer, whereas Paley inferred from the presence of design that it was God."

Me: Again, repeatedly, the Court confuses extra-scientific implications of a scientific theory with the theory itself. William Paley would likely think that the Big Bang was a creative act by God, but that does not make the Big Bang theory unscientific. In fact I myself suspect that the Big Bang may have been a supernatural act, but I would not say that science has determined the universe was begun by God—just that science has determined the universe had a beginning. To reach to a conclusion of God or the supernatural requires philosophical and other arguments beyond science.

Scholarly diligence in making proper distinctions should not be impugned as craftiness. I do not "refuse to identify the designer" as the Court accuses. Starting in *Darwin's Black Box* and continuing up through my testimony at trial, I have repeatedly affirmed that I think the designer is God, and repeatedly pointed out that that personal affirmation goes beyond the scientific evidence, and is not part of my scientific program.

Court: (15) "Expert testimony revealed that this inductive argument is not scientific and as admitted by Professor Behe, can never be ruled out. (2:40 (Miller); 22:101 (Behe); 3:99 (Miller))."

Me: Whether the induction is "scientific" of course depends on the definition of science. The induction is based on reasoning from physical evidence, which in my view does make it scientific. As far as design being "never ruled out," as I explained earlier, science never rules anything out as a matter of logic; that is, science can't prove in some absolute sense that something doesn't exist. The task of science is simply to adduce evidence to help support one view or weigh against another.

Court: (16) "Indeed, the assertion that design of biological systems can be inferred from the 'purposeful arrangement of parts' is based upon an analogy to human design.... Professor Behe testified that the strength of the analogy depends upon the degree of similarity entailed in the two propositions; however, if this is the test, ID completely fails."

Me: The Court has switched in the space of a paragraph from calling the argument for ID an "inductive argument" to calling it an "analogy." That is a critical confusion. As I testified, the ID argument is an induction, not an analogy. Inductions do not depend on the degree of similarity of examples within the induction. Examples only have to share one or a subset of relevant properties. For example, the induction that, *ceteris paribus*, black objects become warm in the sunlight holds for a wide range of dissimilar objects. A black automobile and a black rock become warm in the sunlight, even though they have many dissimilarities. The induction holds because they share a similar relevant property, their blackness. The induction that many fragments rushing away from each other indicates a past explosion holds for both firecrackers and the universe (in the Big Bang theory), even though firecrackers and the universe have many, many dissimilarities. Cellular machines and machines in our everyday world share a relevant property—their functional complexity, born of a purposeful arrangement of parts—and so inductive conclusions to design can be drawn on the basis of that shared property. To call an induction into doubt one has to show that dissimilarities make a relevant difference to the property one wishes to explain. Neither the judge nor the Darwinists he uncritically embraces have done that in respect to intelligent design.

Court: (17) "Unlike biological systems, human artifacts do not live and reproduce over time. They are non-replicable, they do not undergo genetic recombination, and they are not driven by natural selection. (1:131-33 (Miller); 23:57-59 (Behe))."

Me: Despite Darwinian claims, none of these factors has ever been shown to account for the molecular machinery of life, so we have no reason to think they affect the induction. (See above.)

Court: (18) "For human artifacts, we know the designer's identity, human, and the mechanism of design, as we have experience based upon empirical evidence that humans can make such things, as well as many other attributes including the designer's abilities, needs, and desires.... Professor Behe's only response to these seemingly insurmountable points of disanalogy was that the inference still works in science fiction movies. (23:73 (Behe))."

Me: Again, the Court confuses an analogy with an induction. Our knowledge of the nature of the designer is not necessary for a conclusion of design based on induction, any more than knowledge of what caused the Big Bang was necessary before we could inductively conclude that the universe had an explosive beginning. The Court objects to my illustration drawn from science fiction movies, but the objection misses the point. The SETI project (Search for Extraterrestrial Intelligence) is an actual, non-fictional project. It is based on our ability to recognize certain effects of non-human, alien intelligence were we to encounter such effects. As I noted, SETI was featured in the science fiction film *Contact* based upon a work by Carl Sagan. The film was fictional, but it accurately depicted SETI's purpose and how they seek to distinguish natural from intelligently designed signals from space.

Court: (19) "This inference to design based upon the appearance of a 'purposeful arrangement of parts' is a completely subjective proposition, determined in the eye of each beholder and his/her viewpoint concerning the complexity of a system."

Me: The court implies that apprehending design is akin to judging if a piece of artwork is attractive—a matter of personal taste. Yet Darwin's theory is widely touted as explaining the strong appearance of design in biology; if such appearance is just a "completely subjective proposition," what is Darwin's theory explaining? The Court neglects to mention that the "completely subjective" appearance of design is—in the view of the adamantly Darwinian evolutionary biologist Richard Dawkins— "overwhelming." I testified to that, to Dawkins' proclamation that "biology is the study of complicated things that give the appearance of having been designed for a purpose," and to other similar statements. I showed the Court a special issue of the journal *Cell* on "Macromolecular Machines," which contained articles with titles such as "Mechanical Devices of the Spliceosome: Motors, Clocks, Springs, and Things." If strong opponents and proponents of design both agree that biology appears designed, then the appearance should not be denigrated by Judge Jones as subjective.

Court: (20) "As Plaintiffs aptly submit to the Court, throughout the entire trial only one piece of evidence generated by Defendants addressed the strength of the ID inference: the argument is less plausible to those for whom God's existence is in question, and is much less plausible for those who deny God's existence. (P-718 at 705)."

Me: As I pointed out in my direct testimony to the Court, the Big Bang theory also was deemed less plausible by some scientists who disliked its supposed extra-scientific implications. I showed the Court an editorial in the prestigious journal *Nature* that carried the title "Down with the Big Bang," and called the Big Bang a "philosophically unacceptable" theory which gave succor to "Creationists." Because real people—including scientists—do not base all of their judgments on strictly scientific reasoning, various scientific theories can be more or less appealing based on their supposed extra-scientific implications. It is unfair to suggest ID is unique in that regard.

CONCLUSION

THE COURT's reasoning in section E-4 is premised on a cramped view of science; the conflation of intelligent design with creationism; an incapacity to distinguish the implications of a theory from the theory itself; a failure to differentiate evolution from Darwinism; and strawman arguments against ID. The Court has accepted the most tendentious and shopworn excuses for Darwinism with great charity and impatiently dismissed evidence-based arguments for design.

All of that is regrettable, but in the end does not impact the realities of biology, which are not amenable to adjudication. On the day after the judge's opinion, December 21, 2005, as before, the cell is run by amazingly complex, functional machinery that in any other context would immediately be recognized as designed. On December 21, 2005, as before, there are no non-design explanations for the molecular machinery of life, only wishful speculations and Just-So stories.

109. EVEN DILBERT HAS TO DO HIS HOMEWORK

This chapter was originally written for inclusion in *Darwin Devolves*, but was cut due to issues of space. It addresses the Dover Trial of 2005—a time when the idea of Intelligent Design was at the center of a media and political frenzy. The glimpse the chapter provides of the intense feelings that ID can generate makes clear why everyone has to do their own homework on controversial issues and make up their own mind.

THE PUBLICATION OF *DARWIN'S BLACK BOX* IN 1996 ARGUABLY started the broad discussion of the modern scientific argument for intelligent design. Although of course the notion of design is at least as ancient as the early Greek philosophers, and although incisive books skeptical of Darwinism and friendly to design were written before *Darwin's Black Box*,[40] the book made an explicit argument for purposeful design based on the latest science—the molecular machinery of life. The book was reviewed in a wide variety of publications, including academic journals, trade bulletins, and magazines of general interest. The *New York Times*, *Wall Street Journal*, *Nature*, *Christianity Today*, *Skeptic*, *Chronicle of Higher Education*, *Philosophy of Science*, *Quarterly Review of Biology*—even *Aboard*, the in-flight magazine of the Bolivian national airlines (!)—all ran reviews or discussions. (Of course not all reviews were, shall we say, completely positive.)

A COLLISION OF IDEAS AND POLITICS

OVER THE ensuing years the ideas in the book gradually made their way into the public consciousness, and some people decided to use those ideas for their own purposes. In 2004 the school board of the tiny town of Dover, Pennsylvania, near the state capital of Harrisburg, voted to have a (surprisingly poorly written) statement concerning evolution read to high school students in biology class.[41] The statement was frankly skeptical of Darwin's theory, and informed the students that there was a book in the school library—*Of Pandas and People*—that discussed intelligent design.[42]

All hell broke loose. The ACLU sued on behalf of several parents. Reporters from all over the globe descended on Harrisburg, for the federal trial, as did a great-great-grandson of Charles Darwin and assorted oddballs. The judge basked in the publicity, granting press interviews even while the trial was in progress, interviews where he let the world know he was brushing up on the issues by watching *Inherit the Wind*—the fictionalized, highly tendentious 1960 movie based on the 1925 Scopes Monkey trial.[43]

It transpired that some members of the school board were biblical creationists—a segment of society that is rarely portrayed with sympathy by the press—and had taken up a collection in church to purchase *Of Pandas and People* for the school library. It wasn't hard for the plaintiffs' attorneys to show that the creationists had a religious motive for their action, which is disallowed for government officials who administer American public schools. The judge ruled for the plaintiffs.

Fair enough. It's no more appropriate for that school board to urge students to read a book explicitly because they think it supports their theistic views than for another school board to assign *The Blind Watchmaker* because they agree with Richard Dawkins's atheism.

But the ruling went much further than simply determining that the school board acted out of inappropriate motives. The judge, a man named John Jones, issued an expansive opinion declaring that the very idea of intelligent design itself was not scientific, that it was in fact necessarily religious, and that design arguments offered by expert witnesses for the defense did not stand up to criticism. The publicity-amenable judge was lionized for his ruling by all the right people, including the staff at *Time* magazine, who honored Jones with a place on its 2006 *Time 100* list of "the 100 men and women whose power, talent or moral example is transforming our world."

As an expert witness for the defense, I myself was on the stand with a lawyer in my face for three straight six-hour days, so it was hard for me to gauge the general public's perception of the affair. Nonetheless, one likely common view is that of Scott Adams, creator of the *Dilbert* comic about the eponymous brilliant engineer trapped in a brainless corporate world. Adams maintains a blog wherein he writes with a philosophical bent about all manner of things. In one post in 2006 he touched on intelligent design and the Dover trial:

> Intelligent Design was put to the test in the Dover trial and failed miserably in convincing a judge it should be considered science. If you read the judge's opinion, he heard from both sides (as few people ever have) and it wasn't even a close call. That's good enough for me. Until that verdict, I was having a hard time with the obvious biases on both sides. I considered all sources unreliable.[44]

In other words, probably like many people, Adams had a hard time sorting out competing claims about design in the cacophonous marketplace of ideas. But thankfully a wise judge heard from both sides, carefully weighed the arguments, and issued a thoroughly objective, informed opinion that sober citizens could take as a reliable guide. That's an understandable point of view on Adams's part. Unfortunately (but perhaps appropriately for Dilbert), it's comically mistaken. In fact, as we'll see, there's no evidence to show that the judge—formerly a lawyer and an unsuccessful candidate for Congress, as well as the politically appointed head of the Pennsylvania Liquor Control Board—comprehended any of the expert testimony at the trial, for either the plaintiffs or defendants.

A BIT OF BACKGROUND

To SET the context for evaluating the trial let me go back to the beginning of my public involvement with design. In the early 1990s, as I began thinking about writing the book that became *Darwin's Black Box*, I realized that anyone in my intellectual position had a big problem. Darwinists famously gloss over profound difficulties with glib tales of how even the most sophisticated organs of life might have evolved. Darwin himself set the example with his hand-waving explanation of eye evolution (discussed in Chapter 2 of *Darwin Devolves*). Often dubbed "Just-So Stories" (after Rudyard Kipling's children's book with stories such as "How the Elephant Got His Trunk"), even Darwinists themselves sometimes make fun of such tales.[45]

On the other hand, to show that facile stories can't seriously explain the phenomenal intricacy of the molecular foundation of life, a writer like me has to pull readers into the daunting complexity. Here's how I explained the quandary in the preface of *Darwin's Black Box*:

> Several years ago, Santa Claus gave my oldest son a plastic tricycle for Christmas. Unfortunately, busy man that he is, Santa had no time to take it out of the box and assemble it before heading off. The task fell to Dad. I took the parts out of the box, unfolded the assembly instructions, and sighed. There were six pages of detailed instructions: line up the eight different types of screws, insert two one-and-a-half-inch screws through the handle into the shaft, stick the shaft through the square hole in the body of the bike, and so on. I didn't want to even read the instructions, because I knew they couldn't be skimmed like a newspaper—the whole purpose is in the details....
>
> Unfortunately, much of biochemistry is like an instruction booklet, in the sense that the importance is in the details.... People who suffer with sickle cell anemia, enduring much pain in their shortened lives, know the importance of the little detail that changed one out of 146 amino acid residues in one out of the tens of thousands of proteins in their body. The parents of children who die of Tay-Sachs or cystic fibrosis, or suffer from diabetes or hemophilia, know more than they want to know about the importance of biochemical details.
>
> So, as a writer who wants people to read my work, I face a dilemma: people hate to read details, yet the story of the impact of biochemistry on evolutionary theory rests solely in the details. Therefore, I have to write the kind of book people don't like to read in order to persuade them of the ideas that push me to write. Nonetheless, complexity must

be experienced to be appreciated. So, gentle reader, I beg your patience; there are going to be a lot of details in this book.[46]

If the very point a guy is trying to make is how *complex* a system is, then he can't *simplify* the description, at least not beyond a certain point. Simplistic descriptions of life are fodder for Darwinian Just-So stories, not for illustrating sophisticated engineering. So the implicit pact that an author (me) makes with a reader (you) is that the reader will make an earnest effort to understand how some biological systems work at the molecular level, and in return I will show the reader the unexpectedly grave obstacles those systems present to Darwin's theory. Any person is, of course, perfectly free to decide not to read the argument, or to skim it but not pay attention to the particulars, or even to lose heart while wading through the details and zone out. But of course such folks then have no intellectual standing to judge the argument.

Testimony

WHICH BRINGS us to the Dover trial. I was the lead witness for the defense, appearing right after the plaintiffs' lawyers had finished their several weeks of presentations. One of the biochemical systems I discussed on the stand was one I had not included in *Darwin's Black Box*, something called the "lac operon" (briefly described in Chapter 2 of *Darwin Devolves*), which is responsible for making the right components needed by bacteria at the right time to metabolize the milk sugar lactose. The lac operon has been investigated since the mid-twentieth century and is one of the *simplest* textbook examples of genetic regulation. The reason I selected it is that Brown University biology professor Kenneth Miller (who was the lead witness for the plaintiffs in the Dover trial) had featured it in his 1999 anti-intelligent design book *Finding Darwin's God*. With his signature salesmanship, in the book Miller had emphatically assured his readers that a then-fifteen-year-old laboratory evolution experiment on the system—done well before my own book came out—had already conclusively demonstrated that irreducible complexity was wrong: "By the very same logic applied by Michael Behe to other systems, therefore, we could conclude that the system had been designed. Except we *know* that it was *not* designed. [emphasis Miller's] We know it evolved because we watched it happen right in the laboratory!"[47]

As I responded in a web post,[48] (included in this volume—essay 18), the work Miller recounted in fact showed little about Darwinian evolution, but did clearly demonstrate the need for the intelligence and involvement of the researcher—that is, the absolute need for intelligent design. The researcher, Barry Hall of the University of Rochester, had purposely set up the experiment with contrived conditions explicitly to guide it to the result he was looking for, and he never claimed anything remotely like what Miller claimed for the experiment—quite the opposite. Miller had turned Hall's good, careful, modest work on its head to spin a Darwinian fairy tale.

I thought this would be a grand example to show the judge the complexity of the system, its resistance to Darwinian evolution, and the way Darwinists such as Miller often distort or badly spin research results to favor their theory. But, as I wrote above, it all depends on the willingness of the audience to make an effort to understand. I naively thought that would be pretty much guaranteed at a federal trial. Boy was I wrong.

Here's how a woman named Lauri Lebo, then a reporter for the *York Daily Record* (York is a city in Pennsylvania near Dover), described my testimony in her unsympathetic 2008 book on the trial, *The Devil in Dover: An Insider's Story of Dogma v. Darwin in Small-Town America*:

"This might be hard to explain," Behe said. "[The evidence indicates that either AS-92 and cys-trp 977] are the only acceptable amino acids at those positions, or that all of the single based substitutions that might be on the pathway to other amino acid replacements at those sites are so deleterious that they constitute a deep selective valley that has not been traversed in the two billion years since those proteins emerged from a common ancestor." [Lebo neglected to tell her readers that the above was a passage that I quoted from a paper by Barry Hall, not my own words.]

[Back to my words.] "Now, translated into more common language, that means that a very similar protein could only work if it became even more similar to the beta-galactosidase that it replaced, and if you then also knock out that EBG-galactosidase, no other protein was able to substitute for the beta galactosidase. So the bottom line, the bottom line is that the only thing demonstrated was that you can get tiny changes in preexisting systems, tiny changes in preexisting systems, which of course everybody already had admitted."

Behe's testimony continued like this for hours. Reporters, at first, valiantly tried to follow along. But as Behe continued, their hands, scribbling notes, gradually... slowed, one by one, and finally paused, hovered over notebooks, then, at last defeated, dropped. The writer next to me dozed. Utterly lost, the rest of us in the jury box began to giggle helplessly. Judge Jones kept his face studiously composed and ignored us....

After Behe exhausted his repertoire on the lac operon, [defense attorney Robert] Muise turned to Judge Jones and said, "Your Honor, we're about to move into the blood clotting system, which is really complex." "Really?" Jones said, facetiously. "We've certainly absorbed a lot, haven't we?" "We certainly have, Your Honor," Muise gushed. "This is Biology 2."[49]

Lebo seemingly wanted to give her readers the impression that I had made no effort to make the science understandable to intelligent lay people, so she quoted a passage that in isolation sounds the most obscure. A more typical passage is the following where, after giving some background, I referred to a color PowerPoint slide on a screen in the courtroom that showed all of the components of the system:

> This little thing marked Y codes for something called a permease. Now, a permease it turns out is a protein whose job it is to allow the lactose to enter the bacterial cell. The bacterial cell is surrounded by a membrane which generally acts as a barrier to largish molecules, and there's this specialized protein, this specialized machine called a permease which, when lactose is around, grabs the lactose from outside the cell, turns it around, and allows it to enter to the inside of the cell.[50]

If a person is unwilling to try to follow along with a description even at that basic level, then they have little tolerance, or even respect, for actual science—as opposed to amusing Just-So stories. A person who won't invest any more effort than that needed to, say, close her eyes and imagine a fish with stubby legs crawling onto a prehistoric beach, will never comprehend the argument for design from biochemistry. Apparently, most reporters covering the trial weren't willing to make that effort to understand Biology 2 (as Robert Muise put it), including Lebo herself, even after several years to think about it.

THE POINTY-HAIRED BOSS

BUT HOW about the judge? Was *he* willing to make the effort to learn Biology 2? After all, the reporters didn't have the same somber obligation that he did. Perhaps at least a sense of duty would make him more attentive than the media. The quote above, however, where Lebo characterizes the judge as "facetiously" responding to the prospect of having to learn about even more biological complexity, does not bode well. And Lebo was not the only one with such an impression. Two expert participants for the plaintiffs' side in the trial later wrote, "Defense attorney Robert Muise led [Behe] through a tortuously detailed explanation of the bacterial flagellum and numerous other molecular systems that appeared to make the judge's eyes glaze over."[51] Yet biochemical details, such as the teeny-tiny ones that lead to deadly diseases, are crucial to understanding the vacuity of Darwinian claims.

To better appreciate what the judge himself did or didn't understand as he sat silently on the bench, let's take a peek inside some of the peculiar workings of our legal system. I'm no lawyer, so I was surprised to find out that after the trial, attorneys for both sides each give the judge a document outlining in detail exactly how they want him to rule. Only after receiving those documents does the judge issue his opinion....

What was striking is that in numerous cases, Jones simply *copied the text the plaintiffs' lawyers gave him.*[52] Whenever the topic concerned the testimony of any of the academic witnesses—whether scientists, philosophers, or theologians, whether for the plaintiffs or for the defendants— the very same language from the lawyers' document opposing intelligent design was inserted into the opinion with copy-and-paste efficiency, sometimes very lightly copyedited. Tellingly, when the opinion shifted to mundane matters, such as school board meetings or local newspaper editorials, the judge used his own voice, apparently perfectly comfortable writing for himself on those topics.

Now, let's ask ourselves why lifting material from somebody else is a bad idea. Why are reporters and politicians disgraced if they're caught doing so? Perhaps more to the point, why are students at all levels taught it's very wrong to plagiarize the work of another person? One reason is that a teacher wants to see if a student understands a topic, and is able to restate arguments in such a way that indicates comprehension. After all, even Dilbert's clueless Pointy-haired Boss could copy from a book on a difficult topic such as, say, quantum mechanics or Aristotle's *Metaphysics*. But copying the text surely doesn't show that the clueless boss understands the material. Apparently the legal system exempts itself from the standards that the rest of us follow.[53]

The dilemma, however, remains. If a judge simply copies a text on a complicated matter, there's no evidence to show he understands it. In such a case the losing side may justifiably suspect that it didn't really get its day in court—that the judge's choice of text to copy had little to do with the inherent logic of the arguments. What's more, as we'll see below, in the particular case of the Dover trial there's every reason to think that Judge Jones was completely at sea. It is doubtful that the former liquor-store bureaucrat grasped the distinction between such pivotal technical topics as flagellin and prothrombin, gene duplication and point mutation, Thomas Aquinas and David Hume, or random mutation and common descent.

If Dilbert's Pointy-haired Boss were in a similar situation, what might he do? A person who can't make his own distinctions on such specialized matters might defer to assertions of the prestigious scientific establishment, to things like, say, a widely ballyhooed—and then quickly disavowed—wacky claim that the bacterial flagellum evolved stepwise from a single prodigy gene,[54] or a confident affirmation by an eminent Darwinist that mice missing a certain blood clotting factor were normal, even though the mice hemorrhaged to death.[55] (Both of those cases are discussed in the appendix of *Darwin Devolves*.) Even some people who could know better if they made the effort—such as the academics who publicly vouched for those mistaken claims—defer to scientists outside their own immediate fields. After all, if you agree with the smartest people around, that shows you're smart too, right? Or at least not dumb.

Dilbert's Pointy-haired Boss might think like that, but should a judge do so while overseeing a federal case? The question of whether the prestigious scientific establishment might be mistaken or prejudiced was at the core of what Judge Jones was supposed to decide impartially. So, for him to simply cut, paste, and rubber-stamp the establishment view—without critically analyzing it, much less understanding both sides—was an abdication of his responsibility.

WE'RE NOT IN KANSAS ANYMORE

IN A lecture series titled "Difficult Dialogues" held the year after the trial at the University of Kansas, which separately featured several of the participants in the trial (including me), the good judge spoke for a half hour on the topic of the need for everyone to respect the judiciary.[56] Curiously, he declined to answer any questions on the intellectual arguments at Dover, protesting that the trial featured "mind-numbingly" technical presentations. Yet if Jones's mind was numbed by the testimony, how could he have reached a justified decision?

The judge also begged off when questioned in Kansas because "the highly technical scientific testimony... is rapidly becoming a distant memory." There's actually good reason to think that the highly technical scientific testimony became a distant memory for him even while the trial was underway. For example, Jones noted at the University of Kansas that my "cross-examination went on for a very long period of time, as did his direct examination." Yet the opinion—copied virtually word-for-word from the opposing lawyers' document—states:

> Professor Behe was questioned concerning his 1996 claim that science would never find an evolutionary explanation for the immune system.

He was presented with fifty-eight peer-reviewed publications, nine books, and several immunology textbook chapters about the evolution of the immune system; however, he simply insisted that this was still not sufficient evidence of evolution, and that it was not "good enough."[57]

"Simply insisted"? During my "very long" three days on the stand I had discussed *every single paper* on immunology that had previously been raised in the plaintiffs' expert-witness testimony and showed that it was either terminally speculative, concerned only with common descent rather than with Darwin's mechanism, or both. (The above-referenced "fifty-eight peer-reviewed publications [and] nine books" suffered from the same defects.) In fact, as I had testified, none of the papers even contained the phrases "random mutation" or "natural selection," let alone reported experiments testing that process.[58] You'd think that, while copying material for his opinion, the judge might have scratched his head and vaguely recalled what I had said. I guess not. Apparently since the plaintiffs' lawyers (whose job is to make the other side look as bad as possible)[59] didn't write my testimony on the immunology literature into their own document, it didn't appear in the judge's opinion either. "A distant memory," indeed.

What's more, when it did occasionally refer to my testimony, Jones's opinion badly mischaracterized it. The quote above essentially paints a picture of me flailing my arms, crying obnoxiously that all that hard scientific work on immunology—all those fifty-eight papers and nine books—is not "good enough." But in fact those were the words of *the* opposing lawyer cross-examining me, not my words.[60] (I had testified that "They're wonderful articles. They're very interesting. They simply just don't address the question that I pose." In other words, they didn't address the question of Darwin's mechanism.) The lawyers' document stuffed *his* words into my mouth—and the judge copied them.

Now consider the following excerpt from the Dover court decision. The only original writing by Judge Jones in this passage are the italicized "although" and the italicized sentence fragment, which acts as a fatuous link between two substantive (if misleading) arguments lifted from the plaintiffs' lawyers' document.

Science cannot be defined differently for Dover students than it is defined in the scientific community as an affirmative action program, as advocated by Professor Fuller, for a view that has been unable to gain a foothold within the scientific establishment. *Although* ID's failure to meet the ground rules of science is sufficient for the Court to conclude that it is not science, *out of an abundance of caution and in the exercise of completeness, we will analyze additional arguments advanced regarding the concepts of ID and science.*

ID is at bottom premised upon a false dichotomy, namely, that to the extent evolutionary theory is discredited, ID is confirmed. This argument is not brought to this Court anew, and in fact, the same argument, termed "contrived dualism" in McLean, was employed by creationists in the 1980's to support "creation science."[61]

Isn't that precious: "out of an abundance of caution... we will analyze additional arguments." Such snippets sprinkled throughout the opinion could easily mislead sincere people such as Scott Adams—or even *Time*'s 2006 committee who chose the "100 men and women whose power, talent or moral example is transforming our world"—to conclude that the wise judge himself held all the facts of the trial in his head, carefully considered them, and then came to an independent judgment.

For what it's worth, my view is that a judge who is either unwilling or unable to follow crucial technical discussions in a trial should either recuse himself or remain silent about those issues in his written opinion, perhaps deciding the case on other grounds if possible. A judge who can't follow technical testimony but nonetheless pretends to understand it by copying documents written by someone else is a sham and a disgrace.

Take-Home Lesson

By the time the whole thing was finished I had a lot more sympathy for the protagonist of Franz Kafka's novel *The Trial*, and a much sharper understanding of the term *Kafkaesque*: "Marked by surreal distortion." On reflection, I've concluded that it pretty much didn't matter what I said on the stand, nor what any of the other expert witnesses on either side said. The outcome of the case was decided long before the trial began. It was decided when the hoopla started, when the media cast the whole affair in terms of stereotypical heroes and villains, and when the judge consulted old Hollywood films for better perspective. A courtroom is no place to discuss intellectual issues.

That circus left town long ago. The practical point I want to make here is that *Dilbert's* Scott Adams is mistaken. Think of any substantive topic on which you hold a strong point of view, and about which you think a court case was wrongly decided. Did you change your views because of the court's decision? Probably not, and for good reason. Judges aren't philosopher-kings, and they're as likely as anyone else to have blind spots and biases on controversial issues. I think Adams himself would quickly see the absurdity of settling an engineering dispute between Dilbert and his coworker Wally by having the Pointy-Haired Boss—dressed solemnly in judge's robes—choose which of their proposals to copy. No one should simply cede to another party his responsibility to decide an important question, especially on a topic as central as the design of life. There's no substitute for doing your own homework: read widely and think independently.

GAME OVER

COURT PRONOUNCEMENTS and bureaucratic rulings notwithstanding, as an objective matter the question of the role of randomness in evolution is now settled. At this point in science's investigation of life it seems like just a cruel taunt to challenge Darwinian biologists to experimentally justify the ability of random mutation and natural selection to make an irreducibly complex molecular machine. The results are already in. The many arduous studies by leading researchers that I recount in *Darwin Devolves* show that Darwin's mechanism strains mightily to account for even the simplest cooperative molecular feature. The fact that the large majority of even beneficial mutations either degrade genes or outright break them indicates that, while Darwin's mechanism does permit species to adapt to particular environments, that adaptation results in ever-decreasing flexibility, making evolution self-limiting. Darwinian processes consume genetic information as fodder; they don't produce it.

The conclusion is inescapable. Starting with the 1872 book *The Genesis of Species* by the eminent English biologist St. George Mivart, the dissident biologists were right all along: Darwin's mechanism can produce modest changes in already-extant life, but it cannot build much of anything. Some simple genetic changes can yield visually impressive or medically significant results, giving a superficial appearance of significant advance, but modern investigations at the molecular level have unmasked them as pretenders.

Something else must explain the functional complexity of life. And that is the subject of my book *Darwin Devolves*.

EPILOGUE: RESTLESS SCIENCE

THE GREAT THING ABOUT SCIENCE IS NOT THAT IT'S NEVER WRONG, but that it's relentless—always pressing on, never completely satisfied. Throughout the history of science, lots of ideas were consistent with the best evidence of the day but turned out to be incorrect, sometimes embarrassingly so. Phrenology—the study of bumps on people's skulls as an indicator of their mental abilities, which was taken seriously for a while in the 1800s—was laughably wrong. But other ideas were what we might call understandably wrong. For example, the idea of ether—a substance which supposedly filled all of outer space—was consistent with the knowledge that light was a wave which presumably required a physical medium through which to travel. Ether was regarded as a hard fact by all the best physicists of the nineteenth century—until a way was invented to test it. Ether turned out to be imaginary.

Another understandably wrong theory was Isaac Newton's. His laws of motion worked wonderfully for centuries under conditions that could be measured at the time. Then scientists developed the ability to probe extremes of speed and size, and Newton's theory had to give ground to the theories of relativity and quantum mechanics. Newton's laws are still very useful to us in our everyday lives, just as modern sailors can still use earth-centered calculations to navigate the oceans despite knowing that the sun is the center of the solar system. But neither geocentrism nor Newton's laws fully describe nature as it actually is.

The notion of the ether and Newton's laws lasted as long as they did because there had been no way to test them, no way to push experiments past the limits to see if the theories could still keep up. For a long time Darwin's ideas about evolution had the same problem—they couldn't be rigorously tested. After all, in theory random mutation and natural selection take a large number of generations to produce measurable results and require whole populations of organisms—thousands or millions or more, not just a few. What scientist has time to wait so many generations to look for a result? What granting agency will award enough money to monitor millions and millions of animals? None, of course. So in the absence of a way to rigorously test it, Darwin's theory was simply assumed to be right.

But once again new techniques have made better experiments possible, and those old limits to testing Darwin's theory have fallen away. First, the underlying physical basis for evolution was discovered in DNA and the genetic code. Next, scientists began to do evolution experiments with microbes (single-celled creatures and viruses), which replicate rapidly in huge numbers, obviating the onerous need to wait a long time and maintain hordes of larger animals. Finally, rapid and easy ways were invented to determine what changes to DNA were happening in evolution experiments.

That final factor was key. Neither Darwin nor any other nineteenth century scientist knew how genetic information was transmitted. After the 1950s scientists knew how it was transmitted, but had no easy way to detect exactly what changes to DNA were occurring. Only in the past quarter-century has the refinement of DNA sequencing technology (the same technology that has allowed specific mutations in human cancers to be tracked down, opening the vista of specific, individualized treatment) permitted the definite identification of which mutations were causing which changes. The most recent results strongly support the arguments of *Darwin's Black Box* (that the elegant machinery of the cell is a clear indication of intelligent design) and *The Edge of Evolution* (that there is a definite limit to what unintelligent evolutionary processes can achieve). They also, as I detail in *Darwin Devolves*, demonstrate something quite unexpected: Darwin's mechanism of random mutation and natural selection is actually powerfully devolutionary—it works chiefly by squandering genetic information for short-term gain.

Science is relentless, and in its own good time has illuminated the murky depths of life. The trend it has revealed is inexorable: the more and more we know about biology, the less and less Darwin's theory is seen to explain and the deeper into life purposeful design is seen to extend.

ENDNOTES

INTRODUCTION

1. Andreas Wagner, *Arrival of the Fittest: Solving Evolution's Greatest Puzzle* (New York: Current, 2014); James A. Shapiro, *Evolution: A View from the 21st Century* (Upper Saddle River, NJ: FT Press, 2011); Eugene V. Koonin, *The Logic of Chance: The Nature and Origin of Biological Evolution* (Upper Saddle River, NJ: FT Press, 2011); Snait B. Gissis and Eva Jablonka, eds., *Transformations of Lamarckism: From Subtle Fluids to Molecular Biology* (Cambridge, MA: MIT Press, 2011); Massimo Pigliucci and Gerd Müller, *Evolution: The Extended Synthesis* (Cambridge, MA: MIT Press, 2010); Peter Beurton, Raphael Falk, and Hans-Jörg Rheinberger, eds., *The Concept of the Gene in Development and Evolution: Historical and Epistemological Perspectives* (Cambridge, UK: Cambridge University Press, 2008); Michael Lynch, *The Origins of Genome Architecture* (Sunderland, MA: Sinauer, 2007); Denis Noble, *The Music of Life: Biology beyond Genes* (Oxford: Oxford University Press, 2006); Richard A. Watson, *Compositional Evolution: The Impact of Sex, Symbiosis, and Modularity on the Gradualist Framework of Evolution* (Cambridge, MA: MIT Press, 2006); Marc W. Kirschner and John C. Gerhart, *The Plausibility of Life: Great Leaps of Evolution* (New Haven, CT: Yale University Press, 2005); Sean B. Carroll, *Endless Forms Most Beautiful: The New Science of Evo Devo* (New York: W.W. Norton, 2005); Eva Jablonka and Marion J. Lamb, *Evolution in Four Dimensions: Genetic, Epigenetic, Behavioral, and Symbolic Variation in the History of Life* (Cambridge, MA: MIT Press, 2005); Mary Jane West-Eberhard, *Developmental Plasticity and Evolution* (Oxford: Oxford University Press, 2003); and Lynn Margulis and Dorion Sagan, *Acquiring Genomes: A Theory of the Origins of Species* (New York: Basic Books, 2002).

2. Kevin Laland et al., "Does Evolutionary Theory Need a Rethink?," *Nature* 514 (October 8, 2014): 161–164, https://www.nature.com/news/does-evolutionary-theory-need-a-rethink-1.16080.

3. John Farley, *The Spontaneous Generation Controversy: From Descartes to Oparin* (Baltimore: Johns Hopkins University Press, 1977), 73.

4. Michael Levine and Eric H. Davidson, "Gene Regulatory Networks for Development," *PNAS* (April 5, 2005): 4936–4942, https://doi.org/10.1073/pnas.0408031102.

5. Bruce Alberts, "The Cell as a Collection of Protein Machines: Preparing the Next Generation of Molecular Biologists," *Cell* 92, no. 3 (February 6, 1998): 291–294, https://doi.org/10.1016/S0092-8674(00)80922-8. (Emphasis in original.)

PART ONE (1–13)

1. Charles Darwin, *The Origin of Species*, 6th ed. [1872], (New York: New York University Press, 1988), 154.

2. The blood clotting cascade is the "Ultimate Rube Goldberg machine" in the sense of being a very sophisticated, irreducibly complex machine, not in the sense of being comically over-engineered and roundabout, as are cartoon Rube Goldberg machines. The blood clotting cascade is necessarily multifaceted because its task demands it be so.

3. "The Great Debate, Part I: Miller & Pennock vs. Dembski & Behe," YouTube, video, 1:08:08, August 20, 2010, https://www.youtube.com/watch?v=CmMVgOTCukQ&t=62s.

4. Franklin Harold, *The Way of the Cell* (Oxford: Oxford UP, 2003), 205. Harold's endnote accompanying that passage references me and my book *Darwin's Black Box* (*DBB*) as guilty of inferring intelligent design, and also recommends James Shapiro's *National Review* response to *DBB*. There Shapiro uses

nearly identical language when he writes, "There are no detailed Darwinian accounts for the evolution of any fundamental biochemical or cellular system, only a variety of wishful speculations." James Shapiro, "In the Details… What?," *National Review* (Sept. 16, 1996), 64.

5. Anne-Claude Gavin et al., "Functional Organization of the Yeast Proteome by Systematic Analysis of Protein Complexes," *Nature* 415 (2002): 141–147.

6. Douglas D. Axe, "Extreme Functional Sensitivity to Conservative Amino Acid Changes on Enzyme Exteriors," *Journal of Molecular Biology* 301 (2000): 585-595.

7. For an in-depth discussion of Darwinian evolution and structures such as disulfide bonds, see Chapter 9 of my 2019 book *Darwin Devolves*.

8. James Shapiro, "In the Details… What?," *National Review* (Sept. 16, 1996), 64. Franklin Harold echoes this point in *The Way of the Cell* (Oxford: Oxford UP, 2003), 205.

9. H. Allen Orr, "Darwin v. Intelligent Design (Again)," *Boston Review* (December 1996/ January 1997), https://web.archive.org/ web/20141006072402/https://bostonreview. net/archives/BR21.6/orr.html.

10. By *biochemistry* I mean all sciences that investigate life at the molecular level, including molecular biology, much of embryology, immunology, genetics, etc.

11. Lucy Shapiro, "The Bacterial Flagellum: From Genetic Network to Complex Architecture," *Cell* 80, no. 4 (February 24, 1995): 525–527, https://doi.org/10.1016/0092-8674(95)90505-7.

12. Once these comments were no longer available at *Boston Review*, I posted them at Access Research Network, 1999, https://web. archive.org/web/20000816142840/http:// www.arn.org/docs/behe/mb_brrespbr.htm.

13. Michael J. Behe, *Darwin's Black Box: The Biochemical Challenge to Evolution* (New York: The Free Press, 1996), 39.

14. Behe, *Darwin's Black Box*, 46–47.

15. Behe, *Darwin's Black Box*, 66.

16. Behe, *Darwin's Black Box*, 196–197.

17. Behe, *Darwin's Black Box*, 205–208.

18. Behe, *Darwin's Black Box*, 29.

19. Behe, *Darwin's Black Box*, 232–233.

20. Behe, *Darwin's Black Box*, 206–207.

21. Thomas H. Bugge et al., "Loss of Fibrinogen Rescues Mice from the Pleiotropic Effects of Plasminogen Deficiency," *Cell* 87 (November 15, 1996): 709–719.

22. Behe, *Darwin's Black Box*, 196–197.

23. Tom Cavalier-Smith, "Book Review: *Darwin's Black Box*," *Trends in Ecology and Evolution* 12, no. 4 (April 1, 1997): 163–163, https://doi.org/10.1016/S0169-5347(97)89790-X.

24. Michael J. Behe, *Darwin's Black Box: The Biochemical Challenge to Evolution* (New York: The Free Press, 1996), 239.

25. Eugenie C. Scott, "Not (Just) in Kansas Anymore," *Science* 288, no. 5467 (May 5, 2000): 813–815, https://doi.org/10.1126/ science.288.5467.813. The letter to the editor I wrote in response was published online at *Science* on July 7, 2000, https://web.archive. org/web/20200911211430/https://science. sciencemag.org/content/288/5467/813/ tab-e-letters. My letter, "Intelligent Design Is Not Creationism," was then excerpted with a response by Scott: *Science* 289, no. 5481 (August 11, 2000): 871.

26. David J. DeRosier, "The Turn of the Screw: The Bacterial Flagellar Motor," *Cell* 93, no. 1 (April 3, 1998): 17–20, https://doi. org/10.1016/S0092-8674(00)81141-1.

27. Niall Shanks and Karl Joplin, "Of Mousetraps and Men: Behe on Biochemistry," *Reports of the National Center for Science Education* 20, no. 1–2 (November 17, 2008), https://web.archive.org/ web/20200911211732/https://ncse.ngo/ mousetraps-and-men-behe-biochemistry.

28. David Berlinski, "Has Darwin Met His Match?," *Commentary* (December 2002), https://web.archive.org/ web/20200911212902/https://www.commentarymagazine.com/articles/commentary-bk/has-darwin-met-his-match/.

29. Jerry A. Coyne, "Seeing and Believing," *The New Republic* (February 3, 2009), https:// web.archive.org/web/20160629234724if_/ https://newrepublic.com/article/63388/ seeing-and-believing.

30. Michael J. Behe, *Darwin's Black Box: The Biochemical Challenge to Evolution* (New York: The Free Press, 1996), 72.

31. Edward Behrman, George Marzluf, and Ronald Bentley, "Evidence from Biochemical Pathways in Favor of Unfinished Evolution Rather Than Intelligent Design," *Journal of Chemical Education* 81, no. 7 (July 1, 2004): 1051–1052.

32. Irving M. Copi, *Introduction to Logic* (New York: Macmillan, 1953), 56.

33. Kenneth R. Miller, "Life's Grand Design," *Technology Review* 97, no. 2 (1994): 24–32, https://web.archive.org/web/20030220084204/http://www.mille-randlevine.com/km/evol/lgd/.

34. W. Wayt Gibbs, "The Unseen Genome: Gems among the Junk," *Scientific American* 289, no. 5 (November 2003): 46–53.

35. Lori A. Snyder et al., "Bacterial Flagellar Diversity and Evolution: Seek Simplicity and Distrust It?," *Trends in Microbiology* 17, no. 1 (January 2009): 1–5.

36. Michael J. Behe, *Darwin's Black Box: The Biochemical Challenge to Evolution* (New York: The Free Press, 1996), 72.

Part Two (14–34)

1. One has to be sophisticated about what is regarded as a "step." One mutational step in a biological organism might seem to have large effects, such as the famous antenna-pedia mutation in fruit flies. Although such a change may impress us, it involves only the rearrangement of existing structures; no new structures are made. When thinking about what's involved in making a new structure it's best to think of how many lines of instructions (analogous to lines of computer code) would be needed to build it. See my discussion of this topic in *Darwin's Black Box: The Biochemical Challenge to Evolution* (New York: The Free Press, 1996), 39–41.

2. John H. McDonald, "A Reducibly Complex Mousetrap," last modified January 13, 2003, https://web.archive.org/web/20200206134721/http://udel.edu/~mcdonald/oldmousetrap.html.

3. Behe, *Darwin's Black Box*, 43.

4. To play the game right, one has to compare the probability of these events happening with the probability of any slight "mutation" happening. To give a flavor of what that might mean, a mutation might involve bending the spring in the middle, changing the size of the platform, changing the tension on the spring, extending the end of a metal piece, and so on. A crude feel for the probabilities of the events can be obtained by examining the precision a feature must have for the trap to work. To get the probability for two or more unselected events, one multiplies the probabilities for each. (An unselected event is one that does not improve function by itself.)

5. The work has been done by Barry Hall of the University of Rochester. He could get selection to replace one piece of the *lac* operon (the β-galactosidase) but had to intelligently intervene to keep the bacteria alive by adding the artificial chemical induce IPTG. The bacteria could not replace two required protein parts deleted simultaneously, showing the severe problem of irreducible complexity. For a review of his work see: Barry G. Hall, "Experimental Evolution of Ebg Enzyme Provides Clues about the Evolution of Catalysis and to Evolutionary Potential," *FEMS Microbiology Letters* 174, no. 1 (May 1, 1999): 1–8, https://doi.org/10.1016/S0378-1097(99)00096-8.

6. Torben Halkier, *Mechanisms in Blood Coagulation Fibrinolysis and the Complement System* (Cambridge, UK: Cambridge University Press, 1992), 104.

7. Kenneth R. Miller, *Finding Darwin's God: A Scientist's Search for Common Ground Between God and Evolution* (New York: Cliff Street Books, 1999).

8. Russell F. Doolittle, "A Delicate Balance," *Boston Review* (February/March 1997): 28–29, https://web.archive.org/web/20170119102129/https://bostonreview.net/archives/BR22.1/doolittle.html.

9. Thomas H. Bugge et al., "Loss of Fibrinogen Rescues Mice from the Pleiotropic Effects of Plasminogen Deficiency," *Cell* 87, no. 4 (November 15, 1996): 709–719, https://doi.org/10.1016/S0092-8674(00)81390-2.

10. Doolittle, "A Delicate Balance."

11. T. T. Suh et al., "Resolution of Spontaneous Bleeding Events but Failure of Pregnancy in Fibrinogen-Deficient Mice," *Genes and Development* 9 (1995): 2020–2033, https://doi.org/10.1101/gad.9.16.2020.

12. Doolittle, "A Delicate Balance."

13. H. Allen Orr, "Darwin v. Intelligent Design (Again)," *Boston Review* (December 1996/ January 1997): 28–31, https://web.archive. org/web/20141006072402/https://boston-review.net/archives/BR21.6/orr.html.

14. Douglas J. Futuyma, "Miracles and Molecules," *Boston Review* (February/March 1997): 29–30, https://web.archive.org/ web/20170302174106/https://bostonreview. net/archives/BR22.1/futuyma.html.

15. Neil W. Blackstone, "Argumentum ad Ignoratium," *Quarterly Review of Biology* 72, no. 4 (December 1997): 445–447, https://doi. org/10.1086/419956.

16. Robert Dorit, "Review: Molecular Evolution and Scientific Inquiry, Misperceived," *American Scientist* 85, no. 5 (September/October 1997): 474–475.

17. National Academy of Sciences, *Science and Creationism: A View from the National Academy of Sciences* (Washington, DC: National Academy Press, 1999).

18. National Academy of Sciences, *Science and Creationism*.

19. Wen-Hsiung Li, *Molecular Evolution* (Sunderland, MA: Sinauer, 1997).

20. Ernst Mayr, *One Long Argument: Charles Darwin and the Genesis of Modern Evolutionary Thought* (Cambridge, MA: Harvard University Press, 1991).

21. Russell F. Doolittle, *The Evolution of Vertebrate Blood Clotting* (Herndon, VA: University Science Books, 2013).

22. Miller, *Finding Darwin's God*, Chapter 5.

23. Miller, *Finding Darwin's God*, 156–157.

24. Charles Scriver et al., *The Metabolic Basis of Inherited Disease* (New York: McGraw-Hill, 1989), 2213.

25. Thomas H. Bugge et al., "Plasminogen Deficiency Causes Severe Thrombosis but is Compatible with Development and Reproduction," *Genes and Development* 9, no. 7 (April 1, 1995): 794–807, https://doi. org/10.1101/gad.9.7.794.

26. Miller, *Finding Darwin's God*, 157.

27. Behe, *Darwin's Black Box*, 86.

28. Halkier, *Mechanisms in Blood Coagulation Fibrinolysis*, 104.

29. Keith Robison, "Darwin's Black Box: Irreducible Complexity or Irreproducible Irreducibility?" Talk Origins (website), December 11, 1996, https://web.archive.org/ web/19990127102929/http://www.talkorigins.org/faqs/behe/review.html.

30. Behe, *Darwin's Black Box*, 39.

31. Russell F. Doolittle, "A Delicate Balance," *Boston Review* (February/March 1997): 28-29, https://web.archive.org/ web/20170119102129/https://bostonreview. net/archives/BR22.1/doolittle.html.

32. Thomas H. Bugge et al., "Loss of Fibrinogen Rescues Mice from the Pleiotropic Effects of Plasminogen Deficiency," *Cell* 87, no. 4 (November 15, 1996): 709-719, https://doi. org/10.1016/S0092-8674(00)81390-2.

33. Michael Ruse, "Answering the Creationists: Where They Go Wrong and What They're Afraid of," *Free Inquiry* 18, no. 2 (Spring 1998): 28.

34. Neil Greenspan, "Not-So-Intelligent Design," *The Scientist* (March 3, 2002).

35. For a summary of Luskin's arguments see Casey Luskin, "Kenneth Miller, Michael Behe, and the Irreducible Complexity of the Blood Clotting Cascade Saga," January 1, 2010, https://web.archive.org/ web/20100406142115/https://www.discovery.org/a/14081/.

36. Russell F. Doolittle, Yong Jiang, and Justin Nand, "Genomic Evidence for a Simpler Clotting Scheme in Jawless Vertebrates," *Journal of Molecular Evolution* 66 (2008):185– 196, https://doi.org/10.1007/s00239-008-9074-8.

37. Kenneth R. Miller, *Finding Darwin's God: A Scientist's Search for Common Ground Between God and Evolution* (New York: Cliff Street Books, 1999).

38. Miller, *Finding Darwin's God*, 145.

39. Miller, *Finding Darwin's God*, 146.

40. Barry G. Hall, "Experimental Evolution of Ebg Enzyme Provides Clues about the Evolution of Catalysis and to Evolutionary Potential," *FEMS Microbiology Letters* 174, no. 1 (May 1, 1999): 1–8, https://doi.org/10.1016/ S0378-1097(99)00096-8.

41. Hall, "Experimental Evolution of Ebg."

42. Barry G. Hall, "Evolution of a Regulated Operon in the Laboratory," *Genetics* 101, no. 3–4 (July 1982): 335–344, https:// web.archive.org/web/20170815151351/

https://www.genetics.org/content/genetics/101/3-4/335.full.pdf.

43. Barry G. Hall, "Evolution on a Petri Dish: The Evolved B-galactosidase System as a Model for Studying Acquisitive Evolution in the Laboratory," in *Evolutionary Biology*, eds. M. K. Hecht, B. Wallace, and G. T. Prance (New York: Plenum Press, 1982), 15: 85–150.

44. John Cairns, "Mutation and Cancer: The Antecedents to Our Studies of Adaptive Mutation," *Genetics* 148, no. 4 (April 1, 1998): 1433–1440, https://web.archive.org/web/20170817175624/https://www.genetics.org/content/genetics/148/4/1433.full.pdf; P. L. Foster, "Mechanisms of Stationary Phase Mutation: A Decade of Adaptive Mutation," *Annual Review of Genetics* 33 (December 1999): 57–88, https://doi.org/10.1146/annurev.genet.33.1.57; Johnjoe McFadden and Jim Al Khalili, "A Quantum Mechanical Model of Adaptive Mutation," *Biosystems* 50, no. 3 (June 1999): 203–211, https://doi.org/10.1016/S0303-2647(99)00004-0; Barry G. Hall, "Adaptive Mutagenesis: A Process that Generates Almost Exclusively Beneficial Mutations," *Genetica* 102–103 (February 1998): 109–125, https://doi.org/10.1023/A:1017015815643; James A. Shapiro, "Genome Organization, Natural Genetic Engineering and Adaptive Mutation," *Trends in Genetics* 13, no. 3 (1997): 98–104, https://doi.org/10.1016/S0168-9525(97)01058-5.

45. Barry G. Hall, "On the Specificity of Adaptive Mutations," *Genetics* 145 (1997): 39–44, https://web.archive.org/web/20190322203920/https://www.genetics.org/content/genetics/145/1/39.full.pdf.

46. Hall, "Evolution of a Regulated Operon in the Laboratory."

47. Hall, "Experimental Evolution of Ebg Enzyme."

48. Hall, "Experimental Evolution of Ebg Enzyme."

49. Hall, "Experimental Evolution of Ebg Enzyme."

50. Hall, "Evolution on a Petri dish."

51. Miller, *Finding Darwin's God*, 146.

52. Hall, "Evolution of a Regulated Operon in the Laboratory."

53. Hall, "Evolution of a Regulated Operon in the Laboratory."

54. Hall, "Evolution of a Regulated Operon in the Laboratory."

55. Miller's prose is often exaggerated and sometimes borders on the bombastic. Perhaps he uses such a relentlessly emphatic style in the hope of overwhelming readers through the sheer force of his words. Perhaps he just has a much-larger-than-average share of self-confidence. Fortunately, in this section on the "acid test," experiments exist to show that his prose is bluster. Let me be blunt: Miller always writes (or speaks) with the utmost confidence, even when experiments show him to be quite wrong. I would caution readers of his work not to be swayed by his tone, whose confidence never wavers even when the evidence does.

56. Miller, *Finding Darwin's God*, 147.

57. Kenneth Miller, "The Mousetrap Analogy; or, Trapped by Design." Miller's original link no longer functions, but his remarks are available at https://web.archive.org/web/20080725085532/https://chem.tufts.edu/answersinscience/Miller-Mousetrap.html.

58. In his essay Miller wrote the following: "Behe argues that MacDonald's four simpler mousetraps do not present a good model of a 'Darwinian process.' Even the simplest mousetrap, Behe argues, requires 'the involvement of intelligence,' and the 'involvement of intelligence at any point in a scenario is fatal.' I agree. And if I or MacDonald or any one else had presented the simpler mousetraps as examples of an evolutionary transition, Behe would be right."

59. James A. Shapiro, "In the Details... What?," *National Review*, September 16, 1996, 62–65, https://web.archive.org/web/20140808233919/https://shapiro.bsd.uchicago.edu/Shapiro.1996.Nat'lReview.pdf.

60. Jerry A. Coyne, "God in the Details," *Nature* 383 (1996): 227–228, https://doi.org/10.1038/383227a0.

61. Tom Cavalier-Smith, "Book Review: *Darwin's Black Box*," *Trends in Ecology and Evolution* 12, no. 4 (April 1, 1997): 162–163, https://doi.org/10.1016/S0169-5347(97)89790-X.

62. Andrew Pomiankowski, "The God of the Tiny Gaps," *New Scientist*, September 14, 1996, https://web.archive.org/web/20160413053953/https://www.newscientist.com/article/mg15120474-100-review-the-god-of-the-tiny-gaps/.

63. Robert Dorit, "Review: Molecular Evolution and Scientific Inquiry, Misperceived," *American Scientist* 85, no. 5 (September/October 1997): 474–475.

64. Bruce Weber, "Irreducible Complexity and the Problem of Biochemical Emergence," *Biology & Philosophy* 14 (1999): 593–605.

65. Peter Atkins, "Review of *Darwin's Black Box*," The Secular Web, 1998, https://web.archive.org/web/19990829062010/https://infidels.org/library/modern/peter_atkins/behe.html.

66. John Catalano, "Publish or Perish: Some Published Works on Biochemical Evolution," The TalkOrigins Archive, accessed August 26, 2020, https://web.archive.org/web/19981207013643/http://www.talkorigins.org/faqs/behe/publish.html.

67. Michael J. Behe, *Darwin's Black Box: The Biochemical Challenge to Evolution* (New York: The Free Press, 1996), 179.

68. Cavalier-Smith, "Book Review: *Darwin's Black Box*."

69. Wen-Hsiung Li, *Molecular Evolution* (Sunderland, MA: Sinauer, 1997).

70. Li, *Molecular Evolution*.

71. See my earlier essay on blood clotting.

72. John H. Gillespie, *The Causes of Molecular Evolution* (New York: Oxford University Press, 1991); Robert K. Selander, Andrew G. Clark, and Thomas S. Whittam, *Evolution at the Molecular Level* (Sunderland, MA: Sinauer Associates, 1991).

73. Austin L. Hughes and Meredith Yeager, "Molecular Evolution of the Vertebrate Immune System," *BioEssays* 19, no. 9 (1997): 777–786, https://web.archive.org/web/20200912154637/https://www.deepdyve.com/lp/wiley/molecular-evolution-of-the-vertebrate-immune-system-iWJvbjjKZs.

74. Takehiro Kusakabe et al., "Evolution of Chordate Actin Genes: Evidence from Genomic Organization and Amino Acid Sequences," *Journal of Molecular Evolution* 44

(1997): 289–298, https://doi.org/10.1007/PL00006146.

75. Behe, *Darwin's Black Box*, 150–151, 206–207.

76. David Ussery, A Biochemist's Response to "The Biochemical Challenge to Evolution." The website is no longer functional nor is the page archived at the Internet Archive, but see David Ussery, "A Biochemist's Response to 'The Biochemical Challenge to Evolution,'" *Bios* 70, no. 1 (March 1999): 40–45.

77. Behe, *Darwin's Black Box*, 114.

78. Ussery, "A Biochemist's Response."

79. In a later version of his review (the website has been updated several times, making it a moving target that is hard to pin down precisely), Ussery did note explicitly that one needed to search abstracts as well as titles to come up with the total of 130 papers. He then noted that a total of just four papers have both words in the title. These papers were not picked up in my search because they either were published after my search was completed in 1995, or because the papers were published before the mid-1980s (which is outside the scope of a CARL search). None of the papers affects the questions discussed in this manuscript.

80. A. Ramos et al., "Outbreak of Nosocomial Diarrhea by *Clostridium difficile* in a Department of Internal Medicine," *Enfermedades Infecciosas Y Microbiologia Clinica* 16 (1998): 66–69.

81. Yutaka Furutani et al., "Evolution of the Trappin Multigene Family in the Suidae," *Journal of Biochemistry* 124, no. 3 (September 1998): 491–502, https://doi.org/10.1093/oxfordjournals.jbchem.a022140.

82. A. Simonsen et al., "Syntaxin-16, a Putative Golgi t-SNARE," *European Journal of Cell Biology* 75 (1998): 223–231.

83. James A. Shapiro, "In the Details... What?"

84. Jerry A. Coyne, "God in the Details."

85. Kenneth R. Miller, *Finding Darwin's God: A Scientist's Search for Common Ground Between God and Evolution* (New York: Cliff Street Books, 1999).

86. Antony M. Dean and G. Brian Golding, "Protein Engineering Reveals Ancient Adaptive Replacements in Isocitrate Dehydroge-

nase," *PNAS* 94, no. 7 (1997): 3104–3109, https://doi.org/10.1073/pnas.94.7.3104.

87. John M. Logsdon, Jr. and W. Ford Doolittle, "Origin of Antifreeze Protein Genes: A Cool Tale in Molecular Evolution," *PNAS* 94, no. 8 (April 15, 1997): 3485–3487, https://doi.org/10.1073/pnas.94.8.3485.

88. Siegfried M. Musser and Sunney I. Chan, "Evolution of the Cytochrome C Oxidase Proton Pump," *Journal of Molecular Evolution* 46, no. 5 (May 1998): 508–520, https://doi.org/10.1007/PL00006332.

89. See my earlier discussion of "mousetrap evolution."

90. Enrique Meléndez-Hevia, Thomas G. Waddell, and Marta Cascante, "The Puzzle of the Krebs Citric Acid Cycle: Assembling the Pieces of Chemically Feasible Reactions, and Opportunism in the Design of Metabolic Pathways During Evolution," *Journal of Molecular Evolution* 43 (September 1996): 293–303, https://doi.org/10.1007/BF02338838.

91. Behe, *Darwin's Black Box*, 141–142, 150–151.

92. Richard H. Thornhill and David W. Ussery, "A Classification of Possible Routes of Darwinian Evolution," *Journal of Theoretical Biology* 203, no. 2 (2000): 111–116, https://doi.org/10.1006/jtbi.2000.1070.

93. Michael J. Behe, *Darwin's Black Box: The Biochemical Challenge to Evolution* (New York: The Free Press, 1996).

94. Paul Davies, *The Fifth Miracle: The Search for the Origin and Meaning of Life* (New York: Simon & Schuster, 1999).

95. Michael Denton, *Nature's Destiny: How the Laws of Biology Reveal Purpose in the Universe* (New York: Free Press, 1998).

96. National Academy of Sciences, *Science and Creationism: A View from the National Academy of Sciences* (Washington, DC: National Academy Press, 1999), 7.

97. National Academy of Sciences, 25.

98. Jerry A. Coyne, "God in the Details," *Nature* 383 (1996): 227–228, https://doi.org/10.1038/383227a0.

99. Russell F. Doolittle, "A Delicate Balance," *Boston Review* (February/March 1997): 28–29, https://web.archive.org/web/20170119102129/https://bostonreview.net/archives/BR22.1/doolittle.html.

100. Kenneth R. Miller, *Finding Darwin's God: A Scientist's Search for Common Ground Between God and Evolution* (New York: Cliff Street Books, 1999).

101. Thomas H. Bugge et al., "Loss of Fibrinogen Rescues Mice from the Pleiotropic Effects of Plasminogen Deficiency," *Cell* 87 (November 15, 1996): 709–19.

102. Barry G. Hall, "Experimental Evolution of Ebg Enzyme Provides Clues about the Evolution of Catalysis and to Evolutionary Potential," *FEMS Microbiology Letters* 174, no. 1 (May 1, 1999): 1–8, https://doi.org/10.1016/S0378-1097(99)00096-8.

103. National Academy of Sciences, 22.

104. Kenneth Miller leads readers of *Finding Darwin's God* into thinking such a process would be very easy. He writes, "If microevolution can redesign one gene in fewer than two hundred generations (which in this case took only thirteen days!), what principles of biochemistry or molecular biology would prevent it from redesigning dozens or hundreds of genes over a few weeks or months to produce a distinctly new species? There are no such principles of course..." (108). Well, then, why doesn't he just take an appropriate bacterial species, knock out the genes for its flagellum, place the bacterium under selective pressure (for mobility, say), and experimentally produce a flagellum—or any equally complex system—in the laboratory? (A flagellum, after all, has only 30–40 genes, not the hundreds Miller claims would be easy for natural selection to rapidly redesign.) If he did that, my claims would be utterly falsified. But he won't even try it because he is grossly exaggerating the prospects of success.

105. Coyne, "God in the Details."

106. Miller, *Finding Darwin's God*, 147.

107. Bugge et al., "Loss of Fibrinogen Rescues Mice."

108. R. Flietstra, "A Response to Michael Behe," *Books & Culture* (September/October 1998): 37–38.

109. H. Allen Orr, "Darwin v. Intelligent Design (Again)," *Boston Review* (December 1996/January 1997): 28–31, https://web.

archive.org/web/20141006072402/https://bostonreview.net/archives/BR21.6/orr.html.

110. Orr, "Darwin v. Intelligent Design."

111. Behe, *Darwin's Black Box*, 39.

112. Behe, *Darwin's Black Box*, 39.

113. P. R. Gross, "The Dissent of Man," *Wall Street Journal*, July 30, 1996, A12.

114. Douglas J. Futuyma, "Miracles and Molecules," *Boston Review* (February/March 1997): 29–30, https://web.archive.org/web/20170302174106/https://bostonreview.net/archives/BR22.1/futuyma.html.

115. David Ussery, "A Biochemist's Response to 'The Biochemical Challenge to Evolution,'" *Bios* 70, no. 1 (March 1999): 40–45.

116. Behe, *Darwin's Black Box*, 72.

117. Bruce Weber, "Irreducible Complexity and the Problem of Biochemical Emergence," *Biology & Philosophy* 14 (1999): 593–605.

118. James A. Shapiro, "A Third Way," *Boston Review* (February/March 1997): 32–33, https://web.archive.org/web/20170922162237/https://bostonreview.net/archives/BR22.1/shapiro.html.

119. Weber, "Irreducible Complexity."

120. Niall Shanks and Karl Joplin, "Of Mousetraps and Men: Behe on Biochemistry," *Reports of the National Center for Science Education* 20, no. 1–2 (November 17, 2008), https://web.archive.org/web/20200911211732/https://ncse.ngo/mousetraps-and-men-behe-biochemistry.

121. Michael J. Behe, "Self-Organization and Irreducibly Complex Systems: A Reply to Shanks and Joplin," *Philosophy of Science* 67 (2000): 155–162.

122. Behe, *Darwin's Black Box*, 203–204. To be clear, I am not saying that the evidence for intelligent design is weak or that I expect intelligent design to be refuted. Not at all. I'm saying intelligent design is falsifiable in principle by future scientific discoveries, but nonetheless, given what we know today, intelligent design stands out as the best explanation for irreducibly complex biological systems, and the case for any purely naturalistic explanation appears increasingly weak with the ongoing progress of molecular biology.

123. Robert T. Pennock, *Tower of Babel: The Evidence Against the New Creationism* (Cambridge, MA: MIT Press, 1999).

124. Michael J. Behe, "The God of Science: The Case for Intelligent Design," *The Weekly Standard*, June 7, 1999, 35–37.

125. That the Big Bang theory has extra-scientific implications can be seen in the reaction of those who do not welcome the implications. For example, in a 1989 editorial in *Nature* with the intriguing title "Down with the Big Bang," John Maddox wrote, "Creationists and those of similar persuasions seeking support for their theories have ample justification in the doctrine of the Big Bang. That, they might say, is when (and how) the Universe was created." John Maddox, "Down with the Big Bang," *Nature* 340 (1989): 425, https://doi.org/10.1038/340425a0.

126. Orr, "Darwin v. Intelligent Design."

127. William Dembski, *The Design Inference: Eliminating Chance Through Small Probabilities* (New York: Cambridge University Press, 1998).

128. William Dembski, *Intelligent Design: The Bridge between Science and Theology* (Downers Grove, IL: Intervarsity Press, 1999).

129. Michael Ruse, "Enough Speculation," *Boston Review* (February/March 1997): 31–32, https://web.archive.org/web/20200912171611/https://bostonreview.net/archives/BR22.1/ruse.html.

130. Douglas J. Futuyma, "Miracles and Molecules."

131. Behe, *Darwin's Black Box*, 245–250.

132. Francis Crick and L. E. Orgel, "Directed Panspermia," *Icarus* 19 (July 1973): 341–346, https://doi.org/10.1016/0019-1035(73)90110-3.

133. Douglas J. Futuyma, *Science on Trial* (New York: Pantheon Books, 1982).

134. Russell F. Doolittle, "A Delicate Balance," *Boston Review* (February/March 1997): 28–29, https://web.archive.org/web/20170119102129/https://bostonreview.net/archives/BR22.1/doolittle.html.

135. Andrew Pomiankowski, "The God of the Tiny Gaps," *New Scientist*, September 14, 1996, https://web.archive.org/web/20160413053953/https://www.newscientist.com/article/mg15120474-100-review-the-god-of-the-tiny-gaps/.

136. Pomiankowski, "The God of the Tiny Gaps."

137. Neil W. Blackstone, "Argumentum ad Ignorantiam," *Quarterly Review of Biology* 72, no. 4 (December 1997): 445–447, https://doi.org/10.1086/419956.

138. Irving M. Copi, *Introduction to Logic* (New York: Macmillan, 1953), 76.

139. Copi, *Introduction to Logic*, 77.

140. Marc Lipsitch, "Fighting an Evolutionary War," *Forward*, October 25, 1996.

141. Paul Draper, "Irreducible Complexity and Darwinian Gradualism: A Reply to Michael J. Behe," *Faith and Philosophy* 19, no. 1 (January 2002): 3–21, https://doi.org/10.5840/faithphil20021912.

142. H. Allen Orr, "Darwin v. Intelligent Design (Again)," *Boston Review* (December 1996/January 1997): 28–31, https://web.archive.org/web/20141006072402/https://bostonreview.net/archives/BR21.6/orr.html.

143. Michael J. Behe, "The Sterility of Darwinism," *Boston Review* (February/March 1997): 24. While *Boston Review* has reliably archived my critics' remarks, they removed mine. However, a reproduction is available at https://web.archive.org/web/20100406134917/https://www.discovery.org/a/66/.

144. Orr, "Darwin v. Intelligent Design."

145. Orr, "Darwin v. Intelligent Design."

146. Orr, "Darwin v. Intelligent Design."

147. Behe, "The Sterility of Darwinism."

148. Orr, "Darwin v. Intelligent Design."

149. Charles Darwin, *The Origin of Species* [1859] (New York: Bantam Books, 1999), 155–158.

150. Draper, "Irreducible Complexity."

151. Michael J. Behe, *Darwin Devolves: The New Science about DNA That Challenges Evolution* (New York: HarperOne, 2019), chap. 9.

152. Draper, "Irreducible Complexity."

153. Draper, "Irreducible Complexity."

154. Michael J. Behe, *Darwin's Black Box: The Biochemical Challenge to Evolution* (New York: The Free Press, 1996), 43–35.

155. John H. McDonald, "A Reducibly Complex Mousetrap," last modified January 13, 2003, https://web.archive.org/web/20200206134721/http://udel.edu/~mcdonald/oldmousetrap.html.

156. Michael J. Behe, "A Mousetrap Defended," Discovery Institute, July 31, 2000, https://web.archive.org/web/20090310182942/https://www.discovery.org/a/446/.

157. Darwin, *The Origin of Species*, 158.

158. Draper, "Irreducible Complexity."

159. Michael J. Behe, *Darwin Devolves: The New Science about DNA That Challenges Evolution* (New York: HarperOne, 2019), 283–301.

160. Draper, "Irreducible Complexity."

161. Alvin Plantinga, *Where the Conflict Really Lies: Science, Religion, and Naturalism* (New York: Oxford University Press, 2011), 231.

162. Plantinga, *Where the Conflict Really Lies*, 235.

163. Elliott Sober, "Intelligent Design and Probability Reasoning," *International Journal for Philosophy of Religion* 52, no. 2 (October 2002): 65–80.

164. Plantinga, *Where the Conflict Really Lies*, 235–236.

165. Most recently in *Darwin Devolves*, chap. 10.

166. Michael J. Behe, *Darwin's Black Box: The Biochemical Challenge to Evolution* (New York: The Free Press, 1996), 39.

167. Charles Darwin, *The Origin of Species* [1859].

168. Niall Shanks and Karl H. Joplin, "Redundant Complexity: A Critical Analysis of Intelligent Design in Biochemistry," *Philosophy of Science* 66, no. 2 (June 1999): 268–282, https://doi.org/10.1086/392687.

169. Shanks and Joplin, "Redundant Complexity."

170. Peter Gray and Stephen K. Scott, *Chemical Oscillations and Instabilities: Non-linear Chemical Kinetics* (Oxford: Clarendon Press, 1994).

171. Shanks and Joplin, "Redundant Complexity," 272–273.

172. William Dembski, *The Design Inference: Eliminating Chance Through Small Probabilities* (New York: Cambridge University Press, 1998).

173. Milton T. Stubbs and Wolfram Bode, "Coagulation Factors and Their Inhibitors," *Current Opinion in Structural Biology* 4, no. 6 (1994): 823–832, https://doi.org/10.1016/0959-440X(94)90263-1.

174. Richard J. Field, "A Reaction Periodic in Time and Space," *Journal of Chemical Education* 49, no. 5 (1972): 308–311, https://doi.org/10.1021/ed049p308.

175. Field, "A Reaction Periodic in Time and Space," 308.

176. John J. Tyson, "What Everyone Should Know about the Belousov-Zhabotinsky Reaction," in *Frontiers in Mathematical Biology*, ed. Simon A. Levin (Springer-Verlag: Berlin, 1994), 577.

177. Shanks and Joplin, "Redundant Complexity," 273.

178. Michael Schreckenberg and Dietrich E. Wolf, *Traffic and Granular Flow* (Singapore: Springer, 1998).

179. Albert Goldbeter, *Biochemical Oscillations and Cellular Rhythms: The Molecular Bases of Periodic and Chaotic Behaviour* (Cambridge, UK: Cambridge University Press, 1996), part 1.

180. Goldbeter, *Biochemical Oscillations*, part 3.

181. Shanks and Joplin, "Redundant Complexity," 268.

182. Morimitsu Nishikimi and Kunio Yagi, "Molecular Basis for the Deficiency in Humans of Gulonolactone Oxidase, a Key Enzyme for Ascorbic Acid Biosynthesis," *American Journal of Clinical Nutrition* 54, no. 6 (December 1991): 1203S–1208S, https://doi.org/10.1093/ajcn/54.6.1203s.

183. Thomas Kolter and Konrad Sandhoff, "Glycosphingolipid Degradation and Animal Models of GM2-Gangliosidoses," *Journal of Inherited Metabolic Disease* 21 (August 1998): 548–563, https://doi.org/10.1023/A:1005419122018.

184. Charles R. Scriver et al., *The Metabolic and Molecular Bases of Inherited Disease* (New York: McGraw-Hill, Health Professions Division, 1995).

185. Shanks and Joplin, "Redundant Complexity," 279.

186. Thomas H. Bugge et al., "Loss of Fibrinogen Rescues Mice from the Pleiotropic Effects of Plasminogen Deficiency," *Cell* 87, no. 4 (November 15, 1996): 709–719, https://doi.org/10.1016/S0092-8674(00)81390-2.

187. Thomas H. Bugge et al., "Fatal Embryonic Bleeding Events in Mice Lacking Tissue Factor, the Cell-Associated Initiator of Blood Coagulation," *PNAS* 93, no. 13 (June 25, 1996): 6258–6263, https://doi.org/10.1073/pnas.93.13.6258.

188. William Y. Sun et al., "Prothrombin Deficiency Results in Embryonic and Neonatal Lethality in Mice," *PNAS* 95, no. 13 (June 23, 1998): 7597–7602, https://doi.org/10.1073/pnas.95.13.7597.

189. Behe, *Darwin's Black Box*, chap 4.

190. Stephen Wolfram, *A New Kind of Science* (Champaign, IL: Wolfram Media, 2002).

191. "Forum on 'Intelligent Design' Held at the American Museum of Natural History (April 23, 2002)," https://web.archive.org/web/20200912175308/https://ncse.ngo/forum-intelligent-design-held-american-museum-natural-history-april-23-2002.

192. Shin-Ichi Aizawa, "Flagellar Assembly in *Salmonella typhimurium*," *Molecular Microbiology* 19 (January 1996): 1–5, https://doi.org/10.1046/j.1365-2958.1996.344874.x.

193. Christoph J. Hueck, "Type III Protein Secretion Systems in Bacterial Pathogens of Animals and Plants," *Microbiology and Molecular Biology Reviews* 62 (1998): 379–433, https://doi.org/10.1128/MMBR.62.2.379-433.1998.

194. Sharon Begley, "Evolution Critics Come Under Fire for Flaws in 'Intelligent Design,'" *The Wall Street Journal*, February 13, 2004, B1, https://web.archive.org/web/20200912175933/https://www.wsj.com/articles/SB107656106762928092.

195. Michael J. Behe, *Darwin's Black Box: The Biochemical Challenge to Evolution* (New York: The Free Press, 1996), 66.

196. Behe, *Darwin's Black Box*, 39.

197. Renyi Liu and Howard Ochman, "Stepwise Formation of the Bacterial Flagellar System," *PNAS* 104, no. 17 (April 24, 2007): 7116–7121, https://doi.org/10.1073/pnas.0700266104.

198. Nick Matzke, "Flagellum Evolution Paper Exhibits Canine Qualities," Panda's Thumb (website), April 16, 2007, https://web.archive.org/web/20071025024505/https://pandasthumb.org/archives/2007/04/flagellum-evolu-1.html.

199. Jennifer Cutraro, "A Complex Tail, Simply Told," Science (website), April 17, 2007, https://web.archive.org/

web/20200912180332/https://www.sci-
encemag.org/news/2007/04/complex-tail-
simply-told.

200. Jamie T. Bridgham, Sean M. Carroll, and
Joseph W. Thornton, "Evolution of Hormone-Receptor Complexity by Molecular
Exploitation," *Science* 312, no. 5770 (April
7, 2006): 97–101, https://doi.org/10.1126/
science.1123348.

201. Abigail Clements et al., "The Reducible
Complexity of a Mitochondrial Molecular
Machine," *PNAS* 106, no. 37 (September 15, 2009): 15791–15795, https://doi.
org/10.1073/pnas.0908264106.

202. Rick Trebino, "How to Publish a Scientific
Comment in 1 2 3 Easy Steps," Georgia
Institute of Technology (website), accessed
September 12, 2020, https://web.archive.
org/web/20130503053618/https://frog.gat-
ech.edu/Pubs/How-to-Publish-a-Scientific-
Comment-in-123-Easy-Steps.pdf.

203. Clements et al., "The Reducible Complexity."

204. Michael J. Behe, *Darwin's Black Box: The
Biochemical Challenge to Evolution* (New
York: The Free Press, 1996); Michael J.
Behe, "Self-organization and Irreducibly
Complex Systems: A Reply to Shanks and
Joplin," *Philosophy of Science* 67 (2000):
155–162; Michael J. Behe, "Reply to My
Critics: A Response to Reviews of *Darwin's
Black Box*: The Biochemical Challenge to
Evolution," *Biology and Philosophy* 16, no.
5 (January 2001): 685–709, https://doi.
org/10.1023/A:1012268700496.

205. Clements et al., "The Reducible Complexity."

206. Clements et al., "The Reducible Complexity."

207. Michael J. Behe, *The Edge of Evolution:
The Search for the Limits of Darwinism* (New
York: Free Press, 2007).

208. Herbert S. Wilf and Warren J. Ewens, "There's Plenty of Time for Evolution,"
PNAS 107, no. 52 (December 28, 2010):
22454–22456, https://doi.org/10.1073/
pnas.1016207107.

209. Julius Lukeš et al., "How a Neutral Evolutionary Ratchet Can Build Cellular Complexity," *IUBMB Life* 63 (2011): 528–537,
https://doi.org/10.1002/iub.489.

210. Michael W. Gray et al., "Irremediable
Complexity?," *Science* 330, no. 6006 (November 12, 2010): 920–921, https://doi.
org/10.1126/science.1198594.

211. Gray et al., "Irremediable Complexity?"

212. Lukeš et al., "How a Neutral Evolutionary
Ratchet."

213. Lukeš et al., "How a Neutral Evolutionary
Ratchet."

214. Dave Speijer, "Does Constructive Neutral
Evolution Play an Important Role in the Origin of Cellular Complexity? Making Sense of
the Origins and Uses of Biological Complexity," *Bioessays* 33 (March 7, 2011): 344–349,
https://doi.org/10.1002/bies.201100010.

215. W. Ford Doolittle et al., "Comment on
'Does Constructive Neutral Evolution Play
an Important Role in the Origin of Cellular Complexity?'" *Bioessays* 33 (April 29,
2011): 427–429, https://doi.org/10.1002/
bies.201100039.

216. John W. Drake et al., "Rates of Spontaneous Mutation," *Genetics* 148, no. 4 (April 1,
1998): 1667–1686, https://web.archive.org/
web/20170816202411/https://www.genetics.
org/content/genetics/148/4/1667.full.pdf.

217. Motoo Kimura, *The Neutral Theory of
Molecular Evolution* (Cambridge, UK: Cambridge University Press, 1983).

218. A. Nissim et al., "Antibody Fragments
from a 'Single Pot' Phage Display Library as
Immunochemical Reagents," *EMBO Journal*
13 (February 1994): 692–698, https://doi.
org/10.1002/j.1460-2075.1994.tb06308.x;
A. D. Griffiths, "Isolation of High Affinity
Human Antibodies Directly from Large
Synthetic Repertoires," *EMBO Journal*
13 (July 1994): 3245–3260, https://doi.
org/10.1002/j.1460-2075.1994.tb06626.x;
Geoffery P. Smith et al., "Small Binding Proteins Selected from a Combinatorial Repertoire of Knottins Displayed on Phage," *Journal of Molecular Biology* 277, no. 2 (March
27, 1998): 317–332, https://doi.org/10.1006/
jmbi.1997.1621.

219. Kimura, *The Neutral Theory of Molecular
Evolution*.

220. Lukeš et al., "How a Neutral Evolutionary
Ratchet."

221. Lukeš et al., "How a Neutral Evolutionary
Ratchet."

222. Lukeš et al., "How a Neutral Evolutionary Ratchet."

223. Lukeš et al., "How a Neutral Evolutionary Ratchet."

224. Tiffany B. Taylor, et al., "Evolutionary Resurrection of Flagellar Motility Via Rewiring of the Nitrogen Regulation System," Science 347, no. 6225 (February 27, 2015): 1014–1017, https://doi.org/10.1126/science.1259145.

225. Ruth Williams, "Evolutionary Rewiring: Strong Selective Pressure Can Lead to Rapid and Reproducible Evolution in Bacteria," The Scientist, February 26, 2015, https://web.archive.org/web/20190405164701/https://www.the-scientist.com/daily-news/evolutionary-rewiring-35878.

226. Wei Li and Chung-Dar Lu, "Regulation of Carbon and Nitrogen Utilization by CbrAB and NtrBC Two-Component Systems in Pseudomonas Aeruginosa," Journal of Bacteriology 189 (2007): 5413–5420, https://doi.org/10.1128/JB.00432-07.

227. Songye Chen et al., "Structural Diversity of Bacterial Flagellar Motors," The EMBO Journal 30, no. 14 (2011): 2972–2981.

228. Rowan Hooper, "High-Power Biological Wheels and Motors Imaged for First Time," New Scientist, March 14, 2016, https://web.archive.org/web/20160317131536/https://www.newscientist.com/article/2080642-high-power-biological-wheels-and-motors-imaged-for-first-time/.

229. Morgan Beeby et al., "Diverse High-Torque Bacterial Flagellar Motors Assemble Wider Stator Rings Using a Conserved Protein Scaffold," PNAS 113, no. 13 (March 29, 2016): E1917–1926, https://doi.org/10.1073/pnas.1518952113.

230. "Evolution Machined Biological Gears into Bacterial Motors," Genetic Engineering and Biotechnology News, March 16, 2016, https://web.archive.org/web/20200912183936/https://www.genengnews.com/news/evolution-machined-biological-gears-into-bacterial-motors/.

231. John Avise, Inside the Human Genome: A Case for Non-Intelligent Design (Oxford: Oxford University Press, 2010).

232. Renyi Liu and Howard Ochman, "Stepwise Formation of the Bacterial Flagellar System," PNAS 104, no. 17 (April 24, 2007): 7116–7121, https://doi.org/10.1073/pnas.0700266104.

233. Avise, Inside the Human Genome, 64.

234. Avise, Inside the Human Genome, 64.

Part Three (36–50)

1. Richard Dawkins, "Inferior Design," New York Times, July 1, 2007, https://web.archive.org/web/20190820225942/https://www.nytimes.com/2007/07/01/books/review/Dawkins-t.html.

2. Sean B. Carroll, "God as Genetic Engineer," Science 316, no. 5830 (June 8, 2007): 1427–1428, https://doi.org/10.1126/science.1145104. In quotations from this article, the original internal references have been deleted.

3. Yelena V. Budovskaya et al., "An Evolutionary Proteomics Approach Identifies Substrates of the cAMP-dependent Protein Kinase, PNAS 102, no. 39 (September 27, 2005): 13933–13938, https://doi.org/10.1073/pnas.0501046102.

4. Michael Behe and Sean Carroll, "Addressing Cumulative Selection," Science 318, no. 5848 (October 12, 2007): 196, https://doi.org/10.1126/science.318.5848.196.

5. Behe and Carroll, "Addressing Cumulative Selection."

6. Worachart Sirawaraporn et al., "Antifolate-Resistant Mutants of Plasmodium falciparum Dihydrofolate Reductase," PNAS 94, no. 4 (1997): 1124; Jirundon Yuvaniyama et al., "Insights into Antifolate Resistance from Malarial DHFR-TS Structures," Nature Structural and Molecular Biology 10 (2003): 357; Conner I. Sandefur et al., "Pyrimethamine-Resistant Dihydrofolate Reductase Enzymes of Plasmodium falciparum are not Enzymatically Compromised in vitro," Molecular and Biochemical Parasitology 154, no. 1 (2007): 1–5.

7. K. Hayton and X. Z. Su, "Genetic and Biochemical Aspects of Drug Resistance in Malaria Parasites," Current Drug Targets—Infectious Disorders 4, no. 1 (2004): 1–10, https://doi.org/10.2174/1568005043480925.

8. Shalini Nair et al., "A Selective Sweep Driven by Pyrimethamine Treatment in Southeast

Asian Malaria Parasites," *Molecular Biology and Evolution* 20, no. 9 (September 2003): 1526–1536, https://doi.org/10.1093/molbev/msg162.

9. Hayton and Su, "Genetic and Biochemical."

10. Kenneth R. Miller, "Falling over the Edge," *Nature* 447 (June 22, 2007): 1055–1056, https://doi.org/10.1038/4471055a.

11. Jerry Coyne, "The Great Mutator," *The New Republic* (June 18, 2007): 38–44, https://web.archive.org/web/20090509052125/https://pondside.uchicago.edu/ecol-evol/faculty/Coyne/pdf/Behe,%20New%20Republic.pdf.

12. "Professor Coyne Addresses Michael Behe's Reply to Coyne's Review of Behe's New Book," Talk Reason (website), June 30, 2007, https://web.archive.org/web/20070709021712/http://www.talkreason.org/articles/Coyne.cfm.

13. Jerry A. Coyne, "God in the Details," *Nature* 383 (1996): 227–228, https://doi.org/10.1038/383227a0.

14. Gert Korthof, "Either Design or Common Descent," Was Darwin Wrong? (website), July 22, 2007, https://web.archive.org/web/20150104204516/http://www.wasdarwinwrong.com/korthof86.htm.

15. Lília Perfeito et al., "Adaptive Mutations in Bacteria: High Rate and Small Effects," *Science* 317, no. 5839 (August 10, 2007): 813–815.

16. Abbie Smith, "ERV & HIV versus Behe. Behe Loses," Panda's Thumb (website), August 2, 2007, https://web.archive.org/web/20071124021156/https://pandasthumb.org/archives/2007/08/erv-hiv-versus.html.

17. Isolde Butler et al., "HIV Genetic Diversity: Biological and Public Health Consequences," *Current HIV Research* 5, no. 1 (February 2007): 23–45.

18. Ian Musgrave, "An Open Letter to Michael Behe," Panda's Thumb (website), October 22, 2007, https://web.archive.org/web/20071024090328/https://pandasthumb.org/archives/2007/10/an-open-letter-3.html.

19. Lisa M. Gomez et al., "Vpu-Mediated CD4 Down-regulation and Degradation is Conserved among Highly Divergent SIV(cpz) Strains," *Virology* 335, no. 1 (April 25, 2005): 46–60, https://doi.org/10.1016/j.virol.2005.01.049.

20. Gomez et al., "Vpu-mediated CD4."

21. D. R. Hout et al., "Vpu: A Multifunctional Protein that Enhances the Pathogenesis of Human Immunodeficiency Virus Type 1," *Current HIV Research* 2 (2004): 255–270.

22. Isolde Butler et al., "HIV Genetic Diversity: Biological and Public Health Consequences," *Current HIV Research* 5, no. 1 (February 2007): 23–45.

23. Butler et al., "HIV Genetic Diversity."

24. Hout et al., "Vpu."

25. Maria Eugenia Gonzalez and Luis Carrasco, "Viroporins," *FEBS Letters* 552 (July 17, 2003): 28-34, https://doi.org/10.1016/S0014-5793(03)00780-4; T. Mehnert et al., "Biophysical Characterization of Vpu from HIV-1 Suggests a Channel-pore Dualism," *Proteins* 70, no. 4 (March 2008):1488–1497, https://doi.org/10.1002/prot.21642.

26. Michael J. Behe, *The Edge of Evolution: The Search for the Limits of Darwinism* (New York: Free Press, 2007), 157–158.

27. Behe, *The Edge of Evolution*, 148–149.

28. Miguel A. Andrade, Sean I. O'Donoghue, and Burkhard Rost, "Adaptation of Protein Surfaces to Subcellular Location," *Journal of Molecular Biology* 276, no. 2 (February 20, 1998): 517–525, https://doi.org/10.1006/jmbi.1997.1498.

29. Donald Voet and Judith Voet, *Biochemistry*, 3rd ed. (New York: Wiley & Sons, 2004), 265.

30. Derek J. Smith et al., "Deriving Shape Space Parameters from Immunological Data," *Journal of Theoretical Biology* 189, no. 2 (November 21, 1997):141–150, https://doi.org/10.1006/jtbi.1997.0495.

31. A. B. Champion et al., "Immunological Comparison of Azurins of Known Amino Acid Sequence: Dependence of Cross-reactivity upon Sequence Resemblance," *Journal of Molecular Evolution* 5 (1975): 291–305, https://doi.org/10.1007/BF01732216; G. J. Nossal, "Tolerance and Ways to Break It," *Annals of the New York Academy of Sciences* 690 (July 31, 1993): 34–41.

32. Zhan-Yun Guo et al., "In Vitro Evolution of Amphioxus Insulin-like Peptide to Mamma-

lian Insulin," *Biochemistry* 41, no. 34 (2002): 10603–10607, https://doi.org/10.1021/bi020223x.

33. Rick Durrett and Deena Schmidt, "Waiting for Two Mutations: With Applications to Regulatory Sequence Evolution and the Limits of Darwinian Evolution," *Genetics* 180, no. 3 (November 1, 2008): 1501–1509, https://doi.org/10.1534/genetics.107.082610.

34. Michael J. Behe, "Waiting Longer for Two Mutations," *Genetics* 181 (February 1, 2009): 819–820, available at https://web.archive.org/web/20100406151902/https://www.discovery.org/scripts/viewDB/index.php?command=view&printerFriendly=true&id=9461.

35. Nicholas J. White, "Antimalarial Drug Resistance," *Journal of Clinical Investigation* 113 (April 15, 2004): 1084–1092.

36. Michael Lynch and John S. Conery, "The Evolutionary Fate and Consequences of Duplicate Genes," *Science* 290, no. 5494 (November 10, 2000): 1151–1155.

37. A few years after this paper appeared, Lehigh's math department invited Rick Durrett to give that year's Everett Pitcher Lecture—a high-profile event.

38. Dion J. Whitehead et al., "The Look-Ahead Effect of Phenotypic Mutations," *Biology Direct* 3 (2008): 18, https://doi.org/10.1186/1745-6150-3-18. To see Eugene V. Koonin's review, scroll down to "7.1 Reviewers Report 1," https://doi.org/10.1186/1745-6150-3-18.

39. Joakim Näsvall et al., "Real-time Evolution of New Genes by Innovation, Amplification, and Divergence," *Science* 338, no. 6105 (October 19, 2012): 384–387.

40. Elizabeth Pennisi, "Gene Duplication's Role in Evolution Gets Richer, More Complex," *Science* 338, no. 6105 (October 19, 2012): 316–317.

41. Avelina Espinosa, "Introduction: Protistan Biology, Horizontal Gene Transfer, and Common Descent Uncover Faulty Logic in Intelligent Design," *Journal of Eukaryotic Microbiology* 57, no. 1 (2010): 1–2, https://doi.org/10.1111/j.1550-7408.2009.00459.x.

42. Most of the second paper actually dealt with making a case for macroevolution based on a series of fossil foraminifera. Whether or not one wishes to classify the series as an example of "macroevolution," the accompanying drawings are not likely to persuade skeptics, and in any event are at the anatomical level rather than the molecular level, where, I have always argued, one must look to determine whether an evolutionary event is feasible by Darwinian processes. Guillermo Paz and Avelina Espinosa, "Integrating Horizontal Gene Transfer and Common Descent to Depict Evolution and Contrast it with 'Common Design,'" *Journal of Eukaryotic Microbiology* 57, no. 1 (2010): 11–18.

43. Mark A. Farmer and Andrea Habura, "Using Protistan Examples to Dispel the Myths of Intelligent Design," *Journal of Eukaryotic Microbiology* 57, no. 1 (2010): 3–10, https://doi.org/10.1111/j.1550-7408.2009.00460.x.

44. Viswanathan Lakshmanan et al., "A Critical Role for PfCRT K76T in *Plasmodium falciparum* Verapamil-Reversible Chloroquine Resistance," *The EMBO Journal* 24 (2005): 2294–2305, https://doi.org/10.1038/sj.emboj.7600681.

45. Hongying Jiang et al., "Genome-wide Compensatory Changes Accompany Drug-Selected Mutations in the *Plasmodium falciparum crt* Gene," *PLoS One* 3, no. 6 (June 25, 2008): e2484, https://doi.org/10.1371/journal.pone.0002484.

46. Stephen K. Gire et al., "Genomic Surveillance Elucidates Ebola Virus Origin and Transmission during the 2014 Outbreak," *Science* 345, no 6202 (September 12, 2014): 1369–1372.

PART FOUR (51–59)

1. Jeffrey E. Barrick et al., "Genome Evolution and Adaptation in a Long-term Experiment with *Escherichia coli*," *Nature* 461 (October 18, 2009): 1243–1247, https://doi.org/10.1038/nature08480.

2. Paul D. Sniegowski et al., "Evolution of High Mutation Rates in Experimental Populations of *E. coli*," *Nature* 387 (1997): 703–705, https://doi.org/10.1038/42701.

3. Robert J. Woods et al., "Second-Order Selection for Evolvability in a Large *Escherichia coli* Population," *Science* 331 (2011): 1433–1436; Jeffrey E. Barrick et al., "*Escherichia coli rpoB*

Mutants Have Increased Evolvability in Proportion to Their Fitness Defects," *Molecular Biology and Evolution* 27, no. 6 (June 2010): 1338–1347, https://doi.org/10.1093/molbev/msq024.

4. Michael J. Behe, *The Edge of Evolution: The Search for the Limits of Darwinism* (New York: Free Press, 2007).

5. Woods et al., "Second-Order Selection."

6. Barrick et al., *"Escherichia coli rpoB* Mutants."

7. Michael J. Behe, "Experimental Evolution, Loss-of-Function Mutations, and 'The First Rule of Adaptive Evolution,'" *Quarterly Review of Biology* 85, no. 4 (December 2010): 1–27, https://doi.org/10.1086/656902.

8. Woods et al., "Second-Order Selection."

9. Woods et al., "Second-Order Selection."

10. Woods et al., "Second-Order Selection."

11. Jamie T. Bridgham, Eric A. Ortlund, and Joseph W. Thornton, "An Epistatic Ratchet Constrains the Direction of Glucocorticoid Receptor Evolution," *Nature* 461 (2009): 515–519, https://doi.org/10.1038/nature08249. See my comments on Thornton's work in the next chapter.

12. Justin R. Meyer et al., "Repeatability and Contingency in the Evolution of a Key Innovation in Phage Lambda," *Science* 335 (January 27, 2012): 428–432.

13. Michael J. Behe, "Experimental Evolution, Loss-of-Function Mutations, and 'The First Rule of Adaptive Evolution,'" *Quarterly Review of Biology* 85, no. 4 (December 2010): 1–27, https://doi.org/10.1086/656902.

14. Behe, "Experimental Evolution."

15. Zachary D. Blount, Christina Z. Borland, and Richard E. Lenski, "Historical Contingency and the Evolution of a Key Innovation in an Experimental Population of *Escherichia coli*," *PNAS* 105, no. 23 (June 10, 2008): 7899–7906, https://doi.org/10.1073/pnas.0803151105.

16. Blount, Borland, and Lenski, "Historical Contingency."

17. B. G. Hall, "Chromosomal Mutation for Citrate Utilization by *Escherichia coli* K-12," *Journal of Bacteriology* 151 (1982): 269–273.

18. Klaas Martinus Pos, Peter Dimroth, and Michael Bott, "The *Escherichia coli* Citrate Carrier CitT: A Member of a Novel Eubacterial Transporter Family Related to the 2-oxo-glutarate/malate Translocator from Spinach Chloroplasts," *Journal of Bacteriology* 180 (1998): 4160–4165.

19. Blount, Borland, and Lenski, "Historical Contingency."

20. Michael J. Behe, *The Edge of Evolution: The Search for the Limits of Darwinism* (New York: Free Press, 2007).

21. Blount, Borland, and Lenski, "Historical Contingency."

22. H. Allen Orr, "A Minimum on the Mean Number of Steps Taken in Adaptive Walks," *Journal of Theoretical Biology* 220, no. 2 (January 21, 2003): 241–247, https://doi.org/10.1006/jtbi.2003.3161.

23. Michael J. Behe, "Experimental Evolution, Loss-of-Function Mutations, and 'The First Rule of Adaptive Evolution,'" *Quarterly Review of Biology* 85, no. 4 (December 2010): 1–27, https://doi.org/10.1086/656902.

24. Zachary D. Blount, Christina Z. Borland, and Richard E. Lenski, "Historical Contingency and the Evolution of a Key Innovation in an Experimental Population of *Escherichia coli*," *PNAS* 105, no. 23 (June 10, 2008): 7899–7906, https://doi.org/10.1073/pnas.0803151105.

25. Zachary D. Blount et al., "Genomic Analysis of a Key Innovation in an Experimental *Escherichia coli* Population," *Nature* 489 (2012): 513–518, https://doi.org/10.1038/nature11514.

26. Erik R. Zinser et al., "Bacterial Evolution through the Selective Loss of Beneficial Genes: Trade-Offs in Expression Involving Two Loci," *Genetics* 164, no. 4 (August 1, 2003): 1271–1277.

27. Normunds Licis and Jan van Duin, "Structural Constraints and Mutational Bias in the Evolutionary Restoration of a Severe Deletion in RNA Phage MS2," *Journal of Molecular Evolution* 63 (2006): 314–329, https://doi.org/10.1007/s00239-005-0012-8.

28. The original links to the Venema series at BioLogos were still broken as of July 28, 2020, but the first two posts in the series were available at the web archive Wayback Machine. Dennis Venema, "Behe, Lenski and the 'Edge' of Evolution, Part 1: Just the FCTs, Please," BioLogos (website), April 19, 2012, https://web.archive.org/web/20121027030436/

https://biologos.org/blog/behe-lenski-and-the-edge-of-evolution-part-1; and "Behe, Lenski and the "Edge" of Evolution, Part 2: Gaining a New Function," https://web.archive.org/web/20121026002024/http://biologos.org/blog/behe-lenski-and-the-edge-of-evolution-part-2.

29. R. C. Olsthoorn and Jan van Duin, "Evolutionary Reconstruction of a Hairpin Deleted from the Genome of an RNA Virus," *PNAS* 93, no. 22 (October 29, 1996): 12256–12261, https://doi.org/10.1073/pnas.93.22.12256.

30. Blount et al., "Genomic Analysis."

31. Elizabeth Pennisi, "The Man Who Bottled Evolution," *Science* 342, no. 6160 (November 15, 2013): 790–793.

32. Michael J. Behe, "Experimental Evolution, Loss-of-Function Mutations, and 'The First Rule of Adaptive Evolution,'" *Quarterly Review of Biology* 85, no. 4 (December 2010): 127, https://doi.org/10.1086/656902.

33. Michael J. Wiser, Noah Ribeck, and Richard E. Lenski, "Long-Term Dynamics of Adaptation in Asexual Populations," *Science* 342, no. 6164 (December 13, 2013): 1364–1367.

34. Sébastien Wielgoss et al., "Mutation Rate Dynamics in a Bacterial Population Reflect Tension Between Adaptation and Genetic Load," *PNAS* 110, no. 1 (January 2, 2013): 222–227, https://doi.org/10.1073/pnas.1219574110.

35. Dustin J. Van Hofwegen, Carolyn J. Hovde, and Scott A. Minnich, "Rapid Evolution of Citrate Utilization by *Escherichia coli* by Direct Selection Requires citT and dctA," *Journal of Bacteriology* 198, no. 7 (2016): 1022–34.

36. John R. Roth and Sophie Maisnier-Patin, "Reinterpreting Long-Term Evolution Experiments: Is Delayed Adaptation an Example of Historical Contingency or a Consequence of Intermittent Selection?," *Journal of Bacteriology* 198, no. 7 (2016): 1009–12.

37. Richard E. Lenski, "On the Evolution of Citrate Use," Telliamed Revisited (website), February 20, 2016, https://web.archive.org/web/20161012191010/https://telliamedrevisited.wordpress.com/2016/02/20/on-the-evolution-of-citrate-use/.

38. Richard Lenski, "Evolution in Action: A 50,000-Generation Salute to Charles Darwin," *Microbe* 6, no. 1 (January 2011): 30-33.

39. Elizabeth Pennisi, "The Man Who Bottled Evolution," *Science* 342, no. 6160 (November 15, 2013): 790–793.

40. Zachary D. Blount et al., "Genomic and Phenotypic Evolution of *Escherichia coli* in a Novel Citrate-Only Resource Environment," *eLife* 9 (2020): e55414, https://doi.org/10.1101/2020.01.22.915975.

41. Richard E. Lenski, "Convergence and Divergence in a Long-Term Experiment with Bacteria," *The American Naturalist* 190, no. S1 (August 2017): S57–S68.

42. Michael J. Behe, *Darwin Devolves: The New Science about DNA That Challenges Evolution* (New York: HarperOne, 2019), chap 7.

43. Zachary D. Blount, Christina Z. Borland, and Richard E. Lenski, "Historical Contingency and the Evolution of a Key Innovation in an Experimental Population of *Escherichia coli*," *PNAS* 105, no. 23 (June 10, 2008): 7899–7906, https://doi.org/10.1073/pnas.0803151105.

44. Zachary D. Blount et al., "Genomic Analysis of a Key Innovation in an Experimental *Escherichia coli* Population," *Nature* 489 (2012): 513–518, https://doi.org/10.1038/nature11514.

45. Elizabeth Pennisi, "The Man Who Bottled Evolution," *Science* 342, no. 6160 (November 15, 2013): 790–793.

46. Behe, *Darwin Devolves*, chap. 7.

47. Nathan H. Lents, S. Joshua Swamidass, and Richard E. Lenski, "The End of Evolution?," *Science* 363, no. 6427 (February 8, 2019): 590, https://doi.org/10.1126/science.aaw4056.

48. Behe, *Darwin Devolves*, chap. 7.

49. Blount et al., "Genomic and Phenotypic Evolution of *Escherichia coli*."

Part Five (60–70)

1. Jamie T. Bridgham, Sean M. Carroll, and Joseph W. Thornton, "Evolution of Hormone-Receptor Complexity by Molecular Exploitation," *Science* 312, no. 5770 (April 7, 2006): 97–101, https://doi.org/10.1126/science.1123348.

2. Christoph Adami, "Reducible Complexity," *Science* 312, no. 5770 (April 7, 2006): 61–63.

3. Jamie T. Bridgham, Eric A. Ortlund, and Joseph W. Thornton, "An Epistatic Ratchet Constrains the Direction of Glucocorticoid Receptor Evolution," *Nature* 461 (2009): 515–519, https://doi.org/10.1038/nature08249.

4. Jamie T. Bridgham, Sean M. Carroll, and Joseph W. Thornton, "Evolution of Hormone-Receptor Complexity by Molecular Exploitation," *Science* 312, no. 5770 (April 7, 2006): 97–101, https://doi.org/10.1126/science.1123348.

5. Michael J. Behe, *The Edge of Evolution: The Search for the Limits of Darwinism* (New York: Free Press, 2007).

6. Bridgham, Ortlund, and Thornton, "An Epistatic Ratchet."

7. Bridgham, Ortlund, and Thornton, "An Epistatic Ratchet."

8. Jamie T. Bridgham, Eric A. Ortlund, and Joseph W. Thornton, "An Epistatic Ratchet Constrains the Direction of Glucocorticoid Receptor Evolution," *Nature* 461 (2009): 515–519, https://doi.org/10.1038/nature08249.

9. Jamie T. Bridgham, Sean M. Carroll, and Joseph W. Thornton, "Evolution of Hormone-Receptor Complexity by Molecular Exploitation," *Science* 312, no. 5770 (April 7, 2006): 97–101, https://doi.org/10.1126/science.1123348.

10. Bridgham, Ortlund, and Thornton, "An Epistatic Ratchet."

11. Jamie T. Bridgham, Eric A. Ortlund, and Joseph W. Thornton, "An Epistatic Ratchet Constrains the Direction of Glucocorticoid Receptor Evolution," *Nature* 461 (2009): 515–519, https://doi.org/10.1038/nature08249.

12. Stephen Jay Gould, "Dollo on Dollo's Law: Irreversibility and the Status of Evolutionary Laws," *Journal of the History of Biology* 3 (1970): 189–212, https://doi.org/10.1007/BF00137351.

13. Bridgham, Ortlund, and Thornton, "An Epistatic Ratchet."

14. Carl Zimmer, "The Blind Locksmith Continued: An Update from Joe Thornton," *Discover*, October 15, 2009, https://web.archive.org/web/20200303144648if_/https://www.discovermagazine.com/the-sciences/the-blind-locksmith-continued-an-update-from-joe-thornton.

15. Carl Zimmer, "The Blind Locksmith Continued: An Update from Joe Thornton," *Discover*, October 15, 2009, https://web.archive.org/web/20200303144648if_/https://www.discovermagazine.com/the-sciences/the-blind-locksmith-continued-an-update-from-joe-thornton.

16. Jamie T. Bridgham, Eric A. Ortlund, and Joseph W. Thornton, "An Epistatic Ratchet Constrains the Direction of Glucocorticoid Receptor Evolution," *Nature* 461 (2009): 515–519, https://doi.org/10.1038/nature08249.

17. Jamie T. Bridgham, Sean M. Carroll, and Joseph W. Thornton, "Evolution of Hormone-Receptor Complexity by Molecular Exploitation," *Science* 312, no. 5770 (April 7, 2006): 97–101, https://doi.org/10.1126/science.1123348.

18. Carl Zimmer, "The Blind Locksmith Continued: An Update from Joe Thornton," *Discover*, October 15, 2009, https://web.archive.org/web/20200303144648if_/https://www.discovermagazine.com/the-sciences/the-blind-locksmith-continued-an-update-from-joe-thornton.

19. Susumu Ohno, *Evolution by Gene Duplication* (Berlin, Germany: Springer-Verlag, 1970).

20. Michael J. Behe and David W. Snoke, "Simulating Evolution by Gene Duplication of Protein Features that Require Multiple Amino Acid Residues," *Protein Science* 13 (January 1, 2004): 2651–2664, https://doi.org/10.1110/ps.04802904.

21. Jamie T. Bridgham, Eric A. Ortlund, and Joseph W. Thornton, "An Epistatic Ratchet Constrains the Direction of Glucocorticoid Receptor Evolution," *Nature* 461 (2009): 515–519, https://doi.org/10.1038/nature08249.

22. Jamie T. Bridgham, Sean M. Carroll, and Joseph W. Thornton, "Evolution of Hormone-Receptor Complexity by Molecular Exploitation," *Science* 312, no. 5770 (April 7, 2006): 97–101, https://doi.org/10.1126/science.1123348.

23. Bridgham, Ortlund, and Thornton, "An Epistatic Ratchet."

24. Carl Zimmer, "The Blind Locksmith Continued: An Update from Joe Thornton," *Discover*, October 15, 2009, https://web.archive.org/web/20200303144648if_/https://www.discovermagazine.com/the-sciences/the-blind-locksmith-continued-an-update-from-joe-thornton.

25. Sean Michael Carroll, Eric A. Ortlund, and Joseph W. Thornton, "Mechanisms for the Evolution of a Derived Function in the Ancestral Glucocorticoid Receptor," *PLOS Genetics* 7, no. 6 (June 2011): e1002117.

26. Jamie T. Bridgham, Eric A. Ortlund, and Joseph W. Thornton, "An Epistatic Ratchet Constrains the Direction of Glucocorticoid Receptor Evolution," *Nature* 461 (2009): 515–519, https://doi.org/10.1038/nature08249.

27. Carroll, Ortlund, and Thornton, "Mechanisms."

28. Michael J. Behe, "Experimental Evolution, Loss-of-Function Mutations, and 'The First Rule of Adaptive Evolution,'" *Quarterly Review of Biology* 85, no. 4 (December 2010): 1–27, https://doi.org/10.1086/656902.

29. Gregory C. Finnigan et al., "Evolution of Increased Complexity in a Molecular Machine," *Nature* 481 (January 9, 2012): 360–364, https://doi.org/10.1038/nature10724.

30. Allan Force et al., "Preservation of Duplicate Genes by Complementary, Degenerative Mutations," *Genetics* 151, no. 4 (April 1, 1999): 1531–1545.

31. A. L. Hughes, "Evolution of Adaptive Phenotypic Traits Without Positive Darwinian Selection," *Heredity* 108 (2012): 347–353, https://doi.org/10.1038/hdy.2011.97.

32. Michael W. Gray et al., "Irremediable Complexity?," *Science* 330, no. 6006 (November 12, 2010): 920-921, https://doi.org/10.1126/science.1198594; Julius Lukeš et al., "How a Neutral Evolutionary Ratchet Can Build Cellular Complexity," *IUBMB Life* 63 (2011): 528–537, https://doi.org/10.1002/iub.489.

33. Michael J. Harms and Joseph W. Thornton, "Historical Contingency and Its Biophysical Basis in Glucocorticoid Receptor Evolution," *Nature* 512 (2014): 203–207, https://doi.org/10.1038/nature13410.

34. Ann Gauger, "Playing Roulette with Life," Evolution News and Views, Discovery Institute, June 20, 2014, https://web.archive.org/web/20200912215141/https://evolutionnews.org/2014/06/playing_roulett/.

Part Six (71–82)

1. Robert L. Summers et al., "Diverse Mutational Pathways Converge on Saturable Chloroquine Transport Via the Malaria Parasite's Chloroquine Resistance Transporter," *PNAS* 111, no. 17 (2014): E1759–E1767, https://doi.org/10.1073/pnas.1322965111.

2. Nicholas J. White, "Antimalarial Drug Resistance," *The Journal of Clinical Investigation* 113 (April 15, 2004):1084–1092, https://doi.org/10.1172/JCI21682.

3. Robert L. Summers et al., "Diverse Mutational Pathways Converge on Saturable Chloroquine Transport Via the Malaria Parasite's Chloroquine Resistance Transporter," *PNAS* 111, no. 17 (2014): E1759–E1767, https://doi.org/10.1073/pnas.1322965111.

4. John Maynard Smith, "Natural Selection and the Concept of a Protein Space," *Nature* 225 (February 7, 1970): 563–564, https://doi.org/10.1038/225563a0.

5. H. Allen Orr, "A Minimum on the Mean Number of Steps Taken in Adaptive Walks," *Journal of Theoretical Biology* 220, no. 2 (January 21, 2003): 241–247, https://doi.org/10.1006/jtbi.2003.3161.

6. Nicholas J. White, "Antimalarial Drug Resistance," *The Journal of Clinical Investigation* 113 (April 15, 2004): 1084–1092, https://doi.org/10.1172/JCI21682.

7. K. Hayton and X. Z. Su, "Genetic and Biochemical Aspects of Drug Resistance in Malaria Parasites," *Current Drug Targets—Infectious Disorders* 4, no. 1 (2004): 1–10, https://doi.org/10.2174/1568005043480925.

8. Sean B. Carroll, "God as Genetic Engineer," *Science* 316, no. 5830 (June 8, 2007): 1427–1428, https://doi.org/10.1126/science.1145104.

9. Kenneth R. Miller, "Falling over the Edge," *Nature* 447 (June 22, 2007): 1055–1056, https://doi.org/10.1038/4471055a.

10. Nicholas J. Matzke, "The Edge of Creationism," *Trends in Ecology and Evolution*

22, no. 11 (2007): 566–567, https://doi.org/10.1016/j.tree.2007.09.004.

11. Richard Dawkins, "Inferior Design," *New York Times*, July 1, 2007, https://web.archive.org/web/20190820225942/https://www.nytimes.com/2007/07/01/books/review/Dawkins-t.html.

12. Robert L. Summers et al., "Diverse Mutational Pathways Converge on Saturable Chloroquine Transport via the Malaria Parasite's Chloroquine Resistance Transporter," *PNAS* 111, no. 17 (2014): E1759–E1767, https://doi.org/10.1073/pnas.1322965111.

13. Jason Flannick et al., "Loss-of-Function Mutations in *SLC30A8* Protect Against Type 2 Diabetes," *Nature Genetics* 46 (2014): 357–363, https://doi.org/10.1038/ng.2915.

14. Jacy Crosby et al., "Loss-of-Function Mutations in *APOC3*, Triglycerides, and Coronary Disease," *New England Journal of Medicine* 371 (July 3, 2014): 22–31.

15. M. Promerová et al., "Worldwide Frequency Distribution of the 'Gait Keeper' Mutation in the DMRT3 Gene," *Animal Genetics* 45 (January 21, 2014): 274–282, https://doi.org/10.1111/age.12120.

16. Kenneth M. Olsen, Nicholas J. Kooyers, and Linda A. Small, "Adaptive Gains through Repeated Gene Loss: Parallel Evolution of Cyanogenesis Polymorphisms in the Genus *Trifolium* (Fabaceae)," *Philosophical Transactions of the Royal Society of London B: Biological Sciences* 369 (August 5, 2014), https://doi.org/10.1098/rstb.2013.0347.

17. Alison K. Hottes et al., "Bacterial Adaptation through Loss of Function," *PLOS Genetics* 9, no. 7 (July 2013): e1003617, https://doi.org/10.1371/journal.pgen.1003617.

18. P. Z. Myers and Ken Miller, "Quote-Mined by Casey Luskin," Pharyngula, July 17, 2014, https://web.archive.org/web/20140720042740/https://freethoughtblogs.com/pharyngula/2014/07/17/quote-mined-by-casey-luskin/.

19. Nicholas J. White, "Antimalarial Drug Resistance," *The Journal of Clinical Investigation* 113 (April 15, 2004):1084–1092, https://doi.org/10.1172/JCI21682.

20. John C. Wootton et al., "Genetic Diversity and Chloroquine Selective Sweeps in *Plasmodium falciparum*," *Nature* 418 (July 18,

2002): 320–323, https://doi.org/10.1038/nature00813.

21. White, "Antimalarial Drug Resistance."

22. S. Paget-McNicol and A. Saul, "Mutation Rates in the Dihydrofolate Reductase Gene of *Plasmodium falciparum*," *Parasitology* 122, no. 5 (May 2001): 497–505, https://doi.org/10.1017/S0031182001007739.

23. Nicholas White, "Antimalarial Drug Resistance and Combination Chemotherapy," *Philosophical Transactions of the Royal Society of London B: Biological Sciences* 354 (April 29, 1999): 739–749, https://doi.org/10.1098/rstb.1999.0426.

24. Myers, "Quote-Mined by Casey Luskin!"

25. Viswanathan Lakshmanan et al., "A Critical Role for PfCRT K76T in *Plasmodium falciparum* Verapamil-Reversible Chloroquine Resistance," *The EMBO Journal* 24 (2005): 2294–2305, https://doi.org/10.1038/sj.emboj.7600681.

26. K. Hayton and X. Z. Su, "Genetic and Biochemical Aspects of Drug Resistance in Malaria Parasites," *Current Drug Targets—Infectious Disorders* 4, no. 1 (2004):1–10, https://doi.org/10.2174/1568005043480925; Shalini Nair et al., "A Selective Sweep Driven by Pyrimethamine Treatment in Southeast Asian Malaria Parasites," *Molecular Biology Evolution* 20, no. 9 (September 2003): 1526–1536, https://doi.org/10.1093/molbev/msg162.

27. Robert L. Summers et al., "Diverse Mutational Pathways Converge on Saturable Chloroquine Transport Via the Malaria Parasite's Chloroquine Resistance Transporter," *PNAS* 111, no. 17 (2014): E1759–E1767, https://doi.org/10.1073/pnas.1322965111.

28. Pradipsinh K. Rathod et al., "Variations in Frequencies of Drug Resistance in *Plasmodium falciparum*," *PNAS* 94, no. 17 (August 19, 1997): 9389–9393, https://doi.org/10.1073/pnas.94.17.9389.

29. Robert L. Summers et al., "Diverse Mutational Pathways Converge on Saturable Chloroquine Transport Via the Malaria Parasite's Chloroquine Resistance Transporter," *PNAS* 111, no. 17 (2014): E1759–E1767, https://doi.org/10.1073/pnas.1322965111.

30. Larry Moran, "Taking the Behe Challenge," Sandwalk (website), August 1, 2014, https://

web.archive.org/web/20140815141236/
https://sandwalk.blogspot.com/2014/08/
taking-behe-challenge.html.

31. Nicholas J. White, "Antimalarial Drug Resistance," *The Journal of Clinical Investigation* 113 (April 15, 2004):1084–1092, https://doi.org/10.1172/JCI21682.

32. Larry Moran, "Why I'm Not a Darwinist," Sandwalk (website), November 17, 2006, https://web.archive.org/web/20070422122619/https://sandwalk.blogspot.com/2006/11/why-im-not-darwinist.html.

33. Viswanathan Lakshmanan et al., "A Critical Role for PfCRT K76T in *Plasmodium falciparum* Verapamil-Reversible Chloroquine Resistance," *The EMBO Journal* 24 (2005): 2294–2305, https://doi.org/10.1038/sj.emboj.7600681.

34. Larry Moran, "Taking the Behe Challenge," Sandwalk (website), August 1, 2014, https://web.archive.org/web/20140815141236/https://sandwalk.blogspot.com/2014/08/taking-behe-challenge.html.

35. Nicholas J. White, "Antimalarial Drug Resistance," *The Journal of Clinical Investigation* 113 (April 15, 2004):1084–1092, https://doi.org/10.1172/JCI21682.

36. Robert L. Summers et al., "Diverse Mutational Pathways Converge on Saturable Chloroquine Transport Via the Malaria Parasite's Chloroquine Resistance Transporter," *PNAS* 111, no. 17 (2014): E1759–E1767, https://doi.org/10.1073/pnas.1322965111.

37. Michael J. Behe, "Experimental Evolution, Loss-of-Function Mutations, and 'The First Rule of Adaptive Evolution,'" *Quarterly Review of Biology* 85, no. 4 (December 2010): 1–27, https://doi.org/10.1086/656902.

38. Larry Moran, "CCC's and the Edge of Evolution," Sandwalk (website), August 15, 2004, https://web.archive.org/web/20140818213539/https://sandwalk.blogspot.com/2014/08/cccs-and-edge-of-evolution.html.

39. Robert L. Summers et al., "Diverse Mutational Pathways Converge on Saturable Chloroquine Transport Via the Malaria Parasite's Chloroquine Resistance Transporter," *PNAS* 111, no. 17 (2014): E1759–E1767, https://doi.org/10.1073/pnas.1322965111.

40. Nicholas J. White, "Antimalarial Drug Resistance," *The Journal of Clinical Investigation* 113 (April 15, 2004):1084–1092, https://doi.org/10.1172/JCI21682.

41. Michael J. Behe, "How Many Ways Are There to Win at Sandwalk?," Evolution News and Views, Discovery Institute, August 15, 2014, https://web.archive.org/web/20200914134555/https://evolutionnews.org/2014/08/how_many_ways_a/ (collected as essay 76 above).

42. Robert L. Summers et al., "Diverse Mutational Pathways Converge on Saturable Chloroquine Transport Via the Malaria Parasite's Chloroquine Resistance Transporter," *PNAS* 111, no. 17 (2014): E1759–E1767, https://doi.org/10.1073/pnas.1322965111.

43. Larry Moran, "Understanding Michael Behe," Sandwalk (website), August 22, 2014, https://web.archive.org/web/20180710052155/http://sandwalk.blogspot.com/2014/08/understanding-michael-behe.html.

44. K. Hayton and X. Z. Su, "Genetic and Biochemical Aspects of Drug Resistance in Malaria Parasites," *Current Drug Targets—Infectious Disorders* 4, no. 1 (2004):1–10, https://doi.org/10.2174/1568005043480925.

45. Nicholas J. White, "Antimalarial Drug Resistance," *The Journal of Clinical Investigation* 113 (April 15, 2004):1084-1092, https://doi.org/10.1172/JCI21682.

46. I. M. Hastings, P. G. Bray, and S. A. Ward, "A Requiem for Chloroquine," *Science* 298, no. 5591 (October 4, 2002): 74–75, https://doi.org/10.1126/science.1077573.

47. Viswanathan Lakshmanan et al., "A Critical Role for PfCRT K76T in *Plasmodium falciparum* Verapamil-Reversible Chloroquine Resistance," *The EMBO Journal* 24 (2005): 2294–2305, https://doi.org/10.1038/sj.emboj.7600681.

48. Kenneth Miller, "Edging Towards Irrelevance," MillerandLevine.com, https://web.archive.org/web/20150430020605/http://www.millerandlevine.com/evolution/behe-2014/Behe-2014.pdf.

49. Michael Behe, "An Open Letter to Kenneth Miller and P. Z. Myers," Evolution News and Views, Discovery Institute, July 21, 2014, https://web.archive.org/

web/20190405190734/https://evolution-news.org/2014/07/show_me_the_num/ (also collected as essay 74 above).

50. Robert L. Summers et al., "Diverse Mutational Pathways Converge on Saturable Chloroquine Transport Via the Malaria Parasite's Chloroquine Resistance Transporter," *PNAS* 111, no. 17 (2014): E1759–E1767, https://doi.org/10.1073/pnas.1322965111.

51. Miller, "Edging Towards Irrelevance," 3–6.

52. Miller, "Edging Towards Irrelevance," 4.

53. Miller, "Edging Towards Irrelevance," 3.

54. Robert L. Summers et al., "Diverse Mutational Pathways Converge on Saturable Chloroquine Transport Via the Malaria Parasite's Chloroquine Resistance Transporter," *PNAS* 111, no. 17 (2014): E1759–E1767, https://doi.org/10.1073/pnas.1322965111.

55. Kenneth Miller, "Edging Towards Irrelevance," MillerandLevine.com, https://web.archive.org/web/20150430020605/http://www.millerandlevine.com/evolution/behe-2014/Behe-2014.pdf, 4.

56. Sumiti Vinayak et al., "Prevalence of the K76T Mutation in the *pfcrt* gene of *Plasmodium falciparum* among Chloroquine Responders in India," *Acta Tropica* 87 (2003): 287–293, https://doi.org/10.1016/S0001-706X(03)00021-4.

57. Umar Farooqand and R. C. Mahajan, "Drug Resistance in Malaria," *Journal of Vector Borne Diseases* 41 (September and December 2004): 45–53.

58. James G. Kublin et al., "Reemergence of Chloroquine-Sensitive *Plasmodium falciparum* Malaria after Cessation of Chloroquine use in Malawi," *Journal of Infectious Diseases* 187, no. 12 (June 15, 2003): 1870–1875, https://doi.org/10.1086/375419.

59. Xinhua Wang et al., "Decreased Prevalence of the *Plasmodium falciparum* Chloroquine Resistance Transporter 76T Marker Associated with Cessation of Chloroquine Use against *P. falciparum* Malaria in Hainan, People's Republic of China," *The American Journal of Tropical Medicine and Hygiene* 72, no. 4 (April 1, 2005): 410–414, https://doi.org/10.4269/ajtmh.2005.72.410.

60. Miller, "Edging Towards Irrelevance," 4.

61. Viswanathan Lakshmanan et al., "A Critical Role for PfCRT K76T in *Plasmodium falciparum* Verapamil-Reversible Chloroquine Resistance," *The EMBO Journal* 24 (2005): 2294–2305, https://doi.org/10.1038/sj.emboj.7600681.

62. Kenneth Miller, "Edging Towards Irrelevance," MillerandLevine.com, https://web.archive.org/web/20150430020605/http://www.millerandlevine.com/evolution/behe-2014/Behe-2014.pdf, 4. (emphasis in original)

63. Miller, "Edging Towards Irrelevance," 4.

64. Robert L. Summers et al., "Diverse Mutational Pathways Converge on Saturable Chloroquine Transport Via the Malaria Parasite's Chloroquine Resistance Transporter," *PNAS* 111, no. 17 (2014): E1759–E1767, https://doi.org/10.1073/pnas.1322965111.

65. Kenneth Miller, "Edging Towards Irrelevance," MillerandLevine.com, https://web.archive.org/web/20150430020605/http://www.millerandlevine.com/evolution/behe-2014/Behe-2014.pdf, 6–8. The phrase "Fabricating the Odds" occurs as a hyperlinked title at the bottom of the first page of the multi-page version of that article. That first page is available at https://web.archive.org/web/20150102211159/http://www.millerandlevine.com/evolution/behe-2014/Behe-1.html.

66. Miller, "Edging Towards Irrelevance," 7.

67. Miller, "Edging Towards Irrelevance," 6.

68. Nicholas White, "Antimalarial Drug Resistance and Combination Chemotherapy," *Philosophical Transactions of the Royal Society of London B: Biological Sciences* 354 (April 29, 1999): 739–749, https://doi.org/10.1098/rstb.1999.0426.

69. Robert L. Summers et al., "Diverse Mutational Pathways Converge on Saturable Chloroquine Transport Via the Malaria Parasite's Chloroquine Resistance Transporter," *PNAS* 111, no. 17 (2014): E1759–E1767, https://doi.org/10.1073/pnas.1322965111.

70. M. Jucker and L. C. Walker, "Self-Propagation of Pathogenic Protein Aggregates in Neurodegenerative Diseases," *Nature* 501, no. 7456 (September 5, 2013): 45–51, https://doi.org/10.1038/nature12481.

PART SEVEN (83–99)

1. Charles Darwin, *The Origin of Species* [1859] (New York: Bantam Books, 1999), 84.
2. Rebecca Stott, *Darwin and the Barnacle* (New York: W. W. Norton, 2003), 213.
3. Tables 1 through 5 can be viewed online in Michael J. Behe, "Experimental Evolution, Loss-of-Function Mutations, and 'The First Rule of Adaptive Evolution,'" *Quarterly Review of Biology* 85, no. 4 (December 2010): 1–27, https://doi.org/10.1086/656902.
4. E. C. C. Lin and T. T. Wu, "Functional Divergence of the l-fucose System in Mutants of *Escherichia coli*," in *Microorganisms as Model Systems for Studying Evolution*, ed. R. P. Mortlock (New York: Plenum Press, 1984), 135–164.
5. Joel L. Sachs and James J. Bull, "Experimental Evolution of Conflict Mediation between Genomes," *PNAS* 102, no. 2 (January 11, 2005): 390–395, https://doi.org/10.1073/pnas.0405738102.
6. Robert T. Salier et al., "Sequence Determinants of Folding and Stability for the P22 Arc Repressor Dimer," *FASEB Journal* 10 (1996): 42–48, https://doi.org/10.1096/fasebj.10.1.8566546; J. U. Bowie et al., "Deciphering the Message in Protein Sequences: Tolerance to Amino Acid Substitutions," *Science* 247, no. 4948 (March 16, 1990): 1306–1310, https://doi.org/10.1126/science.2315699; J. U. Bowie and R. T. Sauer, "Identifying Determinants of Folding and Activity for a Protein of Unknown Structure," *PNAS* 86, no. 7 (April 1, 1989): 2152–2156, https://doi.org/10.1073/pnas.86.7.2152; J. F. Reidhaar-Olson and R. T. Sauer, "Combinatorial Cassette Mutagenesis as a Probe of the Informational Content of Protein Sequences," *Science* 241, no. 4861 (July 1, 1988): 53–57, https://doi.org/10.1126/science.3388019; John F. Reidhaar-Olson and Robert T. Sauer, "Functionally Acceptable Substitutions in Two α–helical Regions of Lambda Repressor," *Proteins* 7 (1990): 306–316, https://doi.org/10.1002/prot.340070403; Wendell A. Lim and Robert T. Sauer, "Alternative Packing Arrangements in the Hydrophobic Core of Lambda Repressor," *Nature* 339 (1989): 31–36, https://doi.org/10.1038/339031a0;

D. D. Axe, N. W. Foster, and A. R. Fersht, "Active Barnase Variants with Completely Random Hydrophobic Cores," *PNAS* 93, no. 11 (May 28, 1996): 5590–5594, https://doi.org/10.1073/pnas.93.11.5590; Wanzhi Huang et al., "Amino Acid Sequence Determinants of β-lactamase Structure and Activity," *Journal of Molecular Biology* 258, no. 4 (May 17, 1996): 688–703, https://doi.org/10.1006/jmbi.1996.0279; Jörg Suckow et al., "Genetic Studies of the Lac Repressor XV: 4000 Single Amino Acid Substitutions and Analysis of the Resulting Phenotypes on the Basis of the Protein Structure," *Journal of Molecular Biology* 261, no. 4 (August 30, 1996): 509–523, https://doi.org/10.1006/jmbi.1996.0479.
7. J. J. Bull et al., "Experimental Evolution Yields Hundreds of Mutations in a Functional Viral Genome," *Journal of Molecular Evolution* 57 (2003): 241–248, https://doi.org/10.1007/s00239-003-2470-1.
8. Lília Perfeito et al., "Adaptive Mutations in Bacteria: High Rate and Small Effects," *Science* 317, no. 5839 (August 10, 2007): 813–815.
9. Luigi Luca Cavalli-Sforza, Paolo Menozzi, and Alberto Piazza, *The History and Geography of Human Genes* (Princeton, NJ: Princeton University Press, 1994).
10. Brendan W. Wren, "The *Yersiniae*—A Model Genus to Study the Rapid Evolution of Bacterial Pathogens," *Nature Reviews Microbiology* 1 (2003): 55–64, https://doi.org/10.1038/nrmicro730.
11. E. Carniel, "Evolution of Pathogenic *Yersinia*: Some Lights in the Dark," in *The Genus Yersinia: Entering the Functional Genomic Era*, eds. M. Skurnik et al. (New York: Kluwer Academic/Plenum, 2003), 3–12.
12. Carniel, "Evolution of Pathogenic *Yersinia*"; Patrick S. G. Chain et al., "Complete Genome Sequence of *Yersinia pestis* Strains Antiqua and Nepal 516: Evidence of Gene Reduction in an Emerging Pathogen," *Journal of Bacteriology* 188 (2006): 4453–4463.
13. Carniel, "Evolution of Pathogenic *Yersinia*."
14. T. Ferenci, "The Spread of a Beneficial Mutation in Experimental Bacterial Populations: The Influence of the Environment and Genotype on the Fixation of *rpoS* Mutations,"

Heredity 100 (2008): 446–452, https://doi.
org/10.1038/sj.hdy.6801077.

15. Gregory I. Lang, Andrew W. Murray,
and David Botstein, "The Cost of Gene
Expression Underlies a Fitness Trade-off
in Yeast," *PNAS* 106, no. 14 (April 7, 2009):
5755–5760, https://doi.org/10.1073/
pnas.0901620106.

16. Jonathon R. Stone and Gregory A. Wray,
"Rapid Evolution of *cis*-regulatory Sequences
via Local Point Mutations," *Molecular Biology and Evolution* 18, no. 9 (September 2001):
1764–1770, https://doi.org/10.1093/oxford-
journals.molbev.a003964.

17. Richard E. Lenski, "Phenotypic and Genomic Evolution During a 20,000-Genera-
tion Experiment with the Bacterium *Escherichia coli*," *Plant Breeding Reviews* 24 (2004):
225–265.

18. Stone and Wray, "Rapid Evolution."

19. Jerry A. Coyne, "Behe's New Paper," Why
Evolution Is True (website), 2010, https://
web.archive.org/web/20200914143231/
https://whyevolutionistrue.
com/2010/12/12/behes-new-paper/.

20. Michael J. Behe, "Experimental Evolution,
Loss-of-Function Mutations, and 'The First
Rule of Adaptive Evolution,'" *Quarterly
Review of Biology* 85, no. 4 (December 2010):
1–27, https://doi.org/10.1086/656902.

21. D. Rokyta et al., "Experimental Genomic
Evolution: Extensive Compensation for Loss
of DNA Ligase Activity in a Virus," *Molecular Biology and Evolution* 19, no. 3 (March
2002): 230–238, https://doi.org/10.1093/
oxfordjournals.molbev.a004076.

22. Jerry A. Coyne, "An Experimental Evolu-
tionist Replies to Behe," Why Evolution Is
True (website), 2010, https://web.archive.
org/web/20200914143711/https://whyevo-
lutionistrue.com/2010/12/20/an-experimen-
tal-evolutionist-replies-to-behe/.

23. Michael J. Behe, "Experimental Evolution,
Loss-of-Function Mutations, and 'The First
Rule of Adaptive Evolution,'" *Quarterly
Review of Biology* 85, no. 4 (December 2010):
1–27, https://doi.org/10.1086/656902.

24. D. Rokyta et al., "Experimental Genomic
Evolution: Extensive Compensation for Loss
of DNA Ligase Activity in a Virus," *Molecular Biology and Evolution* 19, no. 3 (March

2002): 230–238, https://doi.org/10.1093/
oxfordjournals.molbev.a004076.

25. Ann K. Gauger et al., "Reductive Evolution Can Prevent Populations from Taking
Simple Adaptive Paths to High Fitness,"
BIO-Complexity 2010, no. 2 (April 30, 2010):
1–9, https://www.bio-complexity.org/ojs/in-
dex.php/main/article/view/BIO-C.2010.2/
BIO-C.2010.2.

26. Michael J. Behe, "Experimental Evolution,
Loss-of-Function Mutations, and 'The First
Rule of Adaptive Evolution,'" *Quarterly
Review of Biology* 85, no. 4 (December 2010):
1–27, https://doi.org/10.1086/656902.

27. Jerry A. Coyne, "New Genes Arise
Quickly," Why Evolution Is True (website), 2010, https://web.archive.org/
web/20200914144218/https://whyevolu-
tionistrue.com/2010/12/21/new-genes-arise-
quickly/.

28. Sidi Chen, Yong E. Zhang, and Manyuan
Long, "New Genes in *Drosophila* Quickly
Become Essential," *Science* 330, no. 6011
(December 17, 2010): 1682–1685.

29. William C. Ratcliff et al., "Experimental
Evolution of Multicellularity," *PNAS* 109, no.
5 (January 31, 2012): 1595–1600, https://doi.
org/10.1073/pnas.1115323109.

30. Kodjo Ayi et al., "Pyruvate Kinase Deficiency and Malaria," *New England Journal of
Medicine* 358 (April 24, 2008): 1805–1810.

31. Sarah A. Tishkoff et al., "Haplotype Diver-
sity and Linkage Disequilibrium at Human
G6PD: Recent Origin of Alleles that Confer
Malarial Resistance," *Science* 293, no. 5529
(July 20, 2001): 455–462.

32. Juliane Kaminski et al., "Evolution of Facial
Muscle Anatomy in Dogs," *PNAS* 116, no.
29 (July 16, 2019):14677–14681, https://doi.
org/10.1073/pnas.1820653116.

33. Dana S. Mosher et al., "A Mutation in the
Myostatin Gene Increases Muscle Mass and
Enhances Racing Performance in Hetero-
zygote Dogs," *PLOS Genetics* 3, no. 5 (May
25, 2007), https://doi.org/10.1371/journal.
pgen.0030079.

34. Nathan H. Lents, S. Joshua Swamidass,
and Richard E. Lenski, "The End of Evolution?," *Science* 363, no. 6427 (February 8,
2019): 590, https://doi.org/10.1126/science.
aaw4056.

35. Nathan H. Lents, S. Joshua Swamidass, and Richard E. Lenski, "The End of Evolution?," *Science* 363, no. 6427 (February 8, 2019): 590, https://doi.org/10.1126/science.aaw4056.

36. Michael J. Behe, *Darwin Devolves: The New Science about DNA That Challenges Evolution* (New York: HarperOne, 2019), 80–81.

37. Behe, *Darwin Devolves*, 225.

38. Behe, *Darwin Devolves*, 214.

39. Michael J. Behe, *Darwin's Black Box: The Biochemical Challenge to Evolution* (New York: The Free Press, 1996), 206–207.

40. Behe, *Darwin Devolves*, 214–215.

41. Kanwaljit S. Dulai et al., "The Evolution of Trichromatic Color Vision by Opsin Gene Duplication in New World and Old World Primates," *Genome Research* 9 (1999): 629.

42. Joakim Näsvall et al., "Real-time Evolution of New Genes by Innovation, Amplification, and Divergence," *Science* 338, no. 6105 (October 19, 2012): 384–387.

43. Michael Behe, "To Traverse a Maze, It Helps to Have a Mind," Evolution News and Views, Discovery Institute, November 7, 2012, https://web.archive.org/web/20190411163159/https://evolutionnews.org/2012/11/to_traverse_a_m_1/.

44. Kenneth R. Miller, "The Flagellum Unspun: The Collapse of Irreducible Complexity," in *Philosophy of Biology: An Anthology*, eds. A. Rosenberg and R. Arp (Chichester, UK: Wiley-Blackwell, 2009), 439–455.

45. Kenneth R. Miller, "The Flagellum Unspun: The Collapse of Irreducible Complexity, in *Debating Design: From Darwin to DNA*, eds. William Dembski and Michael Ruse (Cambridge, UK: Cambridge University Press, 2004), 81–97.

46. Michael J. Behe, "Irreducible Complexity," in *Debating Design: From Darwin to DNA*, eds. William Dembski and Michael Ruse (Cambridge, UK: Cambridge University Press, 2004), 353–370.

47. Michael J. Behe, "Irreducible Complexity: Obstacle to Darwinian Evolution," in *Philosophy of Biology: An Anthology*, eds. A. Rosenberg and R. Arp (Chichester, UK: Wiley-Blackwell, 2009), 427–438.

48. Kenneth R. Miller, *Finding Darwin's God: A Scientist's Search for Common Ground Between God and Evolution* (New York: Cliff Street Books, 1999).

49. Kenneth R. Miller, "The Evolution of Vertebrate Blood Clotting," MillerandLevine.com, https://web.archive.org/web/20020418153118/http://www.millerandlevine.com/km/evol/DI/clot/Clotting.html.

50. Keith Robison, "Darwin's Black Box: Irreducible Complexity or Irreproducible Irreducibility?," Talk Origins (website), December 11, 1996, https://web.archive.org/web/19990127102929/http://www.talkorigins.org/faqs/behe/review.html.

51. Michael J. Behe, "In Defense of the Irreducibility of the Blood Clotting Cascade," Discovery Institute, July 31, 2000, https://web.archive.org/web/20090106144440/https://www.discovery.org/a/442/.

52. Behe, *Darwin's Black Box*, chap. 4.

53. Behe, *Darwin Devolves*, 294–298.

54. Thomas H. Bugge et al., "Loss of Fibrinogen Rescues Mice from the Pleiotropic Effects of Plasminogen Deficiency," *Cell* 87, no. 4 (November 15, 1996): 709–719, https://doi.org/10.1016/S0092-8674(00)81390-2.

55. X. Xu and R. F. Doolittle, "Presence of a Vertebrate Fibrinogen-Like Sequence in an Echinoderm," *PNAS* 87, no. 6 (March 1, 1990): 2097–2101.

56. Behe, *Darwin Devolves*, 298–301.

57. Behe, *Darwin Devolves*, 300.

58. Michael J. Behe, "Waiting Longer for Two Mutations," *Genetics* 181, no. 2 (February 1, 2009): 819–820, https://doi.org/10.1534/genetics.108.098905.

59. Rick Durrett and Deena Schmidt, "Reply to Michael Behe," *Genetics* 181, no. 2 (February 1, 2009): 821–822, https://doi.org/10.1534/genetics.109.100800.

60. Behe, "Waiting Longer for Two Mutations." A summary of the exchange can be found at https://web.archive.org/web/20090324062937/https://www.discovery.org/a/9611/.

61. Michael J. Behe, "A Key Inference of *The Edge of Evolution* Has Now Been Experimentally Confirmed," Evolution News and Views, Discovery Institute, July 14, 2014, https://web.archive.org/web/20190411042402/

https://evolutionnews.org/2014/07/a_key_inference/.

62. Robert L. Summers et al., "Diverse Mutational Pathways Converge on Saturable Chloroquine Transport Via the Malaria Parasite's Chloroquine Resistance Transporter," *PNAS* 111, no. 17 (2014): E1759–E1767, https://doi.org/10.1073/pnas.1322965111.

63. Benjamin H. Good et al., "The Dynamics of Molecular Evolution over 60,000 Generations," *Nature* 551 (2017): 45–50, https://doi.org/10.1038/nature24287.

64. Michael J. Behe, "Experimental Evolution, Loss-of-Function Mutations, and 'The First Rule of Adaptive Evolution,'" *Quarterly Review of Biology* 85, no. 4 (December 2010): 1–27, https://doi.org/10.1086/656902.

65. Z. D. Blount, C. Z. Borland, and R. E. Lenski, "Historical Contingency and the Evolution of a Key Innovation in an Experimental Population of *Escherichia coli*," *PNAS* 105 (June 10, 2008): 7899–7906.

66. Behe, *Darwin Devolves*, 188–190. [Internal references removed.]

67. Behe, *Darwin Devolves*, 191.

68. Alison K. Hottes et al., "Bacterial Adaptation through Loss of Function," *PLOS Genetics* 9, no. 7 (July 2013): e1003617, https://doi.org/10.1371/journal.pgen.1003617.

69. Hannah Stower, "Adaptation by Loss of Function," *Nature Reviews Genetics* 14 (2013): 596, https://doi.org/10.1038/nrg3557.

70. Behe, *Darwin Devolves*, 248–249.

71. Richard E. Lenski, "On Damaged Genes and Polar Bears," Telliamed Revisited (website), February 22, 2019, https://web.archive.org/web/20190401022715/https://telliamedrevisited.wordpress.com/2019/02/22/on-damaged-genes-and-polar-bears/.

72. Lenski, "On Damaged Genes and Polar Bears."

73. Shiping Liu et al., "Population Genomics Reveal Recent Speciation and Rapid Evolutionary Adaptation in Polar Bears," *Cell* 157, no. 4 (May 8, 2014): 785–794. https://doi.org/10.1016/j.cell.2014.03.054.

74. Nathan H. Lents, S. Joshua Swamidass, and Richard E. Lenski, "The End of Evolution?," *Science* 363, no. 6427 (February 8, 2019): 590, https://doi.org/10.1126/science.aaw4056; Nathan Lents and Arthur Hunt,

"Darwin Devolves: Behe Gets Polar Bear Evolution Very Wrong," The Human Evolution Blog, February 12, 2019, https://web.archive.org/web/20190331065352/https://thehumanevolutionblog.com/2019/02/12/behe-polar-bears/; S. Joshua Swamidass, Nathan Lents, and Arthur Hunt, "Behe And The Polar Bear's Fat," Peaceful Science (website), February 12, 2019, https://web.archive.org/web/20190605012711/https://discourse.peacefulscience.org/t/lents-and-hunt-behe-and-the-polar-bears-fat/4473; Michael J. Behe, "Train Wreck of a Review: A Response to Lenski et al. in *Science*," Evolution News and Science Today, Discovery Institute, February 14, 2019, https://web.archive.org/web/20190217205514/https://evolutionnews.org/2019/02/train-wreck-of-a-review-a-response-to-lenski-et-al-in-science/.

75. R. V. Farese et al., "Knockout of the Mouse Apolipoprotein B Gene Results in Embryonic Lethality in Homozygotes and Protection Against Diet-induced Hypercholesterolemia in Heterozygotes," *PNAS* 92, no. 5 (1995):1774–1778, https://doi.org/10.1073/pnas.92.5.1774.

76. Lents and Hunt, "Darwin Devolves: Behe Gets Polar Bear Evolution Very Wrong." (emphasis in original)

77. Michael J. Behe, "Experimental Evolution, Loss-of-Function Mutations, and 'The First Rule of Adaptive Evolution,'" *Quarterly Review of Biology* 85, no. 4 (December 2010): 1–27, https://doi.org/10.1086/656902.

78. Richard E. Lenski, "On Damaged Genes and Polar Bears," Telliamed Revisited (website), February 22, 2019, https://web.archive.org/web/20190401022715/https://telliamedrevisited.wordpress.com/2019/02/22/on-damaged-genes-and-polar-bears/.

79. R. V. Farese et al., "Knockout of the Mouse Apolipoprotein B Gene Results in Embryonic Lethality in Homozygotes and Protection Against Diet-induced Hypercholesterolemia in Heterozygotes," *PNAS* 92, no. 5 (1995):1774–1778, https://doi.org/10.1073/pnas.92.5.1774.

80. Richard E. Lenski, "Does Behe's "First Rule" Really Show that Evolutionary Biology Has a Big Problem?," Telliamed Revisited (website), February 15, 2019, https://

web.archive.org/web/20190331232241/
https://telliamedrevisited.wordpress.
com/2019/02/15/does-behes-first-rule-
really-show-that-evolutionary-biology-has-a-
big-problem/.

81. Richard Carter and Kamini N. Mendis,
"Evolutionary and Historical Aspects of the
Burden of Malaria," *Clinical Microbiology
Reviews* 15, no. 4 (October 2002): 564–594.

82. Michael J. Behe, "Getting There First: An
Evolutionary Rate Advantage for Adaptive
Loss-of-Function Mutations," in *Biologi-
cal Information: New Perspectives*, eds. R. J.
Marks et al. (Singapore: World Scientific,
2013), 450–473.

83. Richard E. Lenski, "Is the LTEE Breaking Bad?," Telliamed Revisited (website),
February 26, 2019, https://web.archive.org/
web/20190331115149/https://telliamedre-
visited.wordpress.com/2019/02/26/is-the-
ltee-breaking-bad/.

84. Michael J. Behe, "A Response to My
Lehigh Colleagues," Discovery Institute,
March 22, 2019, https://web.archive.org/
web/20200914153218/https://www.
discovery.org/a/a-response-to-my-lehigh-
colleagues/.

85. Gregory I. Lang and Amber M. Rice,
"Evolution Unscathed: *Darwin Devolves*
Argues on Weak Reasoning That Unguided
Evolution is a Destructive Force, Incapable of Innovation," *Evolution* 73, no. 4
(2019): 862–868, https://web.archive.org/
web/20191105025046/http://glanglab.com/
papers/Evolution2019.pdf.

86. Lenski, "Is the LTEE Breaking Bad?"

87. Colin Raeside et al., "Large Chromosomal
Rearrangements During a Long-Term Evolution Experiment with *Escherichia coli*," *MBio*
5, no. 5 (2014): e01377–14.

88. Zachary D. Blount, Christina Z. Borland, and Richard E. Lenski, "Historical
Contingency and the Evolution of a Key
Innovation in an Experimental Population of
Escherichia coli," *PNAS* 105, no. 23 (June 10,
2008): 7899–7906, https://doi.org/10.1073/
pnas.0803151105.

89. Michael J. Behe, "Train Wreck of a Review:
A Response to Lenski et al. in *Science*," Evolu-
tion News and Science Today, Discovery
Institute, February 14, 2019, https://web.

archive.org/web/20190217205514/https://
evolutionnews.org/2019/02/train-wreck-
of-a-review-a-response-to-lenski-et-al-in-
science/.

90. Michael J. Behe, "Experimental Evolution,
Loss-of-Function Mutations, and 'The First
Rule of Adaptive Evolution,'" *Quarterly
Review of Biology* 85, no. 4 (December 2010):
1–27, https://doi.org/10.1086/656902.

91. R. V. Farese et al., "Knockout of the Mouse
Apolipoprotein B Gene Results in Embryonic Lethality in Homozygotes and Protection Against Diet-induced Hypercholesterolemia in Heterozygotes," *PNAS* 92, no. 5
(1995):1774–1778, https://doi.org/10.1073/
pnas.92.5.1774.

92. Rohan Maddamsetti et al., "Core Genes
Evolve Rapidly in the Long-Term Evolution
Experiment with *Escherichia coli*," *Genome
Biology and Evolution* 9, no. 4 (April 2017):
1072–1083, https://doi.org/10.1093/gbe/
evx064.

93. Maddamsetti et al., "Core Genes Evolve
Rapidly."

94. Richard E. Lenski, "Evolution Goes Viral!
(And How Real Science Works)," Telliamed
Revisited (website), March 6, 2019, https://
web.archive.org/web/20190331202402/
https://telliamedrevisited.wordpress.
com/2019/03/06/evolution-goes-viral-and-
how-real-science-works/.

95. Justin R. Meyer et al., "Repeatability and
Contingency in the Evolution of a Key Innovation in Phage Lambda," *Science* 335 (January 27, 2012): 428–432.

96. Michael J. Behe, "More from Lenski's
Lab, Still Spinning Furiously," Evolution
News and Views, Discovery Institute,
January 30, 2012, https://web.archive.org/
web/20190422171518/https://evolution-
news.org/2012/01/more_from_lensk/.

97. Alita Burmeister, Richard E. Lenski, and
Justin R. Meyer, "Host Coevolution Alters
the Adaptive Landscape of a Virus," *Proceedings of the Royal Society B: Biological Sciences*
283, no. 1839 (2016): 20161528, https://doi.
org/10.1098/rspb.2016.1528.

98. Michael J. Behe, "Response to Critics,
Part 2: Sean Carroll," Discovery Institute,
June 26, 2007, https://web.archive.org/
web/20190926042634/https://www.dis-

covery.org/a/response-to-critics-part-2-sean-carroll/.

99. Sean B. Carroll, "God as Genetic Engineer," *Science* 316, no. 5830 (June 8, 2007): 1427–1428.

100. Michael J. Behe, "Experimental Evolution, Loss-of-Function Mutations, and 'The First Rule of Adaptive Evolution,'" *Quarterly Review of Biology* 85, no. 4 (December 2010): 1–27, https://doi.org/10.1086/656902.

101. Behe, "More from Lenski's Lab."

102. James A. Shapiro, *Evolution: A View from the 21st Century* (Upper Saddle River, NJ: FT Press, 2011).

103. Gregory I. Lang and Amber M. Rice, "Evolution Unscathed: *Darwin Devolves* Argues on Weak Reasoning That Unguided Evolution is a Destructive Force, Incapable of Innovation," *Evolution* 73, no. 4 (2019): 862–868, https://doi.org/10.1111/evo.13710.

104. Nathan H. Lents, S. Joshua Swamidass, and Richard E. Lenski, "The End of Evolution?," *Science* 363, no. 6427 (February 8, 2019): 590, https://doi.org/10.1126/science.aaw4056.

105. Nathan Lents and Arthur Hunt, "Darwin Devolves: Behe Gets Polar Bear Evolution Very Wrong," The Human Evolution Blog, February 12, 2019, https://web.archive.org/web/20190331065352/https://thehumanevolutionblog.com/2019/02/12/behe-polar-bears/.

106. Michael J. Behe, "Train Wreck of a Review: A Response to Lenski et al. in *Science*," Evolution News and Science Today, Discovery Institute, February 14, 2019, https://web.archive.org/web/20190217205514/https://evolutionnews.org/2019/02/train-wreck-of-a-review-a-response-to-lenski-et-al-in-science/; Michael J. Behe, "Lessons from Polar Bear Studies," Evolution News and Science Today, Discovery Institute, March 6, 2019 https://web.archive.org/web/20190404050456/https://evolutionnews.org/2019/03/lessons-from-polar-bear-studies/; "*Science* Review Offers False Accusations about Chloroquine Resistance," Evolution News and Science Today, Discovery Institute, February 14, 2019, https://web.archive.org/web/20200429033401/https://evolutionnews.org/2019/02/science-review-offers-false-accusations-about-chloroquine-resistance/; "Perplexing: Michael Behe's Critics Falsely Claim He Ignores Exaptation," Evolution News and Science Today, Discovery Institute, February 19, 2019, https://web.archive.org/web/20190415001930/https://evolutionnews.org/2019/02/perplexing-michael-behes-critics-falsely-claim-he-ignores-exaptation/; Ann Gauger, "Näsvall et al. Demonstrates the Effectiveness of Intelligent Design," Evolution News and Science Today, Discovery Institute, February 25, 2019, https://web.archive.org/web/20200429054558/https://evolutionnews.org/2019/02/nasvall-et-al-demonstrates-the-effectiveness-of-intelligent-design/; "Cited to Attack *Darwin Devolves,* Study Devolves on Close Inspection," Evolution News and Science Today, Discovery Institute, February 13, 2019, https://web.archive.org/web/20190414234327/https://evolutionnews.org/2019/02/cited-to-attack-darwin-devolves-study-devolves-on-close-inspection/.

107. Michael J. Behe, "Experimental Evolution, Loss-of-Function Mutations, and 'The First Rule of Adaptive Evolution,'" *Quarterly Review of Biology* 85, no. 4 (December 2010): 1–27, https://doi.org/10.1086/656902.

108. Michael J. Behe, "Lessons from Polar Bear Studies."

109. Vincent J. Lynch et al., "Elephantid Genomes Reveal the Molecular Bases of Woolly Mammoth Adaptations to the Arctic," *Cell Reports* 12, no. 2 (2015): 217–228, https://doi.org/10.1016/j.celrep.2015.06.027.

110. Richard Carter and Kamini N. Mendis, "Evolutionary and Historical Aspects of the Burden of Malaria," *Clinical Microbiology Reviews* 15, no. 4 (October 2002): 564–594.

111. Richard Dawkins, "Inferior Design," *New York Times,* July 1, 2007, https://web.archive.org/web/20190820225942/https://www.nytimes.com/2007/07/01/books/review/Dawkins-t.html.

112. Jeffrey J. Schoenebeck and Elaine A. Ostrander, "Insights into Morphology and Disease from the Dog Genome Project," *Annual Review of Cell and Developmental Biology* 30 (October 2014): 535–560, https://doi.org/10.1146/annurev-cellbio-100913-012927.

113. Michael J. Behe, "Richard Lenski and Citrate Hype—Now Deflated," Evolution News and Views, Discovery Institute, May 12, 2016, https://web.archive.org/web/20170309115016/https://evolutionnews.org/2016/05/richard_lenski/; "Lenski's Long-Term Evolution Experiment: 25 Years and Counting," Evolution News and Views, Discovery Institute, November 21, 2103, https://web.archive.org/web/20200528185138/https://evolutionnews.org/2013/11/richard_lenskis/; "More from Lenski's Lab, Still Spinning Furiously," Evolution News and Views, Discovery Institute, January 30, 2012, https://web.archive.org/web/20190422171518/https://evolutionnews.org/2012/01/more_from_lensk/; "Richard Lenski, 'Evolvability,' and Tortuous Darwinian Pathways," Evolution News and Views, Discovery Institute, April 18, 2011, https://web.archive.org/web/20200914161051/https://evolutionnews.org/2011/04/richard_lenski_evolvability_an/; "New Work by Richard Lenski," Evolution News and Views, Discovery Institute, October 21, 2009, https://web.archive.org/web/20190906171057/https://evolutionnews.org/2009/10/new_work_by_richard_lenski/.

114. Peter T. Boag and Peter R. Grant, "Intense Natural Selection in a Population of Darwin's Finches (Geospizinae) in the Galápagos," Science 214, no. 4516 (October 2, 1981): 82–85.

115. Richard E. Lenski, "Convergence and Divergence in a Long-Term Experiment with Bacteria," The American Naturalist 190, no. S1 (August 2017): S57–S68.

116. Zachary D. Blount, Christina Z. Borland, and Richard E. Lenski, "Historical Contingency and the Evolution of a Key Innovation in an Experimental Population of Escherichia coli," PNAS 105, no. 23 (June 10, 2008): 7899–7906, https://doi.org/10.1073/pnas.0803151105.

117. Michael J. Behe, "Rose-Colored Glasses: Lenski, Citrate, and BioLogos," Evolution News and Views, Discovery Institute, November 13, 2012, https://web.archive.org/web/20190210015010/https://evolutionnews.org/2012/11/rose-colored_gl/.

118. Zachary D. Blount et al., "Genomic Analysis of a Key Innovation in an Experimental Escherichia coli Population," Nature 489 (2012): 513–518, https://doi.org/10.1038/nature11514.

119. Erik M. Quandt et al., "Fine-Tuning Citrate Synthase Flux Potentiates and Refines Metabolic Innovation in the Lenski Evolution Experiment," eLife 4 (October 14, 2015): e09696, https://doi.org/10.7554/eLife.09696.001.

120. Robert L. Summers et al., "Diverse Mutational Pathways Converge on Saturable Chloroquine Transport Via the Malaria Parasite's Chloroquine Resistance Transporter," PNAS 111, no. 17 (2014): E1759–E1767, https://doi.org/10.1073/pnas.1322965111.

121. Lenski, "Convergence and Divergence."

122. Lenski, "Convergence and Divergence."

123. Sébastien Wielgoss et al., "Mutation Rate Dynamics in a Bacterial Population Reflect Tension Between Adaptation and Genetic Load," PNAS 110, no. 1 (January 2, 2013): 222–227, https://doi.org/10.1073/pnas.1219574110.

124. Gregory I. Lang and Amber M. Rice, "Evolution Unscathed: Darwin Devolves Argues on Weak Reasoning That Unguided Evolution is a Destructive Force, Incapable of Innovation," Evolution 73, no. 4 (2019): 862–868, https://doi.org/10.1111/evo.13710.

125. Daniel G. MacArthur et al., "A Systematic Survey of Loss-of-Function Variants in Human Protein-Coding Genes," Science 335, no. 6070 (February 17, 2012): 823–828.

126. Michael Lynch, "Mutation and Human Exceptionalism: Our Future Genetic Load," Genetics 202, no. 3 (March 1, 2016): 869–875, https://doi.org/10.1534/genetics.115.180471.

127. Kendra J. Lipinski et al., "High Spontaneous Rate of Gene Duplication in Caenorhabditis elegans," Current Biology 21, no. 4 (February 22, 2011): 306–310, https://doi.org/10.1016/j.cub.2011.01.026.

128. Eugene V. Koonin, "Horizontal Gene Transfer: Essentiality and Evolvability in Prokaryotes, and Roles in Evolutionary Transitions," F1000Research 5 (July 25, 2016): 1805.

129. William F. Martin, "Too Much Eukaryote LGT," *BioEssays* 39, no. 12 (October 2017): 1700115, https://doi.org/10.1002/bies.201700115.

130. J. B. Walsh, "How Often Do Duplicated Genes Evolve New Functions?," *Genetics* 139, no. 1 (January 1, 1995): 421–428.

131. Hideki Innan and Fyodor Kondrashov, "The Evolution of Gene Duplications: Classifying and Distinguishing Between Models," *Nature Reviews Genetics* 11 (2010): 97, https://doi.org/10.1038/nrg2689.

132. Michael J. Behe, "Train Wreck of a Review: A Response to Lenski et al. in *Science*," Evolution News and Science Today, Discovery Institute, February 14, 2019, https://web.archive.org/web/20190217205514/https://evolutionnews.org/2019/02/train-wreck-of-a-review-a-response-to-lenski-et-al-in-science/.

133. Xing-Xing Shen et al., "Tempo and Mode of Genome Evolution in the Budding Yeast Subphylum," *Cell* 175, no. 6 (November 29, 2018): 1533–1545.E20, https://doi.org/10.1016/j.cell.2018.10.023.

134. Yuri I. Wolf and Eugene V. Koonin, "Genome Reduction as the Dominant Mode of Evolution," *BioEssays* 35, no. 9 (2013): 829–837, https://doi.org/10.1002/bies.201300037; Ricard Albalat and Cristian Cañestro, "Evolution by Gene Loss," *Nature Reviews Genetics* 17 (April 18, 2016): 379, https://doi.org/10.1038/nrg.2016.39.

135. Virag Sharma et al., "A Genomics Approach Reveals Insights into the Importance of Gene Losses for Mammalian Adaptations," *Nature Communications* 9, no. 1 (March 23, 2018): 1215, https://doi.org/10.1038/s41467-018-03667-1.

136. Zev N. Kronenberg et al., "High-Resolution Comparative Analysis of Great Ape Genomes," *Science* 360, no. 6393 (June 8, 2018): eaar6343.

137. Sangeet Lamichhaney et al., "Evolution of Darwin's Finches and Their Beaks Revealed by Genome Sequencing," *Nature* 518, no. 7539 (February 11, 2015): 371–375, https://doi.org/10.1038/nature14181.

138. David Brawand et al., "The Genomic Substrate for Adaptive Radiation in African Cichlid Fish," *Nature* 513, no. 7518 (September 3, 2014): 375–381, https://doi.org/10.1038/nature13726.

139. Lamichhaney et al., "Evolution of Darwin's Finches."

140. Stephen Jay Gould, *The Structure of Evolutionary Theory* (Cambridge: Harvard University Press, 2002).

141. Robert M. Zink and Hernán Vázquez-Miranda, "Species Limits and Phylogenomic Relationships of Darwin's Finches Remain Unresolved: Potential Consequences of a Volatile Ecological Setting," *Systematic Biology* 68, no. 2 (March 2018): 347–357, https://doi.org/10.1093/sysbio/syy073.

142. Lamichhaney et al., "Evolution of Darwin's Finches."

143. Gregory I. Lang and Amber M. Rice, "Evolution Unscathed: *Darwin Devolves* Argues on Weak Reasoning That Unguided Evolution is a Destructive Force, Incapable of Innovation," *Evolution* 73, no. 4 (2019): 862–868, https://doi.org/10.1111/evo.13710.

144. "Machine," The Free Dictionary, https://www.thefreedictionary.com/machine.

145. Tanya Lewis, "Creature with Interlocking Gears on Legs Discovered," Live Science, September 12, 2013, https://web.archive.org/web/20131003034557/https://www.livescience.com/39577-insects-with-leg-gears-discovered.html.

146. John Hewitt, "Fiber Optic Light Pipes in the Retina Do Much More Than Simple Image Transfer," Phys.org, July 21, 2014, https://web.archive.org/web/20140724004616/http://m.phys.org/news/2014-07-fiber-optic-pipes-retina-simple.html.

147. See, for example, J. Scott Turner, *Purpose and Desire: What Makes Something "Alive" and Why Modern Darwinism Has Failed to Explain It* (San Francisco: HarperOne, 2017).

148. Paul D. Boyer, "The ATP Synthase—A Splendid Molecular Machine," *Annual Review of Biochemistry* 66, no. 1 (July 1997): 717–749, https://doi.org/10.1146/annurev.biochem.66.1.717.

149. D. Voges, P. Zwickl, and W. Baumeister, "The 26S Proteasome: A Molecular Machine Designed for Controlled Proteolysis," *Annual Review of Biochemistry* 68, no. 1 (July 1999): 1015–1068, https://doi.org/10.1146/annurev.biochem.68.1.1015.

150. "The Nobel Prize in Chemistry 2016," press release, Nobel Prize (website), October 5, 2016, https://web.archive.org/web/20180908132711if_/https://www.nobelprize.org/prizes/chemistry/2016/press-release/.

151. Michael Flannery, "Has Ken Miller Refuted Irreducible Complexity with a Tie Clip?," Evolution News and Views, Discovery Institute, October 6, 2014, https://web.archive.org/web/20190421170444/https://evolutionnews.org/2014/10/has_ken_miller/.

152. Michael J. Behe, "Reply to My Critics: A Response to Reviews of Darwin's Black Box: The Biochemical Challenge to Evolution," Biology and Philosophy 16, no. 5 (January 2001): 685–709, https://doi.org/10.1023/A:1012268700496.

153. Niall Shanks and Karl H. Joplin, "Redundant Complexity: A Critical Analysis of Intelligent Design in Biochemistry," Philosophy of Science 66, no. 2 (June 1999): 268–282, https://doi.org/10.1086/392687; Behe, "Reply to My Critics"; Michael J. Behe, "Irreducible Complexity," in Debating Design: From Darwin to DNA, eds. William Dembski and Michael Ruse (Cambridge, UK: Cambridge University Press, 2004), 353–370; Michael J. Behe, "Waiting Longer for Two Mutations," Genetics 181, no. 2 (February 1, 2009): 819–820, https://doi.org/10.1534/genetics.108.098905; Michael J. Behe, "In Defense of the Irreducibility of the Blood Clotting Cascade," Discovery Institute, July 31, 2000, https://web.archive.org/web/20090106144440/https://www.discovery.org/a/442/; Michael J. Behe, "A Mousetrap Defended," Discovery Institute, July 31, 2000, https://web.archive.org/web/20090310182942/https://www.discovery.org/a/446/; Michael J. Behe, "Philosophical Objections to Intelligent Design," Discovery Institute, July 31, 2000, https://web.archive.org/web/20100113213247/https://www.discovery.org/a/445/.

154. Michael J. Behe, "A Blind Man Carrying a Legless Man Can Safely Cross the Street: Experimentally Confirming the Limits to Darwinian Evolution," Evolution News and Views, Discovery Institute, January 11, 2012, https://web.archive.org/web/20190422185541/https://evolutionnews.org/2012/01/a_blind_man_car/.

155. Michael J. Behe, "More from Lenski's Lab, Still Spinning Furiously," Evolution News and Views, Discovery Institute, January 30, 2012, https://web.archive.org/web/20190422171518/https://evolutionnews.org/2012/01/more_from_lensk/.

156. Michael Denton, Evolution: A Theory in Crisis (Bethesda, MD: Adler & Adler, 1986).

157. James Clerk Maxwell, "Ether, or Æther," Encyclopædia Britannica, 9th ed., vol. 8 (Edinburgh: Encyclopedia Brittanica, 1878), 568–572, available at https://web.archive.org/web/20200914190013/https://en.wikisource.org/wiki/Encyclop%C3%A6dia_Britannica,_Ninth_Edition/Ether_%282.%29.

158. Anil Ananthaswamy, "How the Belief in Beauty Has Triggered a Crisis in Physics," Nature 558, no. 7709 (June 12, 2018): 186–187.

159. Sabine Hossenfelder, "Particle Physicists Excited over Discovery of Nothing in Particular," BackReaction, March 14, 2019, https://web.archive.org/web/20190316165211/http://backreaction.blogspot.com/2019/03/particle-physicists-excited-over.html.

160. David Stove, Darwinian Fairytales: Selfish Genes, Errors of Heredity, and Other Fables of Evolution (New York: Encounter Books, 2006); Thomas Nagel, Mind and Cosmos: Why the Materialist Neo-Darwinian Conception of Nature is Almost Certainly False (New York: Oxford University Press, 2012); James Alan Shapiro, Evolution: A View from the 21st Century (Upper Saddle River, NJ: FT Press Science, 2011); Kevin Laland et al., "Does Evolutionary Theory Need a Rethink?," Nature 514, no. 7521 (October 8, 2014): 161–164, https://web.archive.org/web/20141010002926/https://www.nature.com/news/does-evolutionary-theory-need-a-rethink-1.16080; Lynn Margulis and Dorion Sagan, Acquiring Genomes: A Theory of the Origins of Species (New York: Basic Books, 2002); Michael R. Dietrich, "Richard Goldschmidt: Hopeful Monsters and Other 'Heresies,'" Nature Reviews Genetics 4, no. 1 (January 1, 2003): 68–74, https://doi.org/10.1038/nrg979; Alfred Russel Wallace, The World of Life: A Manifestation of Creative

Power, Directive Mind and Ultimate Purpose (London: Chapman and Hall, 1910).

161. Franklin M. Harold, *The Way of the Cell* (Oxford: Oxford University Press, 2001), 205.

162. Jerry A. Coyne, "God in the Details," *Nature* 383 (1996): 227–228, https://doi.org/10.1038/383227a0.

163. Jerry Coyne, "Intelligent Design Gets Even Dumber," *The Washington Post*, March 8, 2019, https://web.archive.org/web/20190401030546if_/https://www.washingtonpost.com/outlook/intelligent-design-gets-even-dumber/2019/03/08/7a8e72dc-289e-11e9-b2fc-721718903bfc_story.html?utm_term=.4e8a7bb722fb.

PART EIGHT (100–109)

1. Michael Richardson, quoted in Elizabeth Pennisi, "Haeckel's Embryos: Fraud Rediscovered," *Science* 277, no. 5331 (1997): 1435. See also M. K. Richardson et al., "There Is No Highly Conserved Embryonic Stage in the Vertebrates: Implications for Current Theories of Evolution and Development," *Anatomy and Embryology* 196 (1997): 91–106. The authors note, "Our survey seriously undermines the credibility of Haeckel's drawings."

2. National Science Education Standards (Washington DC: National Academy Press, 1996).

3. National Academy of Sciences, *Science and Creationism: A View from the National Academy of Sciences*: 2nd Edition (Washington, DC: National Academy Press, 1999), https://web.archive.org/web/20151006021402if_/https://www.nap.edu/read/6024/chapter/1; National Academy of Sciences, *Teaching About Evolution and the Nature of Science* (Washington, DC: National Academy Press, 1998), https://web.archive.org/web/20150918203353if_/https://www.nap.edu/read/5787/chapter/1.

4. *Science and Creationism*, 27.

5. *Science and Creationism*, 27.

6. *Teaching About Evolution and the Nature of Science*, 55.

7. *Science and Creationism*, 7.

8. *Science and Creationism*, 8.

9. *Science and Creationism*, 8.

10. *Science and Creationism*, 21–22.

11. *Science and Creationism*, 22.

12. Michael J. Behe, *Darwin's Black Box: The Biochemical Challenge to Evolution* (New York: The Free Press, 1996).

13. *Science and Creationism*, 21.

14. *Science and Creationism*, 6.

15. Paul Davies, *The Fifth Miracle: The Search for the Origin and Meaning of Life* (New York: Simon & Schuster, 1999), 17.

16. *Science and Creationism*, 7.

17. Tim Wong et al., "Evolution of the Bacterial Flagellum," *Microbe* 2, no. 7 (July 2007): 335–340, https://web.archive.org/web/20180116121528/https://www.asmscience.org/content/journal/microbe/10.1128/microbe.2.335.1.

18. Richard E. Lenski, "Phenotypic and Genomic Evolution During a 20,000-Generation Experiment with the Bacterium *Escherichia coli*," *Plant Breeding Reviews* 24 (2004): 225–265.

19. Richard Carter and Kamini N. Mendis, "Evolutionary and Historical Aspects of the Burden of Malaria," *Clinical Microbiology Reviews* 15, no. 4 (October 2002): 564–594.

20. Ronald Numbers, quoted in Gwen Evans, "Reason or Faith? Darwin Expert Reflects," University of Wisconsin–Madison News, February 3, 2009, https://web.archive.org/web/20151216031720/http://news.wisc.edu/reason-or-faith-darwin-expert-reflects/.

21. Kevin Padian and Nicholas Matzke, "Darwin, Dover, 'Intelligent Design' and Textbooks," *Biochemical Journal* 417, no. 1 (January 2009): 29–42, https://doi.org/10.1042/bj20081534.

22. Michael J. Behe, *The Edge of Evolution: The Search for the Limits of Darwinism* (New York: Free Press, 2007), 231.

23. Padian and Matzke, "Darwin, Dover."

24. National Academy of Sciences, *Science and Creationism: A View from the National Academy of Sciences* (Washington, DC: National Academy Press, 1999), 7, https://web.archive.org/web/20151006021402if_/https://www.nap.edu/read/6024/chapter/1.

25. Kitzmiller v. Dover Area School District, 400 F. Supp. 2d 707 (M.D. PA 2005),

Plaintiffs' Finding of Facts, Case 4:04-cv-02688-JEJ Document 334-1, Filed November 23, 2005, https://web.archive.org/web/20120127154731/www.pamd.uscourts.gov/kitzmiller/04cv2688-334.pdf, 42.

26. Kitzmiller v. Dover Area School District, 400 F. Supp. 2d 707 (M.D. PA 2005), https://web.archive.org/web/20200914195851/https://www.pamd.uscourts.gov/sites/pamd/files/opinions/04v2688d.pdf, 80.

27. Padian and Matzke, "Darwin, Dover."

28. Eugene V. Koonin, "The Cosmological Model of Eternal Inflation and the Transition from Chance to Biological Evolution in the History of Life," *Biology Direct* 2 (May 31, 2007): 15, https://doi.org/10.1186/1745-6150-2-15.

29. John Travis, "On the Origin of the Immune System," *Science* 324, no. 5927 (May 1, 2009): 580–582.

30. Jan Klein and Nikolas Nikolaidis, "The Descent of the Antibody-Based Immune System by Gradual Evolution," *PNAS* 102, no. 1 (January 4, 2005): 169–174, https://doi.org/10.1073/pnas.0408480102.

31. Shirley M. Tilghman, "Strange Bedfellows: Science, Politics, and Religion," President's Page, *Princeton Alumni Weekly*, January 25, 2006, https://web.archive.org/web/20060913035151/https://www.princeton.edu/~paw/archive_new/PAW05-06/07-0125/prezpage.html. That Tilghman has me in view is evident from this passage: "Advocates such as Michael Behe, a professor of physical chemistry at Lehigh University, declare that 'natural selection can only choose among systems that are already working, so the existence in nature of irreducibly complex biological systems poses a powerful challenge to Darwinian theory.'"

32. See the original version of my essay at Evolution News and Views to see the archival photograph and the doctored version, https://web.archive.org/web/20200914200401/https://evolutionnews.org/2009/08/bloggingheads_tv_and_me/. For the source of the photographs, and historical background on the disappearing figure beside Stalin, see Brian Dillon, "The Revelation of Erasure," Tate Etc., September 1, 2006, https://

web.archive.org/web/20200914200728/https://www.tate.org.uk/tate-etc/issue-8-autumn-2006/revelation-erasure.

33. Robert Wright and George Johnson, "Science Saturday [various intelligent design segments]," Bloggingheads TV, videos, varying lengths, September 5, 2009, https://web.archive.org/web/20130426015918/https://bloggingheads.tv/videos/2247.

34. Kitzmiller v. Dover Area School District, 400 F. Supp. 2d 707 (M.D. PA 2005), https://web.archive.org/web/20200914195851/https://www.pamd.uscourts.gov/sites/pamd/files/opinions/04v2688d.pdf.

35. Kitzmiller v. Dover Area School District, 64–89.

36. Michael J. Behe, *Darwin's Black Box: The Biochemical Challenge to Evolution* (New York: The Free Press, 1996), 39.

37. Behe, *Darwin's Black Box*, 40.

38. H. Allen Orr, "Darwin v. Intelligent Design (Again)," *Boston Review* (December 1996/January 1997): 28–31, https://web.archive.org/web/20141006072402/https://bostonreview.net/archives/BR21.6/orr.html.

39. The systems I discussed in *Darwin's Black Box* in 1996 were selected because they were well studied and well understood at the time. For further systems and other considerations that present grave difficulties for random mutation and natural selection as the drivers of the unfolding of life, see my later books, *The Edge of Evolution* and *Darwin Devolves*.

40. See, for example, Michael Denton, *Evolution: A Theory in Crisis* (Bethesda, MD: Adler & Adler, 1986); Phillip E. Johnson, *Darwin on Trial* (Washington, DC: Regnery Gateway, 1991).

41. The entirety of the statement is the following: "Because Darwin's Theory is a theory, it is still being tested as new evidence is discovered. The Theory is not a fact. Gaps in the Theory exist for which there is no evidence. A theory is defined as a well-tested explanation that unifies a broad range of observations. Intelligent design is an explanation of the origin of life that differs from Darwin's view. The reference book, *Of Pandas and People* is available for students to see if they would like to explore this view in an effort to gain

an understanding of what intelligent design actually involves. As is true with any theory, students are encouraged to keep an open mind." Of such molehills are federal cases made. The statement and the complaint can be viewed here: https://web.archive.org/web/20101113160212/https://www.aclu.org/files/evolution/legal/complaint.pdf.

42. Percival Davis and Dean H. Kenyon, *Of Pandas and People: The Central Question of Biological Origins* (Dallas: Haughton, 1993).

43. Amy Worden, "Bad Frog Beer to 'Intelligent Design': The Controversial Ex-Pa. Liquor Board Chief Is Now U.S. Judge in the Closely Watched Trial," *Philadelphia Inquirer*, October 16, 2005. Worden writes: "Sitting in the barren visiting judge's chambers in Harrisburg (he usually hears cases in Williamsport), Jones ponders a question about the movie *Inherit the Wind*, starring Spencer Tracy as lawyer William Jennings Bryan and Gene Kelly as columnist H.L. Mencken. He said he saw the film—based on the landmark Scopes trial that tested a Tennessee law forbidding the teaching of evolution—years ago. Jones says he plans to watch the movie again, soon. 'It would help put things in historical context,' he said. 'I don't know if it would be helpful to the decision I have to make.' Then he laughs. 'You know,' he said, 'nobody ever remembers who played the judge in that movie.'"

44. Scott Adams, Dilbert Blog, March 2006.

45. "The World's First-Ever Festival of Bad Ad Hoc Hypotheses," https://web.archive.org/web/20130417085349/http://bahfest.com/.

46. Michael J. Behe, *Darwin's Black Box: The Biochemical Challenge to Evolution* (New York: The Free Press, 1996), xi–xii.

47. Kenneth R. Miller, *Finding Darwin's God: A Scientist's Search for Common Ground Between God and Evolution* (New York: Cliff Street Books, 1999), 146.

48. Michael J. Behe, "'A True Acid Test': Response to Ken Miller," Discovery Institute, July 31, 2000, https://web.archive.org/web/20100113213242/https://www.discovery.org/a/441/.

49. Lauri Lebo, *The Devil in Dover: An Insider's Story of Dogma v. Darwin in Small-Town*

America (New York: The New Press, 2008), 150–151.

50. Kitzmiller v. Dover Area School District, Case Number 4:04-CV-02688, Transcript of Civil Bench Trial Proceedings Trial Day Ten, Afternoon Session, https://web.archive.org/web/20181113210131/https://www.aclupa.org/files/7613/1404/6696/Day10PM.pdf.

51. Kevin Padian and Nicholas Matzke, "Darwin, Dover, 'Intelligent Design' and Textbooks," *Biochemical Journal* 417, no. 1 (January 2009): 29–42, https://doi.org/10.1042/bj20081534.

52. For a detailed comparison of the court opinion versus the plaintiffs' lawyers' brief, see "'Masterful' Ruling on Intelligent Design Was Copied from ACLU," Discovery Institute, December 12, 2016, https://web.archive.org/web/20080517110113/https://www.discovery.org/a/3828/.

53. Indeed, in the legal system it is not considered "plagiarism" for a judge to copy parts of the brief of a party, but I am told that extensive copying is strongly frowned upon by the legal community, as well it should be. For more on the subject, see Casey Luskin, "Backgrounder on the Significance of Judicial Copying," Evolution News and Views, Discovery Institute, December 13, 2006, https://web.archive.org/web/20190407050811/https://evolutionnews.org/2006/12/media_backgrounder_on_kitzmill/.

54. Renyi Liu and Howard Ochman, "Stepwise Formation of the Bacterial Flagellar System," *PNAS* 104, no. 17 (April 24, 2007): 7116–7121, https://doi.org/10.1073/pnas.0700266104.

55. Thomas H. Bugge et al., "Loss of Fibrinogen Rescues Mice from the Pleiotropic Effects of Plasminogen Deficiency," *Cell* 87, no. 4 (November 15, 1996): 709–719, https://doi.org/10.1016/S0092-8674(00)81390-2.

56. John Jones, "Judicial Independence and Kitzmiller v. Dover, et al.," (lecture, Difficult Dialogues, the University of Kansas, Lawrence, KS, September 26, 2006), https://web.archive.org/web/20140819224306/http://archive.news.ku.edu/2006/july/31/difficult-dialogues.shtml.

57. Kitzmiller v. Dover Area School District, 400 F. Supp. 2d 707.

58. Kitzmiller v. Dover Area School District, Case Number 4:04-CV-02688, Transcript of Civil Bench Trial Proceedings Trial Day Eleven, Morning Session, https://web. archive.org/web/20181113210136/https:// www.aclupa.org/files/8013/1404/6696/ Day11AM.pdf.

59. In what at the time I took as a dumb courtroom stunt, the opposing lawyers dumped a pile of textbooks and folders of papers on immunology on a table close in front of me, apparently in a bit of kabuki theater to act out the moral, "Look at all of this scholarship that opposes the heretic." The fact that none of the papers showed how Darwin's mechanism could build an immune system was irrelevant to the drama. It reminded me of the scene in the old 1940s movie *Miracle on 34th Street*, where the lawyer for a man who claimed to be Santa Claus had the post office deliver bags of letters addressed to the North Pole to the courtroom and pile them on the judge's desk, arguing that if the federal post office thought the defendant was Santa Claus, then who was the judge to disagree? The judge—a local politician up for re-election— was happy to have the newspapers report that he ruled for Santa Claus. In my naiveté I thought part of the job of a real judge was to roll his eyes at such shenanigans and focus on the legal and intellectual issues. But Judge Jones seems to be a fan of old movies.

60. The attorney tried to bait me, asking me several times, "Is that your position today that these articles aren't good enough, you need to see a step-by-step description?" and variations on that. Richard Dawkins helped spread the false quote to millions of people. Richard Dawkins, *The God Delusion* (New York: Houghton Mifflin, 2006).

61. Kitzmiller v. Dover Area School District, 400 F. Supp. 2d 707 (M.D. PA 2005), https://web.archive.org/ web/20200914195851/https://www. pamd.uscourts.gov/sites/pamd/files/ opinions/04v2688d.pdf.

FIGURE CREDITS

PART 1

Figure 1.1. "Mousetrap Rat Trap." Illustration by Christian Dorn (Commongt), 2020, Pixabay. Pixabay license. Modified by Amanda Witt. Also in the frontpiece.

Figure 1.2. "Eukaryotic Cilium Diagram." Illustration by Mariana Ruiz Villarreal (Lady of Hats), 2007, Wikimedia Commons. Public domain.

Figure 1.4. Bacterial flagellum motor (also used as the frontispiece). Illustration by Joseph Condeelis/Light Productions.

PART 2

Figures 2.1-4. Illustrations by John McDonald, 2003, University of Delaware (website), https://web.archive.org/web/20200206134721/http://udel.edu/~mcdonald/oldmousetrap.html. Used with permission. Modified by Brian Gage.

PART 7

Figure 7.1. Planthopper gears. "Interactive Gears in the Hind Legs of Issus Coleoptratus from Cambridge." Photo by Malcolm Burrows and Gregory, 2013, Wikimedia Commons, https://commons.wikimedia.org/wiki/File:Interactive_gears_in_the_hind_legs_of_Issus_coleoptratus_from_Cambridge_gears-3.jpg. CC-BY-SA 3.0 license.

INDEX

Made in the USA
Monee, IL
18 January 2024

51994396R00323